OFFSHORE WIND ENERGY GENERATION

OFFSHORE WIND ENERGY GENERATION

CONTROL, PROTECTION, AND INTEGRATION TO ELECTRICAL SYSTEMS

Olimpo Anaya-Lara
University of Strathclyde, Glasgow, UK

David Campos-Gaona
Morelia Institute of Technology, Mexico

Edgar Moreno-Goytia
Morelia Institute of Technology, Mexico

Grain Adam
University of Strathclyde, Glasgow, UK

Library of Congress Cataloging-in-Publication Data

Anaya-Lara, Olimpo.
    Offshore wind energy generation : control, protection, and integration to electrical systems / by Olimpo Anaya-Lara, David Campos-Gaona, Edgar Lenymirko Moreno-Goytia, Grain Philip Adam.
        1 online resource.
    Includes bibliographical references and index.
    Description based on print version record and CIP data provided by publisher; resource not viewed.
    ISBN 978-1-118-70153-9 (Adobe PDF) – ISBN 978-1-118-70171-3 (ePub) – ISBN 978-1-118-53962-0 (cloth)   1. Wind power plants.   2. Offshore electric power plants.   3. Wind energy conversion systems. I. Campos-Gaona, David.   II. Moreno-Goytia, Edgar Lenymirko.   III. Adam, Grain Philip.   IV. Title.
    TK1541
    621.31′213609162–dc23

                                                                                                    2013048398

A catalogue record for this book is available from the British Library.

ISBN: 978-1-118-53962-0

Set in 10/12pt Times by Aptara Inc., New Delhi, India
Printed and bound in Malaysia by Vivar Printing Sdn Bhd

1   2014

Contents

Preface

The motivation for this book is the rapid growth of offshore wind energy systems and the implications this has on power system operation, control and protection. Developments on wind turbine technology and power electronic converters along with new control approaches have enabled offshore wind energy systems performance to be improved. The authors identified the need for a book that covers fundamental and up-to-date issues on this dynamic topic suitable for beginners or advanced readership. The contents offer information on technology trends for offshore wind energy systems, detailed modelling of variable-speed wind generator technologies and easy-to-use grid integration examples. The textbook is useful to final year undergraduate and postgraduate students, and also practising engineers and scientists in the wind industry with research interest in aspects of wind generator technology and electrical systems for grid integration.

The book is organised in seven chapters and two appendices. Chapter 1 reviews wind turbine basics and discusses challenges of offshore wind farm connection and grid code compliance. Chapter 2 covers in detail the operation in normal and abnormal conditions of DFIG wind generators while Chapter 3 focuses on Fully-Rated Converter technologies. In Chapter 4 electrical collectors and offshore transmission schemes are covered (including multi-terminal dc transmission). Chapter 5 describes technical challenges arising as a result of integrating offshore wind farms into the power grid and provides various case studies. Chapter 6 discusses protection aspects of offshore wind energy systems. Chapter 7 reviews emerging technologies for offshore wind integration, including energy storage and condition monitoring. Topologies, control and operation of voltage source converters are covered in Appendix A. Appendix B presents a number of worked-out examples well suited for university students.

The text presented in this book draws together material on electrical systems of offshore wind farms from many sources such as graduate courses that the authors have taught over many years at universities in the UK, a large number of technical papers published by the IEEE and IET, and research programmes with which they have been closely associated such as the EPSRC-funded SUPERGEN Wind Technologies and the ETI Helmwind project. The authors would like to thank Professor William Leithead for his strong and continued support on how these research programmes are conducted. Special thanks are given to Mr John O. Tande and Professor Kjetil Uhlen for their support and cooperation through the Norwegian Research Centre for Offshore Wind Technology (NOWITECH). The authors wish to thank Dr Giddani O. A. Kalcon who provided very valuable input on VSC-HVDC offshore transmission in Chapter 4 and Dr Nolan Caliao who gave permission to include material from his PhD thesis in Chapter 3. The authors would also like to acknowledge Dr Gustavo Quinonez-Varela and

Dr Ryan Tumilty for the useful discussions during the preparation of the manuscript, Mr William Ross, Mr Alexander Giles, Mr Edward Corr and Mr Philip Morris, who assisted in proofreading the manuscript and Mr Victor Velazquez Cortes and Ms Kamila Nieradzinska who assisted in the preparation of drawings.

Olimpo Anaya-Lara, David Campos-Gaona, Edgar Moreno-Goytia, Grain Adam
2014

About the Authors

Olimpo Anaya-Lara is a Reader in the Institute for Energy and Environment at the University of Strathclyde, UK. Over the course of his career, he has successfully undertaken research on power electronic equipment, control systems design, and stability and control of power systems with increased wind energy penetration. Dr Anaya-Lara is a key participant to the Wind Integration Sub-Programme of the European Energy Research Alliance (EERA) Joint Programme Wind (JP Wind), leading Strathclyde's involvement and contribution to this Sub-Programme. He was appointed to the post of Visiting Professor in Wind Energy at NTNU, Trondheim, Norway funded by Det Norske Veritas (2010–2011). He was a member of the International Energy Annexes XXI Dynamic models of wind farms for power system studies and XXIII Offshore wind energy technology development. He is currently a member of the IEEE and IET, and has published three technical books, as well as over 140 papers in international journals and conference proceedings.

David Campos-Gaona received his PhD from Instituto Tecnológico de Morelia, México. He is currently a research associate in the same institution. During his academic career he has authored and co-authored papers for refereed journals and congresses and worked as a reviewer for IEEE and Wiley transaction papers. His research interests include power electronic equipment control for wind power integration and HVDC transmission systems.

Edgar Moreno-Goytia is a Professor in the Posgrado en Ingeniería Eléctrica, Instituto Tecnológico de Morelia, México. His current research interests include the development of power electronics-based technologies applied to power grids (HVDC, FACTS and electronic power transformers), development of dc grids, and grid integration of large wind farms system integration and other renewables. He has led various research projects and has published more than 20 papers in international journal and conference proceedings. He is a member of the IEEE and IET.

Grain Adam received first-class BSc and MSc degrees in electrical machines and power systems from Sudan University of Science and Technology, Khartoum, Sudan, in 1998 and 2002, respectively, and a PhD degree in power electronics from Strathclyde University, Glasgow, UK, in 2007. He is currently with the Department of Electronic and Electrical Engineering, Strathclyde University, and his research interests are multilevel inverters, electrical machines and power systems control and stability.

Acronyms and Symbols

ac	Alternating current
AFC	Active flow control
AVR	Automatic voltage regulator
CB	Circuit breaker
CC	Current control
CCC	Capacitor-commutated converter
CIA	Constant-ignition angle
CM	Condition monitoring
CSC	Current source converter
dc	Direct current
DFIG	Doubly-fed induction generator
DG	Distributed generation
EMF	Electromotive force
ESCR	Effective short-circuit ratio
ESR	Equivalent series resistance
FACTs	Flexible alternating current transmission system
FC	Flying capacitor
FCL	Fault-current limiter
FRC	Fully-rated converter
FRT	Fault ride-through
FSIG	Fixed-speed induction generator
GIL	Gas-insulated line
GIT	Gas-insulated transformer
GPS	Global positioning system
GSC	Grid-side converter
GTOs	Gate turn-off thyristor
GW	Giga-watt
HP	Horse power
HTS	High-temperature superconducting
HTSC	High-temperature superconducting cables
HV	High voltage
HVAC	High-voltage alternating current
HVDC	High-voltage direct current
IG	Induction generator

IGBT	Insulated-gate bipolar transistor
IMC	Internal model control
IPC	Individual pitch control
LCC	Line-commutated converters
LVRT	Low-voltage ride-through
MIMO	Multiple-input multiple-output
MTDC	Multi-terminal dc
MVA	Mega volt-ampere
MW	Mega watt
NIST	National Institute of Standards and Technology
NSC	Network-side converter
ODE	Ordinary differential equation
O&M	Operation & Maintenance
PCC	Point-of common coupling
PD	Phase disposition
PDC	Power system oscillations damping controller
POD	Phase opposition disposition
PI	Proportional Integral
PLL	Phase lock loop
PM	Permanent Magnet
PMSG	Permanent Magnet synchronous generator
PMU	Phasor measurement unit
PMW	Pulse width modulation
PSS	Power system stabiliser
pu	Per unit
RF	Radio frequency
RMS	Root-mean square
rpm	Revolutions per minute
RSC	Rotor-side converter
SCADA	Supervisory control and data acquisition
SCIG	Squirrel-cage induction generator
SCR	Silicon-controlled-rectifier
SISO	Single input single output
SMES	Superconducting Magnetic Energy Storage
STATCOM	Static synchronous compensator
SVC	Static var compensator
TCR	Thyristor-controlled reactor
TSC	Thyristor-switched capacitor
TSO	Transmission system operator
UHF	Ultra high frequency
VAr	Volt-ampere reactive
VCO	Voltage-controlled oscillator
VDCOL	voltage-dependent current-order limit
VPP	Virtual power plant
VSC	Voltage-source converter
WAMS	Wide-area measurement system

WT	Wind turbine
WTG	Wind turbine generator
XLPE	Cross-linked polyethylene

Symbols Used in Chapter 1

P_{air}	Power in the airflow
ρ	Air density
A	Swept area of rotor, m^2
v	Upwind free wind speed, m/s
C_p	Power coefficient
$P_{\text{wind turbine}}$	Power transferred to the wind turbine rotor
λ	Tip-speed ratio
ω	Rotational speed of rotor
R	Radius to tip of rotor
V_m	Mean annual site wind speed
V_{dc}	Direct voltage

Symbols Used in Chapter 2

v_{as}, v_{bs}, v_{cs}	Stator a b c voltages
r_{as}, r_{bs}, r_{cs}	Stator a b c windings resistance
i_{as}, i_{bs}, i_{cs}	Stator a b c currents
$\psi_{as}, \psi_{bs}, \psi_{cs}$	Stator a b c fluxes
v_{ar}, v_{br}, v_{cr}	Rotor a b c voltages
$r_{ar}r_{br}, r_{cr}$	Rotor a b c windings resistance
i_{ar}, i_{br}, i_{cr}	Rotor a b c currents
$\psi_{ar}, \psi_{br}, \psi_{cr}$	Rotor a b c fluxes
$[L_{IG}]$	Induction generator inductance matrix
L_{ms}	Stator magnetising inductance
L_{ls}	Stator leakage inductance
L_m	Magnetising inductance
L_{lr}	Rotor leakage inductance
N_s	Effective stator windings turns
N_r	Effective rotor windings turns
μ_0	Permeability of free space
r_a	Radius of the induction generator air gap annulus
l	Effective length of the machine (i.e. the effective length of the pole area)
θ_{dq}	Angle between the d axis of the rotating frame and stator phase a of the induction generator
ω_{dq}	Angular speed of the $dq0$ rotating frame
i_{ds}, i_{qs}, i_{0s}	d q 0 components of stator current
v_{ds}, v_{qs}, v_{0s}	d q 0 components of stator voltage
$\psi_{ds}, \psi_{qs}, \psi_{0s}$	d q 0 components of stator flux

i_{dr}, i_{qr}, i_{0r}	$d\,q\,0$ components of rotor current
v_{dr}, v_{qr}, v_{0r}	$d\,q\,0$ components of rotor voltage
$\psi_{dr}, \psi_{qr}, \psi_{0r}$	$d\,q\,0$ components of rotor flux
S_{dq0}	Induction generator instantaneous power in $d\,q\,0$
T_e	Electromagnetic torque
T_{shaft}	Torque from the shaft of the mechanical system
J_g	Generator mass moment of inertia
ω_t	Turbine angular speed
J_t	Turbine mass moment of inertia
$T_{torsion}$	Shaft elasticity
$T_{damping}$	Shaft damping torque
θ_t	Turbine rotor angle
θ_r	Induction generator rotor angle
K_{tot}	Shaft torsion constant
D	Shaft damping constant
S_{B2B}	Back-to-back converter power rating
v_{dc}	Back-to-back converter dc voltage level
r_{cb}	Crowbar resistance
$Q_1, Q_2, Q_3, Q_4, Q_5, Q_6$	VSC IGBTs
$d_1, d_2, d_3, d_4, d_5, d_6$	VSC Diodes
Q_s	Induction generator reactive power
ω_s	Synchronous speed
ω_r	Rotor speed
J	Turbine and generator added moment of inertia
C	dc capacitance
$'$	Quantity referred to the stator
r	Resistance between the VSC and the grid
L	Inductance between the VSC and the grid
pwm	PMW signal
d_a, d_b, d_c	a, b, c, duty cycle
m_a, m_b, m_c	a, b, c, modulator signals
W	Capacitor energy
L_s	Stator inductance
L_r	Rotor inductance
P_{gsc}	Grid-side converter active power
Q_{gsc}	Grid-side converter reactive power
$P_x(s)$	Plant
$K_x(s)$	PI controller
Kp_x, Ki_x	Proportional and integral gains
$\ell_x(s)$	Open loop gain
$B_x(s)$	Closed-loop transfer function
τ_x	Closed-loop time constant
G_x	Inner feedback loop gain
$M_x(s)$	Plant with inner feedback loop gain
$\rightarrow$	Space vector notation
T_s	Stator time constant

T_r	Rotor time constant
u	Voltage dip magnitude
t_0^-, t_0^+	Instant before and after the fault happening
$\vec{v_1}, \vec{v_2}, \vec{v_0}$	Positive, negative and zero sequence components of ac voltage
s_ω^+, s_ω^-	Positive and negative sequence slip speed
T_e^+, T_e^-	Positive and negative sequence electromagnetic torque
$s_{\omega max}$	Induction generator maximum speed
P_r	Rotor active power
P_{ag}	Power across the airgap
$P'(s)$	Plant model
$G(s)$	IMC controller
$L(s)$	IMC filter
$M_x(s)$	Model of plant with inner feedback loop gain
α_x	IMC Filter bandwidth
$F_x(s)$	Two-degrees-of-freedom IMC controller configured as a classical controller
l_m	Model uncertainty
η	Controller performance
ϖ	Normalised control system input

Symbols Used in Chapter 3

Over _	Per unit quantity
b	Base quantity
ϕ_s	Stator magnetic field
ϕ_r	Rotor magnetic field
i_{ds}, i_{qs}	Stator currents in d and q-axis
v_{ds}, v_{qs}	Stator voltages in d and q-axis
ψ_{ds}, ψ_{qs}	Stator flux linkage in d and q-axis
T_e	Electromagnetic torque
T_m	Mechanical torque
P_e	Electrical power
P_m	Mechanical power
Q	Reactive power
ω_b	Base synchronous speed
ω_s	Synchronous speed
ω_r	Rotor speed
J	Inertia constant
H	Per unit inertia constant
K	Shaft stiffness
f	System frequency
C	Capacitance
i_f	Field current
i_{kd}, i_{kq1}, i_{kq2}	Damper winding d and q-axis currents
L_{lkd}, L_{lkq}	Leakage inductance of damper windings in d and q-axis

L_{md}, L_{mq}	Mutual inductance in d and q-axis
L_{lf}	Leakage inductance of the field coil
L_{ls}	Leakage inductance of the stator coil
r_s	Stator resistance
r_f	Field winding resistance
r_{kd}, r_{kq1}, r_{kq2}	Resistance of damper d and q-axis coils
v_{fd}	Field voltage
v_{kd}, v_{kq1}, v_{kq2}	Damper winding voltages in d and q-axis
ψ_f	Field flux linkage
$\psi_{kd}, \psi_{kq1}, \psi_{kq2}$	Damper winding flux linkage in d and q-axis
δ_r	Rotor angle
C_s	Synchronising power coefficient
C_d	Damping power coefficient
i_{dr}, i_{qr}	Rotor currents in d and q-axis
v_{dr}, v_{qr}	Rotor voltages in d and q-axis
ψ_{dr}, ψ_{qr}	Rotor flux linkage in d and q-axis
e_d, e_q	Voltage behind the reactance in d and q-axis
L_m	Mutual inductance between stator and rotor windings
X_m	Magnetising reactance
L_r, L_s	Rotor and stator self inductance
X_r, X_s	Rotor and stator reactance
L_{lr}	Rotor leakage inductance
L_{ls}	Stator leakage inductance
r_r	Rotor resistance
r_s	Stator resistance
s	Slip of an induction generator
p	Number of poles

Symbols Used in Chapter 4

S	Three-phase symmetrical short-circuit level
P_{dc}	Rated dc power
Q_C	Three-phase fundamental MVAr at rated P_{dc}
γ	Extinction angle
α	Delay angle
I_m	Current margin
τ	Dc capacitor time constant
C	Capacitance
X_T	Transformer leakage reactance
R_T	Transformer winding resistance
X_F	Smoothing reactor reactance
V_{dc}	VSC converter dc voltage
R_{dc}	dc link resistance
L_{dc}	dc link inductance
δ	Phase angle difference between the ac system and the VSC converter

θ_s	Phase angle difference between the ac system voltage and the ac current flowing from or into the converter
θ_c	Phase angle difference between the VSC converter voltage and the ac current
x1	Sending end
x2	Receiving end
P	Active Power
v_{sa}, v_{sb}, v_{sc}	ac source three phase voltages
i_{sa}, i_{sb}, i_{sc}	Three phase currents flowing between the AC source and the VSC
v_{ca}, v_{cb}, v_{cc}	Three phase voltages at the VSC terminals
i_{ca}, i_{cb}, i_{cc}	Three phase currents at the VSC terminals
L_t	Inductance of the coupling transformer
R_F	Resistance of the coupling reactor
L_F	Inductance of the coupling reactor
I_{dc}	dc link current
i_{sd}, i_{sq}	d and q currents flowing between the AC source and the VSC
v_{sd}, v_{sq}	d and q AC source voltages
ω	Synchronous speed
M	Modulation index
θ	Converter phase angle
Q	Reactive power
B	TCR susceptance
Q_{SVC}	SVC power firing angle
X_{SVC}	SVC effective reactance
X_{TCR}	TCR fundamental frequency equivalent reactance
X_L	TCR inductor reactance
σ	TCR Conduction angle
V_{VSC}	Statcom voltage
δ	Statcom angle

Symbols Used in Chapter 5

GS-VSC	Grid-side converter
B_G	ac voltage at PCC
B_{WF1}	Wind farm ac voltage
P_g	Wind farm active power
Q_g	Wind farm reactive power
P_g^*	Wind farm active power reference
Q_g^*	Reactive power reference
G	ac grid
WF-VSC	Wind farm-side converter
B_{WF1}	Wind farm ac voltage
VSC	Voltage source converter
V_{dc}	VSC converter dc voltage
P	VSC active power

Symbols Used in Chapter 6

I_{pickup}	Over current protection pickup current
I_{max}	Maximum load current
$I_{sc\,min}$	Smallest short circuit current
v_{dc_sc}	Equivalent dc voltage in a HVDC link during a pole-to-pole fault
i_{dc_sc}	Equivalent dc current in a HVDC link during a pole-to-pole fault
C, L_{dc}, r_{dc}	Equivalent capacitance, inductance and resistance in a HVDC link during a pole to pole fault
v'_{dc_sc}	Equivalent dc voltage in a HVDC link during a pole-to-earth fault
i'_{dc_sc}	Equivalent dc current in a HVDC link during a pole-to-earth fault
C', L'_{dc}, r'_{dc}	Equivalent capacitance, inductance and resistance in a HVDC link during a pole-to-earth fault
r_f	Fault resistance in a HVDC link during a pole-to-earth fault

1

Offshore Wind Energy Systems

1.1 Background

With construction restrictions inhibiting the deployment of wind turbines onshore, offshore installations are more attractive (e.g. in the UK) (The Crown State, 2011). By mid-2012, offshore wind power installed globally was 4620 MW, representing about 2% of the total installed wind power capacity. Over 90% of the offshore wind turbines currently installed across the globe are situated in the North, Baltic and Irish Seas, along with the English Channel. Most of the rest is in two demonstration projects off China's coast. According to the more ambitious projections, a total of 80 GW of offshore wind power could be installed worldwide by 2020, with three quarters of this in Europe (GWEC, 2013).

All current offshore wind installations are relatively close to shore, using well-known onshore wind turbine technology. However, new offshore wind sites located far from shore have been identified, with clusters of wind farms appearing at favourable locations for wind power extraction, like in the UK Dogger Bank and German Bight (Figure 1.1) (European Union, 2011). The depths of the waters at these sites are in excess of 30 m.

1.2 Typical Subsystems

The typical subsystems in an offshore wind farm are shown in Figure 1.2. At first glance, it comprises the same elements of an onshore wind farm. However, the environment in which a turbine operates allows a distinction to be made. Considering that the nature of the sea state will act to prohibit accessibility of wind turbines for repair, there is a greater need for offshore wind turbines to be reliable and not require regular repair. This requirement means that the designs and controllers of offshore wind turbines differ from those seen with onshore wind turbines. This is to ensure that performance is maximised whilst minimising cost (German Energy Agency, 2010).

Offshore Wind Energy Generation: Control, Protection, and Integration to Electrical Systems, First Edition.
Olimpo Anaya-Lara, David Campos-Gaona, Edgar Moreno-Goytia and Grain Adam.
© 2014 John Wiley & Sons, Ltd. Published 2014 by John Wiley & Sons, Ltd.
Companion Website: www.wiley.com/go/offshore_wind_energy_generation

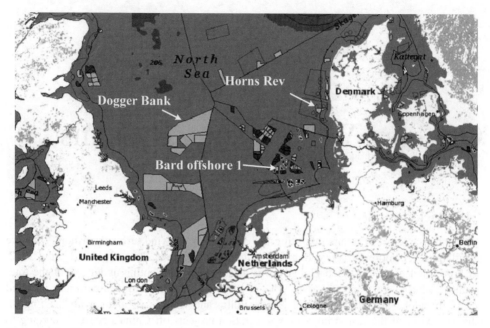

Figure 1.1 Europe's offshore wind farms in operation, construction and planning (Source: www.4coffshore.com/offshorewind).

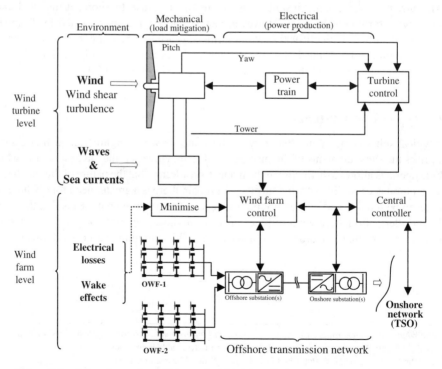

Figure 1.2   Subsystems of an offshore wind farm installation (Anaya-Lara *et al.*, 2013).

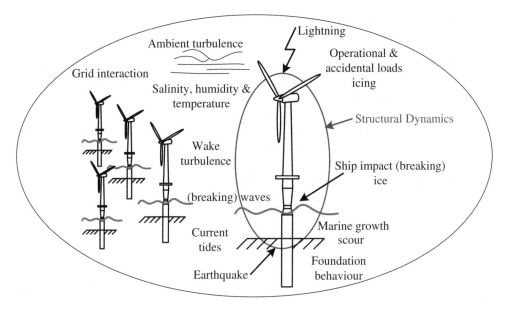

Figure 1.3 Impacts on a bottom-fixed wind turbine (Fischer, 2006).

In the offshore environment, loads are induced by wind, waves, sea currents, and in some cases, floating ice (Figure 1.3), introducing new and difficult challenges for offshore wind turbine design and analysis. Accurate estimation and proper combination of these loads are essential to the turbine and associated controllers design process. Offshore wind turbines have different foundations to onshore wind turbines. The foundations are subjected to hydrodynamic loads. This hydrodynamic loading will inevitably exhibit some form of coupling to the aerodynamic loading seen by the rotor, nacelle and tower. This is an additional problem that must be considered when designing offshore wind turbines. Ideally, the total system composed of rotor/nacelle, tower, substructure and foundation should be analysed using an integrated model (Nielsen, 2006). Development of novel wind turbine concepts optimised for operation in rough offshore conditions along with better O&M strategies is crucial. In addition, turbine control philosophy must be consistent and address the turbine as a whole dynamic element, bearing in mind trade-offs in terms of mechanical performance and power output efficiency (Anaya-Lara *et al.*, 2013).

At the wind farm level, the array layout and electrical collectors must be designed on a site-specific basis to achieve a good balance between electrical losses and wake effects. For power system studies, it is typical to represent the wind farm by an aggregated machine model (and controller). However, more detailed wind farm representations are required to take full advantage of control capabilities, exploring further coordinated turbine control and operation to achieve a better array design. Full exploitation of the great potential offered by offshore wind farms will require the development of reliable and cost-effective offshore grids for collection of power, and its transmission and connection to the onshore network whilst complying with the grid codes. It is anticipated that power electronic equipment (e.g. HVDC and FACTs), and their enhanced control features, will be fundamental in addressing these objectives.

1.3 Wind Turbine Technology

1.3.1 Basics

Wind turbines produce electricity by using the power of the wind to drive an electrical generator (Fox *et al.*, 2007; Anaya-Lara *et al.*, 2009). Wind passes over the blades generating lift and exerting a turning force. The rotating blades turn a shaft that goes into a gearbox, which increases the rotational speed to that which is appropriate for the generator. The generator uses magnetic fields to convert the rotational energy into electrical energy. The power output goes to a transformer, which steps up the generator terminal voltage to the appropriate voltage level for the power collection system.

A wind turbine extracts kinetic energy from the swept area of the blades (Figure 1.4).

The power in the airflow is given by (Burton *et al.*, 2001; Manwell *et al.*, 2002):

$$P_{\text{air}} = \frac{1}{2}\rho A v^3 \tag{1.1}$$

where ρ is the air density, A is the swept area of the rotor in m^2, and v is the upwind free wind speed in m/s. The power transferred to the wind turbine rotor is reduced by the power coefficient, C_{p}:

$$P_{\text{wind turbine}} = C_{\text{p}}P_{\text{air}} = \frac{1}{2}\rho A v^3 C_p \tag{1.2}$$

A maximum value of C_{p} is defined by the Betz limit, which states that a turbine can never extract more than 59.3% of the power from an air stream. In practice, wind turbine rotors have maximum C_{p} values in the range 25–45%. It is also conventional to define a tip-speed ratio, λ, as

$$\lambda = \frac{\omega R}{v} \tag{1.3}$$

where ω is the rotational speed of the rotor and R is the radius to tip of the rotor.

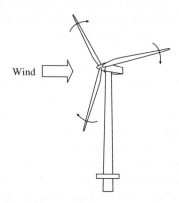

Wind

Figure 1.4 Horizontal-axis wind turbine.

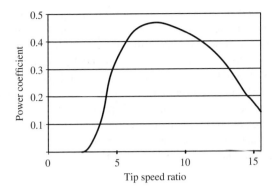

Figure 1.5   Illustration of power coefficient/tip-speed ratio curve, C_p/λ.

The tip-speed ratio, λ, and the power coefficient, C_p, are dimensionless and so can be used to describe the performance of any size of wind turbine rotor. Figure 1.5 shows that the maximum power coefficient is only achieved at a single tip-speed ratio. The implication of this is that fixed rotational speed wind turbines could only operate at maximum efficiency for one wind speed. Therefore, one argument for operating a wind turbine at variable rotational speed is that it is possible to operate at maximum C_p over a range of wind speeds.

The power output of a wind turbine at various wind speeds is conventionally described by its power curve. The power curve gives the steady-state electrical power output as a function of the wind speed at the hub height. An example of a power curve for a 2-MW wind turbine is given in Figure 1.6.

The power curve has three key points on the velocity scale:

- Cut-in wind speed – the minimum wind speed at which the machine will deliver useful power.
- Rated wind speed – the wind speed at which rated power is obtained.
- Cut-out wind speed – the maximum wind speed at which the turbine is allowed to deliver power (usually limited by engineering loads and safety constraints).

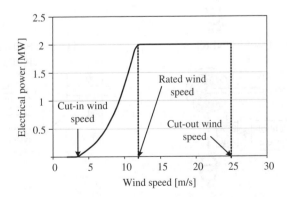

Figure 1.6   Power curve for a 2-MW wind turbine.

Below the cut-in speed of about 4–5 m/s, the wind speed is too low for useful energy production, so the wind turbine remains shut down. When the wind speed is above this value, the wind turbine begins to produce energy; the power output increases following a broadly cubic relationship with wind speed (although modified by the variation in C_p) until rated wind speed is reached at about 11–12 m/s. Above rated wind speed, the aerodynamic rotor is arranged to limit the mechanical power extracted from the wind and so reduce the mechanical loads on the drive train. Then at very high wind speeds, typically above 25 m/s, the turbine is shut down. The choice of cut-in, rated and cut-out wind speed is made by the wind turbine designer who, for typical wind conditions, will try to balance obtaining maximum energy extraction with controlling the mechanical loads (Anaya-Lara *et al.*, 2009).

1.3.2   Architectures

Figure 1.7 shows the main wind turbine generator concepts which are divided into fixed-speed wind turbines (type A), and variable-speed wind turbines (types B, C and D) (Tande *et al.*, 2007; Fox *et al.*, 2007).

1.3.2.1   Fixed-Speed Wind Turbines

A fixed-speed wind turbine (Type A in Figure 1.7) employs a three-phase squirrel-cage induction generator (SCIG) driven by the turbine via a gearbox and directly connected to the grid through a step-up transformer. Thus, the induction generator will provide an almost

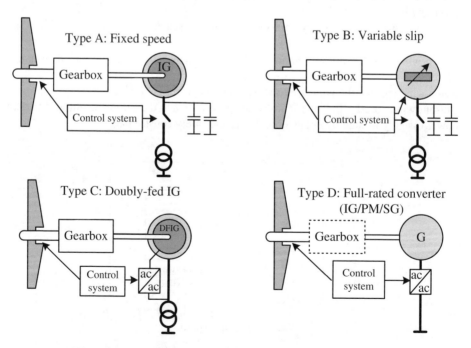

Figure 1.7   Overview of wind turbine concepts (Tande *et al.*, 2007).

constant rotational speed, that is only varying by the slip of the generator (typically about 1%). The reactive power consumption of the induction generator is provided via a capacitors bank, whereas a soft-starter limits the inrush current to the induction generator during start-up. At wind speeds above rated, the output power is limited by natural aerodynamic stall or by active pitching of the blades before the wind turbine is stopped at cut-out wind speed. Modern fixed-speed wind turbines are commonly equipped with capacitors that are connected in steps using power electronic switches for fast reactive power compensation control. A Static VAr Compensator (SVC) can be applied either for controlling the reactive power exchange to a certain value (e.g. zero for unity power factor), or for contributing to voltage control with droop settings just as any other utility-scaled power plant (Tande *et al.*, 2007).

1.3.2.2 Variable-Speed Wind Turbines

Variable-speed operation offers increased efficiency and enhanced power control. The variable-speed operation is achieved either by controlling the rotor resistance of the induction generator, that is slip control (Type B in Figure 1.7), or by a power electronic frequency converter between the generator and the grid (Types C and D in Figure 1.7). The variable slip concept yields a speed range of about 10%, whereas the use of a frequency converter opens for larger speed variations. All variable-speed concepts are expected to yield quite small power fluctuations, especially during operation above rated wind speed. They are also expected to offer smooth start-up.

In regards to power quality, the basic difference between the three variable-speed concepts is that Type B does not have a power electronic converter and thus reactive power capabilities as a fixed-speed wind turbine, whereas Types C and D each have a converter that offers dynamic reactive power control. The reactive power capability of Types C and D may differ as the Doubly-Fed Induction Generator (DFIG) concept of Type C uses a converter rated typically about 30% of the generator and not 100% as is the case for the Fully-Rated Converter (FRC) wind turbine of the Type D concept. The network-side converters of all major wind turbine suppliers offering Types C or D concepts are voltage source converters (VSCs), allowing independent control of active and reactive power (within the apparent power rating of the converter). The converters are based on fast-switching transistors, for example insulated-gate bipolar transistors (IGBTs); consequently, they are not expected to cause harmonic currents that may significantly distort the voltage waveform (Tande *et al.*, 2007).

1.3.3 Offshore Wind Turbine Technology Status

Currently installed offshore wind turbines are adapted from standard onshore wind turbine designs with significant upgrades to account for sea conditions. These modifications include strengthening the tower to handle the added loading from waves, along with pressurized nacelles and environmental control to keep corrosive sea spray away from critical drive train and electrical components.

Offshore turbine power capacity is greater than standard onshore wind turbines, currently ranging from 2 to 5 MW (Figure 1.8). The current generation of offshore wind turbines typically are three-bladed horizontal-axis, yaw-controlled, active blade-pitch-to-feather controlled, upwind rotors, which are nominally 80 m to approximately 130 m in diameter (E.ON Climate

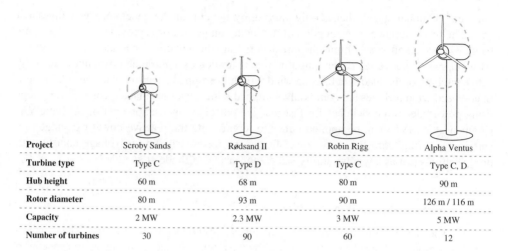

Project	Scroby Sands	Rødsand II	Robin Rigg	Alpha Ventus
Turbine type	Type C	Type D	Type C	Type C, D
Hub height	60 m	68 m	80 m	90 m
Rotor diameter	80 m	93 m	90 m	126 m / 116 m
Capacity	2 MW	2.3 MW	3 MW	5 MW
Number of turbines	30	90	60	12

Figure 1.8   Offshore wind turbine development (E.ON Climate and Renewables, 2012).

and Renewables, 2012). Offshore wind turbines are generally larger because there are fewer constraints on component and assembly equipment transportation, which limit land-based machine size. In addition, larger turbines can extract more total energy for a given project site area than smaller turbines (Dolan *et al.*, 2009). A critical issue in developing very-large wind turbines is that the physical scaling laws do not allow some of the components to be increased in size without a change in the fundamental technology.

In onshore wind turbines, the drive train is typically designed around a modular, fixed-ratio, three-stage, gearbox speed with planetary stages on the low-speed side and helical stages on the high-speed side. Offshore towers are shorter than onshore ones for the same output because wind shear (the change in wind velocity resulting from the change in elevation) is lower offshore, which reduces the energy capture potential of increasing tower height (Dolan *et al.*, 2009).

1.4   Offshore Transmission Networks

Recent research has produced a number of proposals for a future integrated European transmission network (Supergrid), where large quantities of new offshore wind farms in the North Sea are connected to each other and to major load centres (Figure 1.9). In the long term, benefits of the operation of renewable energy sources are predicted to arise from better interconnection. Energy derived from renewable sources such as wind is inherently variable; that is, their energy output is variable and often difficult to accurately predict. If wind farms from different geographical regions are connected to a single transmission system, in addition to inputs from solar and hydroelectric schemes, then the total output from the renewable sources is expected to become 'smoother' and easier to predict.

It is anticipated that the development of the Supergrid will involve complex interconnections integrating wind farms into clusters and wind power plants. By way of example, Figure 1.10 presents a schematic of a generic ideal scenario interconnecting wind generation, and oil and gas platforms with various onshore grids (Anaya-Lara *et al.*, 2011).

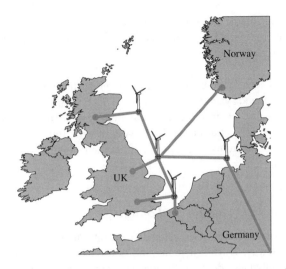

Figure 1.9 Phase 1 Supergrid (Source: www.friendsofthesupergrid.eu).

1.5 Impact on Power System Operation

There are significant differences between wind power and conventional synchronous generation (Slootweg, 2003):

- Wind turbines use different, often converter-based, generating systems compared with those used in synchronous generation.

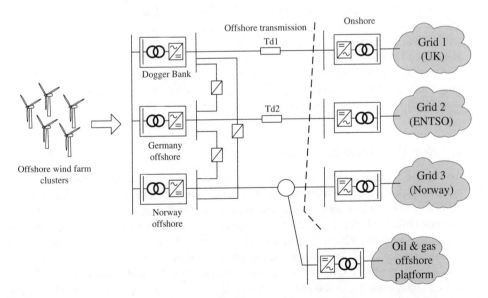

Figure 1.10 Generic offshore power network (Td stands for transmission distance) (Anaya-Lara *et al.*, 2011).

- The wind is not controllable and fluctuates stochastically.
- The typical size of individual wind turbines is much smaller than that of a conventional utility synchronous generator.

Due to these differences, wind energy systems interact differently with the power system and may have both local and system-wide impacts on operation. Local impacts, such as busbar voltages and power quality, occur in the electrical vicinity of a wind turbine or wind farm, and can be attributed to a specific turbine or farm. System wide impacts, such as power system dynamics and stability and frequency support, affect the behaviour of the power system as a whole (UCTE, 2004).

1.5.1   Power System Dynamics and Stability

Squirrel-cage induction generators used in fixed-speed turbines can cause local voltage collapse after rotor speed runaway. During a fault (and consequent grid voltage depression), they accelerate due to the imbalance between the mechanical power from the wind and the electrical power that can be supplied to the grid. When the fault is cleared, these machines absorb reactive power, further depressing the network voltage. If the voltage does not recover quickly enough, the wind turbines continue to accelerate and to consume larger amounts of reactive power. This eventually leads to voltage and rotor speed instability. In contrast to synchronous generators, whose exciters increase reactive power output during low network voltages and thus support voltage recovery after a fault, squirrel-cage induction generators tend to impede voltage recovery (Anaya-Lara et al., 2009).

With variable-speed wind turbines, the sensitivity of the power electronics to over-currents caused by the network voltage depressions can have serious consequences for the stability of the power system. If the penetration level of variable-speed wind turbines in the system is high and they disconnect at relatively small voltage reduction, a voltage drop over a wide geographic area can lead to a large generation deficit. Such a voltage drop could, for instance, be caused by a fault in the transmission grid. To prevent this, Grid Companies and Transmission System Operators require that wind turbines have a Fault Ride-Through capability and withstand voltage drops of certain magnitudes and durations without tripping. This prevents the disconnection of a large amount of wind power in the event of a remote network fault.

1.5.2   Reactive Power and Voltage Support

The voltage on a transmission network is determined mainly by the interaction of reactive power flows with the reactive inductances of such network. Fixed-speed induction generators absorb reactive power to maintain their magnetic field and have no direct control over their reactive power flow. Therefore, in the case of fixed-speed induction generators, the only way to support the network voltage is to reduce the reactive power drawn from the network by using shunt compensators. Variable-speed wind turbines have the capability of reactive power control and may be able to support the voltage of the network. In many situations, the reactive power and voltage control at the point of connection of the wind farm is achieved by using reactive power compensation equipment such as static VAr compensators (SVCs) or static compensators (STATCOMs).

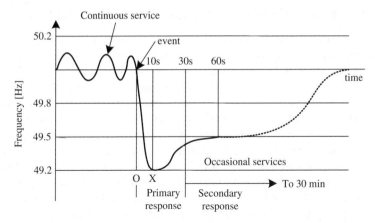

Figure 1.11   Frequency control in England and Wales (Erinmez *et al.*, 1999).

1.5.3   Frequency Support

With the projected increase in wind generation, a potential concern for transmission system operators is the capability of wind farms to provide dynamic frequency support in the event of sudden changes in power network frequency (see Figure 1.11). To provide frequency support from a generation unit, the generator power must increase or decrease as the system frequency changes. Thus, in order to respond to low network frequency, it is necessary to de-load the wind turbine leaving a margin for power increase. A fixed-speed wind turbine can be de-loaded if the pitch angle is controlled such that a fraction of the power that could be extracted from wind will be "spilled". A variable-speed wind turbine can be de-loaded by operating it away from the maximum power extraction curve, thus leaving a margin for frequency control (Anaya-Lara *et al.*, 2009).

1.5.4   Wind Turbine Inertial Response

A FSIG wind turbine acts in a similar manner to a synchronous machine when a sudden change in frequency occurs. For a drop in frequency the machine starts decelerating. This results in the conversion of kinetic energy of the machine to electrical energy, thus giving a power surge. The inverse is true for an increase in system frequency. In the case of a DFIG wind turbine, equipped with conventional controls, the control system operates to apply a restraining torque to the rotor according to a pre-determined curve against rotor speed. This is decoupled from the power system frequency so there is no contribution to the system inertia.

With a large number of DFIG and/or FRC wind turbines connected to the network, the angular momentum of the system will be reduced and the frequency may drop very rapidly during the phase OX in Figure 1.11. Therefore, it is important to reinstate the effect of the machine inertia within these wind turbines. It is possible to emulate the inertia response by manipulating their control actions. The emulated inertia response provided by these generators is referred to as fast primary response (also called 'virtual' or 'synthetic' inertia).

1.6   Grid Code Regulations for the Connection of Wind Generation

Grid connection codes define the requirements for the connection of generation and loads to an electrical network, which ensure efficient, safe and economic operation of the transmission and/or distribution systems (Anaya-Lara *et al.*, 2009). Grid Codes specify the mandatory minimum technical requirements that a power plant should fulfil and additional support that may be called on to maintain the second-by-second power balance and maintain the required level of quality and security of the electricity supply. The additional services that a power plant should provide are normally agreed between the transmission system operator and the power plant operator through market mechanisms. The connection codes normally focus on the point of connection between the Public Electricity System and the new generation. This is very important for wind farm connections, as the Grid Codes demand requirements at the point of connection of the wind farm not at the individual wind turbine generator terminals.

Grid Codes specify the levels and time period of the output power of a generating plant that should be maintained within the specified values of grid frequency and grid voltages. Typically this requirement is defined as shown in Figure 1.12 where the values of voltage, V_1 to V_4, and frequency, f_1 to f_4, differ from country to country. Grid Codes also specify the steady-state operational region of a power plant in terms of active and reactive power requirements. The definition of the operational region differs from country to country. For example, Figure 1.13 shows the operational regions as specified in the Great Britain and Ireland Grid Codes.

Traditionally, wind turbine generators were tripped off once the voltage at their terminals reduced to less than 20% retained voltage. However, with the penetration of wind generation increasing, Grid Codes now generally demand Fault Ride-Through (FRT) – or Low-Voltage Ride-Through (LVRT) – capability for wind turbines connected to transmission networks. Figure 1.14 shows a plot illustrating the general shape of voltage tolerance that most grid operators demand. When reduced system voltage occurs following a network fault, generator tripping is only permitted when the voltage is sufficiently low and for a time that puts it in the shaded area indicated in Figure 1.14.

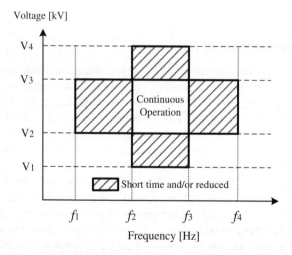

Figure 1.12   Typical shape of continuous and reduced output regions (after GB and Irish Grid Codes).

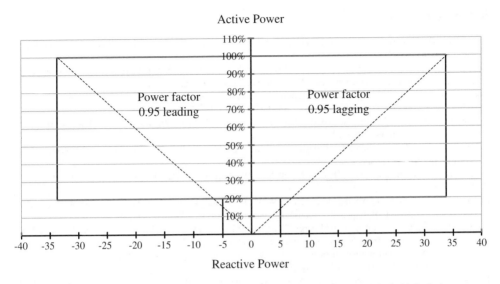

Figure 1.13 Typical steady-state operating region (after GB and Irish Grid Codes).

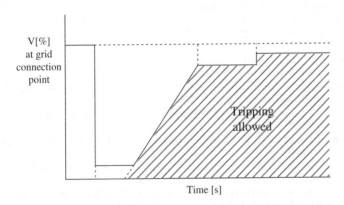

Figure 1.14 Typical shape of Fault Ride-Through capability plot (after GB and Irish Grid Codes).

The large-scale deployment of offshore wind turbines in the North Sea will potentially involve various wind turbine providers, introducing different turbine designs, with varying specifications and performance characteristics. It is envisaged that control requirements and dynamic performance of these future offshore wind power systems, with such a variety of technology and complex grid arrangements, may be significantly different from conventional and comparatively simpler existing power networks. Consequently, the establishment of suitable Grid Codes satisfying so many variables is a difficult challenge that needs to be addressed.

Acknowledgements

The material in Sections 1.3.1, 1.5 and 1.6 of this chapter is adapted from that originally published in Anaya-Lara *et al.*, 2009.

References

Anaya-Lara, O., Jenkins, N., Ekanayake, J. *et al.* (2009) *Wind Generation Systems – Modelling and Control*, Wiley, ISBN 0-470-71433-6.

Anaya-Lara, O., Kalcon, G., Adam, G. *et al.* (2011) North Sea Offshore Networks Basic Connection Schemes: Dynamic Performance Assessment, EPE Wind Chapter 2011, Trondheim, Norway.

Anaya-Lara, O., Tande, J.O., Uhlen, K. and Adaramola, M. (2013) Control Challenges and Possibilities for Offshore Wind Farms, SINTEF Report, TR A7258, ISBN: 978-82-594-3604-7.

Burton, T., Sharpe, D., Jenkins, N. and Bossanyi, E. (2001) *Wind Energy Handbook*, John Wiley & Sons, Ltd, Chichester, ISBN 10: 0471489972.

Dolan, D., Jha, A., Gur, T. *et al.* (2009) *Comparative Study of Offshore Wind Turbine Generators (OWTG) Standards*, MMI Engineering, Oakland, CA.

E.ON Climate and Renewables (2012) E.ON Offshore Wind Energy Factbook, www.eon.com, last accessed: 4 September 2013.

Erinmez, I.A., Bickers, D.O., Wood, G.F. and Hung, W.W. (1999) NGC Experience with frequency control in England and Wales – Provision of frequency response by generator. IEEE PES Winter Meeting.

European Union (2011) European Wind Energy Technology Platform (TPWind): Strategic Research Agenda/Market Development Strategy from 2008 to 2030, www.windplatform.eu, last accessed: 17 December 2013.

Fischer, T.A. (2006) Load mitigation of an offshore wind turbine by optimisation of aerodynamic damping and vibration control. MSc. Thesis, DTU, July, 2006.

Fox, B., Flynn, D., Bryans, L. *et al.* (2007) Wind Power Integration: Connection and System Operational Aspects. IET Power and Energy Series 50, ISBN: 10: 0863414494.

German Energy Agency (2010) DENA Study, Planning of the Grid Integration of Wind Energy in Germany Onshore and Offshore up to the year 2020, www.dena.de, last accessed: 17 December 2013.

GWEC (2013) Global Offshore: Current Status and Future Prospects, www.gwec.net, last accessed: 4 September 2013.

Manwell, J.F., McGowan, J.G. and Rogers, A.L. (2002) *Wind Energy Explained: Theory, Design and Application*, John Wiley & Sons, Ltd., Chichester, ISBN: 10: 0471499722.

Nielsen, F.G. (2006) Specialist Committee V.4, "Ocean wind and wave energy utilisation," 16th International Ship and Offshore Structures Congress, Southampton, UK.

Slootweg, J.G. (2003) Wind power: Modelling and impacts on power system dynamics. PhD Thesis, Technical University of Delft.

Tande, J.O., Di Marzio, G. and Uhlen, K. (2007) System requirements for wind power plants, SINTEF TR A6586, available at www.sintef.no/wind.

The Crown State (2011) Round 3 Offshore Wind Farm Connection Study, Version 1, National Grid, The Crown Estate, www.thecrownstate.co.uk, last accessed: 20 February 2013.

Union for the Co-ordination of Transmission of Electricity, UCTE (2004) Integrating wind power in the European power systems: prerequisites for successful and organic growth. UCTE Position Paper.

2

DFIG Wind Turbine

2.1 Introduction

The 2009 technology report of the European Wind Energy Association surveys 95 different models and 29 different brands of wind turbines above 1 MW showing that the large majority (72 models) use some kind of variable-speed technology (Paul Gardner, 2009). Variable-speed wind turbines have the ability to track the maximum power for the prevailing wind speed and the possibility of reducing loads in the mechanical structure by damping, to some extent, the torque peaks generated by intermittent wind gusts.

The Doubly Fed Induction Generator (DFIG) wind turbine has been the dominant variable-speed technology in the market for models above 1.5 MW. The stator windings are connected directly to the ac grid through the turbine transformer, whilst the rotor windings are connected to the ac grid via slip-rings and a back-to-back (B2B) power electronic converter. The power exchange between this converter and the ac grid depends on the operating point of the machine and the desired control over the speed and reactive power of the DFIG. The B2B converter partially decouples the network electrical frequency from the rotor mechanical frequency, enabling variable-speed operation of the wind turbine. A DFIG typically provides ±30% speed range around the synchronous speed, assuming the B2B converter is rated at around 30% the power rating of the wind turbine (WT). Figure 2.1 shows a schematic diagram of a DFIG and main components.

The main components of a DFIG wind turbine are described below (Manwell *et al.*, 2010, Nelson, 2009).

2.1.1 Induction Generator (IG)

The IG is of the wound-rotor type with slip-rings. Usually, the stator-to-rotor winding turns ratio is adjusted to attain a lower rotor current magnitude suitable for the current ratings of the B2B converter.

Offshore Wind Energy Generation: Control, Protection, and Integration to Electrical Systems, First Edition.
Olimpo Anaya-Lara, David Campos-Gaona, Edgar Moreno-Goytia and Grain Adam.
© 2014 John Wiley & Sons, Ltd. Published 2014 by John Wiley & Sons, Ltd.
Companion Website: www.wiley.com/go/offshore_wind_energy_generation

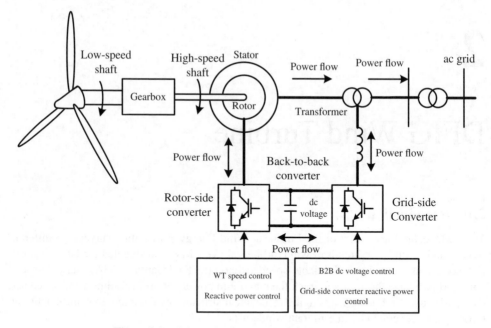

Figure 2.1   Schematic diagram of a DFIG wind turbine.

2.1.2   Back-to-Back Converter

The B2B converter consists of two three-phase Voltage Source Converters (VSCs) that is the rotor-side converter (RSC) and the grid-side converter (GSC), linked via a common dc bus. The RSC is connected to the rotor windings and the GSC is connected to the ac grid via a coupling reactor. The RSC controls the wind turbine speed and reactive power consumption by modifying the rotor currents. The GSC mainly keeps a steady dc voltage level on the B2B dc bus (which is of primary importance for the correct operation of the RSC) and provides, to some extent, reactive power support to the ac grid. The architecture of both converters allows bidirectional power flow making the B2B converter a rotor slip-power recovery device (i.e. recovers the power generated in the rotor windings that otherwise would have been lost as heat in the windings) for the IG.

2.1.3   Gearbox

The gearbox steps up the speed of the low-speed rotor to a higher speed suitable for the electrical generator. The gearbox speed ratio depends on various factors such as rated wind speed and number of poles of the generator, amongst others. By way of example, a 5-MW wind turbine with 11 m/s rated wind speed has a gearbox ratio of about 1:90 (Jonkman, 2006).

2.1.4   Crowbar Protection

The crowbar protects the B2B converter from over currents induced in the rotor circuit during an ac grid fault. The crowbar protection disconnects the RSC and shorts circuit the rotor windings to a bank of resistors. The resistance is chosen large enough to reduce the transient currents back to a safe level in a short period of time. Modern crowbar protection circuits use power electronic devices such as IGBTs or GTOs.

The use of the crowbar protection in DFIG wind turbines is necessary to comply with the Fault Ride-Through requirements defined in Grid Codes. Without the crowbar protection the DFIG cannot remain connected to the ac grid in the event of a fault because the RSC cannot withstand the large transient currents of the rotor. Also, the B2B dc circuit has limited capacity to withstand the sudden inrush of energy coming from the rotor. However, when the crowbar protection is used, the RSC is protected from the initial transient rotor currents, and the additional rotor resistance allows the IG to transiently over speed without reaching a critical operating point or consuming high amounts of reactive power. Therefore, the DFIG can remain connected to the ac grid and ride through the fault.

2.1.5 Turbine Transformer

The transformer is typically located at the base of the wind turbine tower. The low-voltage side levels are usually between 575 and 690 V at 50/60 Hz (although modern offshore turbines are now using higher voltages). These voltages are transformed to the voltage level of the wind farm collector system, for example 33 kV.

2.2 DFIG Architecture and Mathematical Modelling

2.2.1 IG in the abc Reference Frame

The equivalent circuit of the IG is shown in Figure 2.2. The generator is represented by six circuits: three for the stator and three for the rotor. Each circuit of the IG is composed of a self-inductance; and five mutual inductances. For modelling purposes we use in this chapter the motor convention to represent power and currents.

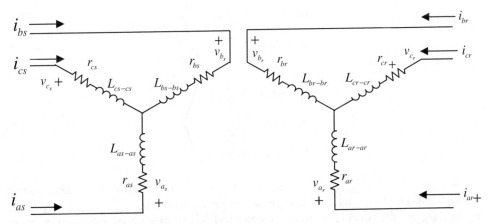

Figure 2.2 Equivalent circuit of an IG.

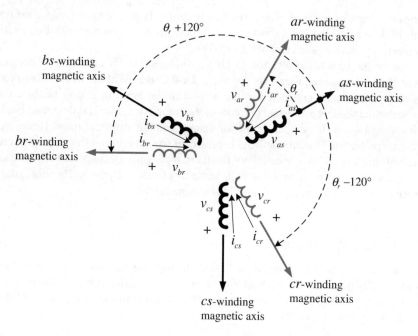

Figure 2.3   Magnetic axes of an IG.

The rotor and stator self-inductances and the mutual inductances between the rotor and stator phases do not depend on the rotor position. Only the mutual fluxes between the rotor and stator are dependant of the rotor position as shown in Figure 2.3.

The voltage equations of the IG in the *abc* domain are given below:

$$
\left.
\begin{aligned}
v_{as} &= r_{as}i_{as} + \frac{d}{dt}\psi_{as} \\
v_{bs} &= r_{bs}i_{bs} + \frac{d}{dt}\psi_{bs} \\
v_{cs} &= r_{cs}i_{cs} + \frac{d}{dt}\psi_{cs}
\end{aligned}
\right\}
\tag{2.1}
$$

$$
\left.
\begin{aligned}
v_{ar} &= r_{ar}i_{ar} + \frac{d}{dt}\psi_{ar} \\
v_{br} &= r_{br}i_{br} + \frac{d}{dt}\psi_{br} \\
v_{cr} &= r_{cr}i_{cr} + \frac{d}{dt}\psi_{cr}
\end{aligned}
\right\}
\tag{2.2}
$$

where $v_{as}, v_{bs}, v_{cs}, \ r_{as}, r_{bs}, r_{cs}, i_{as}, i_{bs}, i_{cs}, \ \psi_{as}, \psi_{bs}, \psi_{cs}$ are the voltage, resistance, current and flux linkage of the stator *a, b, c* phases, respectively. $v_{ar}, v_{br}, v_{cr} \ r_{ar}r_{br}, r_{cr}, i_{ar}, i_{br}, i_{cr}$ $\psi_{ar}, \psi_{br}, \psi_{cr}$ are the voltage, resistance, current and flux linkage of the rotor *a, b, c* phases respectively.

If the stator and rotor three-phase circuits are balanced then the following is true:

$$\left.\begin{matrix} v_{as} = v_{bs} = v_{cs} = v_s \\ r_{as} = r_{bs} = r_{cs} = r_s \\ i_{as} = i_{bs} = i_{cs} = i_s \\ v_{ar} = v_{br} = v_{cr} = v_r \\ r_{ar} = r_{br} = r_{cr} = r_r \\ i_{ar} = i_{br} = i_{cr} = i_r \end{matrix}\right\} \tag{2.3}$$

The equation that defines the flux linkages of the IG is:

$$\left[\psi_{as}, \psi_{bs}, \psi_{cs}, \psi_{ar}, \psi_{br}, \psi_{cr}\right]^T = L_{IG} \left[i_{a_s}, i_{b_s}, i_{c_s}, i_{a_r}, i_{b_r}, i_{c_r}\right]^T \tag{2.4}$$

where L_{IG} is the IG inductance matrix, defined below:

$$[L_{IG}] = \begin{pmatrix} L_{as-as} & L_{as-bs} & L_{as-cs} & L_{as-ar} & L_{as-br} & L_{as-cr} \\ L_{as-bs} & L_{bs-bs} & L_{bs-cs} & L_{bs-ar} & L_{bs-br} & L_{bs-cr} \\ L_{as-cs} & L_{bs-cs} & L_{cs-cs} & L_{cs-ar} & L_{cs-br} & L_{cs-cr} \\ L_{as-ar} & L_{bs-ar} & L_{cs-as} & L_{as-ar} & L_{ar-br} & L_{ar-cr} \\ L_{as-br} & L_{bs-br} & L_{cs-br} & L_{ar-br} & L_{br-br} & L_{br-cr} \\ L_{as-cr} & L_{bs-cs} & L_{cs-cr} & L_{ar-cr} & L_{br-cr} & L_{cr-cr} \end{pmatrix} \tag{2.5}$$

where the subscript of L in the matrix indicates the self-inductance or the mutual inductance between the stator and rotor phases. For example the inductance L_{as-cr} is the mutual inductance between the stator phase a and the rotor phase c. In order to calculate the inductance matrix in Eq. (2.5) it is necessary to compute the flux linkages between the windings of the IG.

The computation of the flux linkages of the IG using the construction parameters of the windings, (i.e. spatial distribution, number of turns, core material, air-gap width, machine length along with magnetic equations) is out of the scope of this book; however, it is valuable to show the final results of the magnetic equations that define such flux linkages because they allow the self- and mutual inductances in Eq. (2.5) to be defined.

The IG total flux linkage of the stator winding in phase a linked by the differential flux crossing the air gap, considering a sinusoidal winding distribution on an iron core, is defined as follows (Novotny and Lipo, 1996):

$$\psi_{as} = \mu_0 N_{as}^2 i_{as} \left(\frac{rl}{g}\right) \left(\frac{\pi}{4}\right) \tag{2.6}$$

where μ_0 is the permeability of free space ($4\pi \times 10^{-7} Wb/(A \cdot m)$), N_{as} is the effective number of turns of the stator winding in phase a, r is the radius of the air gap annulus, l is the effective length of the machine (i.e. the effective length of the pole area) and g is the air-gap length respectively.

Thus, the magnetizing inductance associated to the stator winding in phase a, given a flux linkage ψ_{as}, (also referred as the self-inductance of the stator phase a) is

$$L_{mas} = \frac{\psi_{as}}{i_{as}} = \mu_0 N_{as}^2 \left(\frac{rl}{g}\right)\left(\frac{\pi}{4}\right) \tag{2.7}$$

The remaining windings of the IG are also linked by the current circulating in the stator winding in phase a, producing a mutual flux linkage.

If the three-phase circuits of the stator are assumed balanced with identical effective winding turns N_s, then

$$L_{mas} = L_{mbs} = L_{mcs} = L_{ms} \tag{2.8}$$

The remaining stator and rotor windings of the IG are also linked by the current circulating in the stator winding in phase a, producing a mutual flux linkage. For the case of the stator winding in phase b, the total flux linkages due to a current in the stator winding in phase a are defined as

$$\psi_{as-bs} = \mu_0 N_{as} N_{bs} i_{as} \left(\frac{rl}{g}\right)\left(\frac{\pi}{4}\right)\cos(\beta) \tag{2.9}$$

where N_{bs} is the number of effective turns of the stator winding in phase b, and β is the magnetic axis separation, in degrees, of the stator winding in phase b from the magnetic axis of stator winding in phase a. Thus, the magnetizing inductance associated to the stator winding in phase b, given a flux linkage ψ_{as-bs}, with $\beta = 120°$, (also referred as mutual inductance between the stator a and b phases) is

$$L_{mas-bs} = \frac{\psi_{as-bs}}{i_{as}} = -\mu_0 N_{as} N_{bs}\left(\frac{rl}{g}\right)\left(\frac{\pi}{8}\right) \tag{2.10}$$

In addition, since the magnitude of the currents and the number of turns of the windings in the stator phases are the same, and since $\cos(120°) = \cos(-120°)$, then the mutual inductances between the stator phases a, b, and c due to the currents in the stator phases a, b and c respectively, can be calculated from (2.9), as

$$L_{mas-mbs-mcs} = L_{mbs-mas-mcs} = L_{mcs-mbs-mas} = -\frac{L_{ms}}{2} \tag{2.11}$$

Given that the phase circuits in the rotor are also balanced with identical effective winding turns N_r, the self- and mutual inductances between the rotor phases can be readily calculated from (2.7) and (2.10) (if changing N_{as} by N_r), this is

$$L_{mar} = L_{mbr} = L_{mcr} = L_{mr} = \mu_0 N_r^2 \left(\frac{rl}{g}\right)\left(\frac{\pi}{4}\right) = \left(\frac{N_r}{N_s}\right)^2 L_{mr} \tag{2.12}$$

$$L_{mar-mbr-mcr} = L_{mbr-mar-mcr} = L_{mcr-mbr-mar} = -\frac{L_{mr}}{2} = -\left(\frac{N_r}{N_s}\right)^2 \frac{L_{ms}}{2} \tag{2.13}$$

Notice in (2.12) and (2.13) that the rotor's self- and mutual inductances can be written in terms of the stator mutual inductance if the turns-ratio N_r/N_s is taken into account; this is useful to reduce the number of variables in the inductance matrix.

The next step is to obtain the expression for the flux linkage between the stator winding in phase a and the rotor winding in phase a, for a current circulating in the stator winding in phase a. Making use of (2.9) and taking into account the separation θ_r (in degrees), between the magnetic axis of the stator winding in phase a and the magnetic axis of the rotor winding in phase a, then the flux linkage ψ_{as-ar}, is given by the following expression:

$$\psi_{as-ar} = \mu_0 N_s N_r i_{as} \left(\frac{rl}{g}\right)\left(\frac{\pi}{4}\right)\cos(\theta_r) \tag{2.14}$$

Thus, the magnetising inductance associated to the rotor winding in phase a given a flux linkage of ψ_{as-ar} is

$$L_{as-ar} = \mu_0 N_s N_r \left(\frac{rl}{g}\right)\left(\frac{\pi}{4}\right)\cos(\theta_r) = \left(\frac{N_r}{N_s}\right) L_{ms}\cos(\theta_r) \tag{2.15}$$

Knowing that the three-phase rotor and stator circuits are balanced, it can be proven that the mutual inductances between the stator windings in phases a, b and c with their rotor counterparts and vice-versa are

$$L_{as-ar} = L_{bs-br} = L_{cs-cr} = L_{ar-as} = L_{br-bs} = L_{cr-cs} = \left(\frac{N_r}{N_s}\right) L_{ms}\cos(\theta_r) \tag{2.16}$$

The remaining mutual inductances between the stator and rotor phases, those which are not calculated using Eq. (2.16), are a function of both the rotor position and the angle between the magnetic axis of the windings in stator and rotor phases; these inductances turn out to be:

$$L_{as-br} = L_{bs-cr} = L_{cs-ar} = L_{ar-bs} = L_{br-cs} = L_{cr-as} = \left(\frac{N_r}{N_s}\right) L_{ms}\cos(\theta_r + 120°) \tag{2.17}$$

$$L_{as-cr} = L_{bs-ar} = L_{cs-br} = L_{ar-cs} = L_{br-as} = L_{cr-bs} = \left(\frac{N_r}{N_s}\right) L_{ms}\cos(\theta_r - 120°) \tag{2.18}$$

Finally, the flux that does not link the turns of the different windings of the IG has to be taken into account. This flux instead closes around the stator slot itself and/or the air-gap and/or the ends of the machine. This flux is called the leakage flux and the inductance it produces, called leakage inductance, is only reflected as series inductance on each winding of the stator and the rotor phases. Thus, the total inductance of the stator and rotor windings in the abc phases are given by the following expression:

$$\begin{aligned} L_{as-as} &= L_{bs-bs} = L_{cs-cs} = L_{ls} + L_{ms} \\ L_{ar-ar} &= L_{br-br} = L_{cr-cr} = L_{lr} + \left(\frac{N_r}{N_s}\right)^2 L_{ms} \end{aligned} \tag{2.19}$$

where L_{ls} and L_{lr} are the stator and rotor leakage inductance produced by the leakage flux in the stator and rotor phases respectively.

Thus, using (2.7), (2.8), (2.11)–(2.13), (2.15)–(2.19) L_{IG} can be represented as shown:

$[L_{IG}] =$

$$
\begin{bmatrix}
L_{ls} + L_{ms} & -L_{ms}/2 & -L_{ms}/2 \\
-L_{ms}/2 & L_{ls} + L_{ms} & -L_{ms}/2 \\
-L_{ms}/2 & -L_{ms}/2 & L_{ls} + L_{ms} \\
\left(\dfrac{N_r}{N_s}\right) L_{ms} \cos\theta & \left(\dfrac{N_r}{N_s}\right) L_{ms} \cos\left(\theta_r - 120°\right) & \left(\dfrac{N_r}{N_s}\right) L_{ms} \cos\left(\theta_r + 120°\right) \\
\left(\dfrac{N_r}{N_s}\right) L_{ms} \cos\left(\theta_r + 120°\right) & \left(\dfrac{N_r}{N_s}\right) L_{ms} \cos\theta_r & \left(\dfrac{N_r}{N_s}\right) L_{ms} \cos\left(\theta_r - 120°\right) \\
\left(\dfrac{N_r}{N_s}\right) L_{ms} \cos\left(\theta_r - 120°\right) & \left(\dfrac{N_r}{N_s}\right) L_{ms} \cos\left(\theta_r + 120°\right) & \left(\dfrac{N_r}{N_s}\right) L_{ms} \cos\theta_r
\end{bmatrix}
$$

$$
\begin{bmatrix}
\left(\dfrac{N_r}{N_s}\right) L_{ms} \cos\theta_r & \left(\dfrac{N_r}{N_s}\right) L_{ms} \cos\left(\theta_r + 120°\right) & \left(\dfrac{N_r}{N_s}\right) L_{ms} \cos\left(\theta_r - 120°\right) \\
\left(\dfrac{N_r}{N_s}\right) L_{ms} \cos\left(\theta_r + 120°\right) & \left(\dfrac{N_r}{N_s}\right) L_{ms} \cos\theta & \left(\dfrac{N_r}{N_s}\right) L_{ms} \cos\left(\theta_r + 120°\right) \\
\left(\dfrac{N_r}{N_s}\right) L_{ms} \cos\left(\theta_r + 120°\right) & \left(\dfrac{N_r}{N_s}\right) L_{ms} \cos\left(\theta_r - 120°\right) & \left(\dfrac{N_r}{N_s}\right) L_{ms} \cos\theta \\
L_{lr} + \left(\dfrac{N_r}{N_s}\right)^2 L_{ms} & -\left(\dfrac{N_r}{N_s}\right)^2 L_{ms}/2 & -\left(\dfrac{N_r}{N_s}\right)^2 L_{ms}/2 \\
-\left(\dfrac{N_r}{N_s}\right)^2 L_{ms}/2 & L_{lr} + \left(\dfrac{N_r}{N_s}\right)^2 L_{ms} & -\left(\dfrac{N_r}{N_s}\right)^2 L_{ms}/2 \\
-\left(\dfrac{N_r}{N_s}\right)^2 L_{ms}/2 & -\left(\dfrac{N_r}{N_s}\right)^2 L_{ms}/2 & L_{lr} + \left(\dfrac{N_r}{N_s}\right)^2 L_{ms}
\end{bmatrix}
$$

$$(2.20)$$

Finally, the electrical power at the terminals of the IG is

$$
[I]^T [V] = [I]^T [R] [I] + \frac{d}{dt}\left([L_{IG}] [I]\right) = [I]^T [R] [I] + \frac{\partial}{\partial \theta_r}\left([L_{IG}] [I]\right) \cdot \frac{\partial \theta_r}{\partial t}
$$

$$
+ \frac{\partial}{\partial [I]}\left([L_{IG}] [I]\right) \cdot \frac{\partial [I]}{\partial t} \tag{2.21}
$$

where $[V]$, $[I]$ and $[R]$ are vectors containing the six terms of the voltage, current and resistance of the IG, respectively.

Solving the partial derivatives in (2.21), the following equation is obtained:

$$
[I]^T [V] = [I]^T [R] [I] + \frac{d}{dt}\left(\frac{1}{2}[L_{IG}] [I] [I]^T\right) + \frac{1}{2}[I]^T \frac{d}{d\theta_r}[L_{IG}] [I] \omega_r \tag{2.22}
$$

The first term on the right-hand side of (2.22) represents the winding copper losses; the second term represents the rate-of-change of the energy stored in the magnetic field between windings. The term $1/2 \, [I]^T \, d/d\theta_r \, [L_{IG}] \, [I] \, \omega_r$ represents the rate of energy converted to mechanical work, also known as the electromechanical power. The electromechanical torque T_e is given by dividing the electromechanical power by the mechanical speed of the IG (Boldea et al., 2001), this is:

$$T_e = \frac{p_g}{2} \frac{1}{2} \, [I]^T \, \frac{d}{d\theta_r} \, [L_{IG}] \, [I] \tag{2.23}$$

where p_g is the number of pole-pairs of the IG.

2.2.2   IG in the dq0 Reference Frame

The detailed simulation of an IG in *abc* components requires solving a seventh-order model (this is if the equations describing the dynamics of the rotor speed are omitted). Moreover, the inductance matrix and its derivative need to be solved for the rotor electrical angle on each integration step. This requires a considerable computational effort and does not provide an easy framework to establish the control rules of the DFIG generator.

To reduce the computational effort and to simplify control design and analysis of the IG, a transformation to *dq0* coordinates is used (also named the Park transformation first proposed in 1929 by Robert H. Park). This transforms a three-phase time-domain signal from a stationary coordinate system, *abc*, to a rotating coordinate system of two orthogonal phases *dq*, and a zero sequence component. The *q* axis of the *dq0* coordinate system is chosen to be leading the *d* axis by 90°. The angular speed of the *dq0* coordinate system can be chosen, in case of the IG, to match the synchronous speed or the rotor speed; this allows mathematically 'seeing' any voltage, current or flux vector that is rotating at the same angular speed to that of the *dq0* frame as a constant spatial distribution. Hence, any balanced three-phase ac quantities can be transformed to two dc quantities and a zero-sequence component, simplifying the calculation of the IG control rules and its analysis. The *dq0* transformation also allows the representation of any transient event between phases (balanced or not) (Ong, 1998). Figure 2.4 shows the application of the *dq0* transformation to a three-phase function where θ_{dq} is the

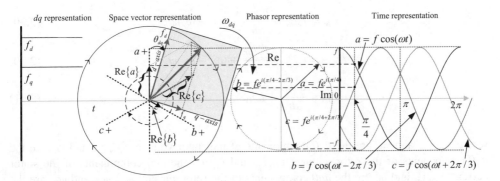

Figure 2.4   Illustration of the *dq0* transformation applied to a three-phase system.

angle between the d axis of the rotating frame and phase a of the three-phase system, ω is the angular speed of the three-phase abc system and ω_{dq} is the angular speed of the $dq0$ rotating frame.

θ_{dq} is a function of ω_{dq} and of the initial value $\theta_{dq}(0)$:

$$\theta_{dq} = \int_0^t \omega_{dq}(t)dt + \theta_{dq}(0) \tag{2.24}$$

A three-phase system is transformed to $dq0$ reference frame by applying the mathematical transformation T_{dq0} given in Eq. (2.25).

$$T_{dq0} = \frac{2}{3} \begin{bmatrix} \cos\theta_{dq} & \cos\left(\theta_{dq} - \dfrac{2\pi}{3}\right) & \cos\left(\theta_{dq} + \dfrac{2\pi}{3}\right) \\ \sin\theta_{dq} & \sin\left(\theta_{dq} - \dfrac{2\pi}{3}\right) & \sin\left(\theta_{dq} + \dfrac{2\pi}{3}\right) \\ \dfrac{1}{2} & \dfrac{1}{2} & \dfrac{1}{2} \end{bmatrix} \tag{2.25}$$

Assuming for convenience that the $dq0$ reference frame rotates at the same angular frequency of the electrical system, that is $\omega_{dq} = \omega_s$, and that $\theta_{dq}(0) = 0$, then the following equations for the voltage and flux linkages of the IG in $dq0$ coordinates can be obtained:

$$v_{ds} = r_s i_{ds} + \frac{d\psi_{ds}}{dt} - \omega_s \psi_{qs} \tag{2.26}$$

$$v_{qs} = r_s i_{qs} + \frac{d\psi_{qs}}{dt} + \omega_s \psi_{ds} \tag{2.27}$$

$$v_{0s} = \frac{d\psi_0}{dt} + r_s i_{0s} \tag{2.28}$$

$$v_{dr} = r_r i_{dr} + \frac{d\psi_{dr}}{dt} - (\omega_s - \omega_r)\psi_{qr} \tag{2.29}$$

$$v_{qr} = r_r i_{qr} + \frac{d\psi_{qr}}{dt} + (\omega_s - \omega_r)\psi_{dr} \tag{2.30}$$

$$v_{0r} = \frac{d\psi_{0r}}{dt} + r_r i_{0r} \tag{2.31}$$

where v_{ds}, v_{qs} and v_{0s} are the $dq0$ components of the stator voltage; v_{dr}, v_{qr} and v_{0r} are the $dq0$ components of the rotor voltage; i_{ds}, i_{qs} and i_{0s} are the $dq0$ components of the stator current; i_{dr}, i_{qr} and i_{0r} are the $dq0$ components of the rotor current; ψ_{ds}, ψ_{qs} and ψ_{0s} are the $dq0$ components of the stator flux; ψ_{dr}, ψ_{qr} and ψ_{0r} are the $dq0$ components of the rotor flux.

A common practice when modelling an IG is to refer all the rotor variables to the stator windings. For the *dq0* quantities of the IG this can be done as follow:

$$
v'_{dr} = \frac{N_s}{N_r}v_{dr}, \; v'_{qr} = \frac{N_s}{N_r}v_{qr}, \; v'_{0r} = \frac{N_s}{N_r}v_{0r}
$$

$$
\psi'_{dr} = \frac{N_s}{N_r}\psi_{dr}, \; \psi'_{qr} = \frac{N_s}{N_r}\psi_{qr}, \; \psi'_{0r} = \frac{N_s}{N_r}\psi_{0r}
$$

$$
i'_{dr} = \frac{N_r}{N_s}i_{dr}, \; i'_{qr} = \frac{N_r}{N_s}i_{qr}, \; i'_{0r} = \frac{N_r}{N_s}i_{0r}
$$

$$
L'_{lr} = \left(\frac{N_s}{N_r}\right)^2 L_{lr}, \; r'_r = \left(\frac{N_s}{N_r}\right)^2 r_r
$$

(2.32)

where the primed quantities denote the rotor values referred to the stator side. In this way, the equations that define the flux linkages of the IG in the *dq0* frame, after applying T_{dq0} to $[L_{IG}]$, are given as

$$
\begin{bmatrix} \psi_{ds} \\ \psi_{qs} \\ \psi_{0s} \\ \psi'_{dr} \\ \psi'_{qr} \\ \psi'_{0r} \end{bmatrix} = \begin{bmatrix} L_{ls}+L_m & 0 & 0 & L_m & 0 & 0 \\ 0 & L_{ls}+L_m & 0 & 0 & L_m & 0 \\ 0 & 0 & L_{ls}+L_m & 0 & 0 & 0 \\ L_m & 0 & 0 & L'_{lr}+L_m & 0 & 0 \\ 0 & L_m & 0 & 0 & L'_{lr}+L_m & 0 \\ 0 & 0 & 0 & 0 & 0 & L'_{lr}+L_m \end{bmatrix} \begin{bmatrix} i_{ds} \\ i_{qs} \\ i_{0s} \\ i'_{dr} \\ i'_{qr} \\ i'_{0r} \end{bmatrix}
$$

(2.33)

where L_m is the magnetising inductance defined as

$$
L_m = \frac{3}{2}L_{ms}
$$

(2.34)

It is important to notice in (2.33) that the IG inductance matrix is now independent of the rotor-position when expressed in *dq0* form. This translates into a substantial reduction in the computation time of the IG model since $[L_{IG}]$ is computed only once and not on each integration step.

The instantaneous power of the IG in *dq* frame is given below:

$$
S_{dq0} = v_{qs}i_{qs} + v_{ds}i_{ds} + v'_{dr}i'_{dr} + v'_{qr}i'_{qr} + v_{0s}i_{0s} + v'_{0r}i'_{0r}
$$

(2.35)

$$
S_{dq0} = \frac{3}{2}\left(i_{qs}\frac{d}{dt}\psi_{qs} + i_{ds}\frac{d}{dt}\psi_{ds} + i'_{qr}\frac{d}{dt}\psi'_{qr} + i'_{dr}\frac{d}{dt}\psi'_{dr} + i'_{0s}\frac{d}{dt}\psi'_{0s} + i'_{0r}\frac{d}{dt}\psi'_{0r}\right)
$$
$$
+ \frac{3}{2}\left(i_{ds}^2 + i_{qs}^2 + i_{0s}^2\right)r_s + \frac{3}{2}\left(i'^2_{dr} + i'^2_{qr} + i'^2_{0r}\right)r'_r + \frac{3}{2}\left(i'_{dr}\psi'_{qr} - i'_{qr}\psi'_{dr}\right)\omega_r
$$
$$
+ \frac{3}{2}\left(-i'_{dr}\psi'_{qr} + i'_{qr}\psi'_{dr} - i_{ds}\psi_{qs} + i_{qs}\psi_{ds}\right)\omega_s
$$

(2.36)

In Eq. (2.36) the $id\psi/dt$ terms represent the rate-of-change of the energy stored in the magnetic field between windings. The $i^2 r$ terms represent the winding copper losses. The $\omega\psi i$ terms represent the electromechanical power. The electromechanical torque T_e is given by dividing the electromechanical power by the mechanical speed of the IG:

$$T_e = \left(\frac{3}{2}\right)\left(\frac{P_g}{2\omega_r}\right)\left[\left(i'_{dr}\psi'_{qr} - i'_{qr}\psi'_{dr}\right)\omega_r + \left(i'_{qr}\psi'_{dr} - i'_{dr}\psi'_{qr} + i_{qs}\psi_{ds} - i_{ds}\psi_{qs}\right)\omega_s\right] \quad (2.37)$$

From (2.33) the following relationship can be obtained:

$$i_{qs}\psi_{ds} - i_{ds}\psi_{qs} = -i'_{qr}\psi'_{dr} + i'_{dr}\psi'_{qr} \quad (2.38)$$

Using (2.38) T_e reduces to

$$T_e = \left(\frac{3}{2}\right)\left(\frac{P_g}{2}\right)\left(i'_{dr}\psi'_{qr} - i'_{qr}\psi'_{dr}\right)$$

$$= \left(\frac{3}{2}\right)\left(\frac{P_g}{2}\right)\left(i_{qs}\psi_{ds} - i_{ds}\psi_{qs}\right) \quad (2.39)$$

$$= \left(\frac{3}{2}\right)\left(\frac{P_g}{2}\right) L_m\left(i'_{dr}i_{qs} - i'_{qr}i_{ds}\right)$$

The derivation of the zero sequence components of the stator and rotor voltages, currents and fluxes was included in order to show the complete application of the $dq0$ transformation to a three-phase system. However, given the type of connection of a DFIG wind turbine (i.e. three-wire delta or wye connection between supply and stator windings and three-wire wye connection between rotor phases), the zero sequence currents are zero by physical constraints – regardless of the phases being balanced or not (Novotny and Lipo, 1996; Ong, 1998). Thus, the sixth order $dq0$ model of the IG can be reduced to a simpler fourth order without losing information. The reduced fourth-order model of the IG used in this book for control design and analysis purposes is given by the following set of equations:

$$v_{ds} = r_s i_{ds} + \frac{d\psi_{ds}}{dt} - \omega_s \psi_{qs} \quad (2.40)$$

$$v_{qs} = r_s i_{qs} + \frac{d\psi_{qs}}{dt} + \omega_s \psi_{ds} \quad (2.41)$$

$$v'_{dr} = r'_r i'_{dr} + \frac{d\psi'_{qr}}{dt} - (\omega_s - \omega_r)\psi'_{qr} \quad (2.42)$$

$$v'_{qr} = r'_r i'_{qr} + \frac{d\psi'_{qr}}{dt} + (\omega_s - \omega_r)\psi'_{dr} \quad (2.43)$$

$$\psi_{qs} = L_{ls} i_{qs} + L_m\left(i_{qs} + i'_{qr}\right) \quad (2.44)$$

$$\psi_{ds} = L_{ls} i_{ds} + L_m\left(i_{ds} + i'_{dr}\right) \quad (2.45)$$

$$\psi'_{qr} = L'_{lr}i'_{qr} + L_m \left(i_{qs} + i'_{qr} \right) \tag{2.46}$$

$$\psi'_{dr} = L'_{lr}i'_{dr} + L_m \left(i_{ds} + i'_{dr} \right) \tag{2.47}$$

$$T_e = \left(\frac{3}{2}\right) \left(\frac{P_g}{2}\right) L_m(i'_{dr}i_{qs} - i'_{qr}i_{ds}) \tag{2.48}$$

2.2.3   Mechanical System

The mechanical system of a DFIG wind turbine is modelled in some literature as a lumped single-mass system that comprises the masses of the turbine and IG rotor producing a first-order model (which is simple to compute). However, the use of a two-mass model that considers the stiffness of the shaft connecting the IG rotor mass to the turbine mass is more accurate. This is because the transient behaviour of a two-mass system can be considerably different from that of a single-mass system if the stiffness of shaft in the two-mass model is not rigid enough (i.e. below 0.3 pu torque/rad (Akhmatov, 2003)).

In the case of the DFIG the stiffness of the shaft is considered low enough to justify the use of a two-mass model (Akhmatov, 2003; SHIMA *et al.*, 2008) as shown in Figure 2.5.

The rotor speed of the induction generator is defined as (Ackermann, 2012):

$$\frac{d\omega_r}{dt} = \frac{1}{J_g} \left(T_e - T_{shaft} \right) \tag{2.49}$$

where ω_r is the angular speed of the IG in radians per second, T_e is the electrical torque produced by the IG in N·m, T_{shaft} is the incoming torque form the shaft connecting the IG with the wind turbine rotor in N·m, and J_g is the IG mass moment of inertia in kg·m^2.

The rotor speed of the turbine ω_t is defined as

$$\frac{d\omega_t}{dt} = \frac{1}{J_t} \left(T_{mech} - T_{shaft} \right) \tag{2.50}$$

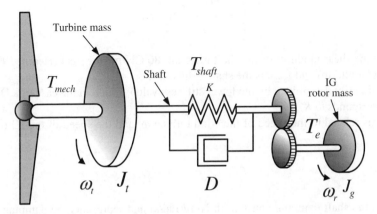

Figure 2.5   Schematic diagram of the mechanical system of a DFIG wind turbine.

where T_{mech} is the mechanical torque produced by the wind turbine in N·m, J_t is the moment of inertia of the turbine mass in kg·m^2.

The shaft torque is given by

$$T_{shaft} = T_{torsion} + T_{damping} \tag{2.51}$$

where $T_{torsion}$ represents the elasticity of the shaft, and $T_{damping}$ represents the damping torque of the shaft.

$T_{torsion}$ is expressed as a function of the angle of the turbine rotor θ_t and the angle of IG rotor θ_r (in radians).

These angles are given as

$$\frac{d\theta_r}{dt} = \omega_r \tag{2.52}$$

$$\frac{d\theta_t}{dt} = \omega_t \tag{2.53}$$

Using θ_r and θ_t the equation for $T_{torsion}$ is

$$T_{torsion} = K_{tot} \left(\theta_t - \theta_r \right) \tag{2.54}$$

where K_{tot} is the torsion constant of the shaft given in N·m/rad; that is, the effective shaft stiffness of the mechanical system. K_{tot} can be calculated as a parallel connection of the shaft stiffness of the IG K_r and turbine rotors K_t in N·m/rad:

$$K_{tot} = K_t || K_r = \left(\frac{1}{K_t} + \frac{1}{K_r} \right)^{-1} \tag{2.55}$$

The stiffness of each shaft is calculated, in turn, using the following equation:

$$K = \frac{\pi G r_{shaft}^4}{2 l_{shaft}} \text{N·m/rad} \tag{2.56}$$

where G is the shear modulus of the shaft material (80 GN/m^2 for steel (Hearn, 1997)), l_{shaft} is the shaft length (m) and r_{shaft} is the shaft radius (m).

K_{tot} is primarily determined by the lower stiffness value, which in the case of a DFIG wind turbine corresponds to K_t.

$T_{damping}$ is related to the speed of the wind turbine ω_t and the speed of the IG rotor ω_r as follows:

$$T_{damping} = D(\omega_t - \omega_r) \tag{2.57}$$

where D is the shaft damping constant in N·m/(rad/s) and represents the damping torque in both the wind turbine and the IG.

2.2.4   Crowbar Protection

When a fault occurs near the DFIG terminals a series of events follow that prompt the need for additional protection for the DFIG B2B converter (Pannell *et al.*, 2010).

- As a consequence of the sudden drop in the stator voltage, transient current peaks are induced (including transient dc components).
- Due to the magnetic coupling between stator and rotor windings, and the law of flux conservation, the transient currents are induced as rotor transient currents that add up to the steady-state currents. Consequently, the transient currents in the rotor can reach up to two or three times their nominal value. These currents are also transmitted to the terminals of the RSC.
- Because of the high RSC currents the dc voltage of the B2B converter increases and may damage the capacitor if it exceeds its design specifications. Also, the capability of the GSC to transfer energy from the dc circuit to the ac grid during the fault reduces and consequently, the extra energy charging the dc capacitor cannot be delivered to the ac network to maintain the dc voltage within limits.
- The RSC is unable to handle the transient currents coming from the rotor because its power handling capacity, S_{B2B}, is a fraction of the nominal power of the DFIG (around 0.3 pu). This makes the RSC the most fault-sensitive component of the DFIG system and is disconnected rapidly to avoid permanent damage to its components. Thus, the crowbar protection monitors i_r and the dc link voltage, v_{dc}, of the B2B. If any of these exceeds a pre-defined limit the crowbar protection is activated. The limit values are usually of <2 pu for i_r' and <1.3 pu for v_{dc} (Petersson, 2005; Pannell *et al.*, 2010).

The crowbar protection disconnects the RSC from the rotor circuits and short circuits the rotor phases using an external crowbar resistance r_{cb}. The value of r_{cb} is chosen to provide an effective damping of i_r' and T_e which could be achieved with an r_{cb} of at least ten times that of r_r'. Some typical values for r_{cb} are in the range of 0.5 to 1.5 pu (Hansen *et al.*, 2007).

The effects of shorting the rotor windings via the r_{cb} are equivalent to changing the rotor voltage Eqs. (2.29) and (2.30) to

$$0 = (r_r' + r_{cb})i_{dr}' + \frac{d\psi_{dr}'}{dt} - (\omega_s - \omega_r)\psi_{qr}' \tag{2.58}$$

$$0 = (r_r' + r_{cb})i_{qr}' + \frac{d\psi_{qr}'}{dt} + (\omega_s - \omega_r)\psi_{dr}' \tag{2.59}$$

The triggering logic for the crowbar protection controller can be explained using the schematic shown in Figure 2.6. The magnitude of the rotor currents and the magnitude of v_{dc} are compared with the thresholds i_{th} and v_{th} respectively, in case any of the two conditions becomes true, the crowbar protection trigger signal activates, and a monostable device produces a continuous crowbar activation signal for a period of time t_{cr} (Pannell *et al.*, 2010; Campos-Gaona *et al.*, 2013).

The electric circuit of the crowbar protection can be constructed using a three-phase full-bridge rectifier and a switching device such as a GTO or IGBT connected in parallel between the rotor and the RSC terminals as shown in Figure 2.7.

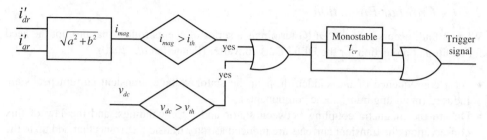

Figure 2.6 Trigger logic of the crowbar protection.

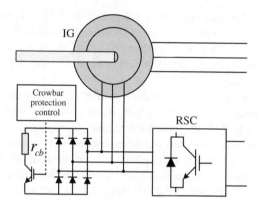

Figure 2.7 Crowbar protection connected in parallel between the rotor and the RSC terminals.

2.2.5   Modelling of the DFIG B2B Power Converter

Figure 2.8 shows a two-level B2B based on IGBTs where Q_x1, d_x1, Q_x2 and d_x2 refer to a specific IGBT and anti-parallel diode number of circuit 1 or circuit 2 respectively. The VSCs can perform four-quadrant operation (i.e. the direction of the power flow can be reversed at any time). Notice that each phase branch is composed of a set of two IGBT-diode devices connected to a dc circuit.

The switching pattern of the IGBTs (along with associated anti-parallel diodes) allows generating a specific voltage at the converter terminals and a constant current flow between

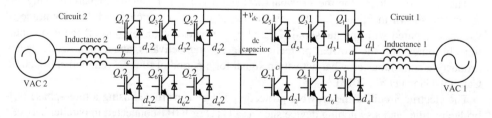

Figure 2.8 An IGBT-based B2B converter.

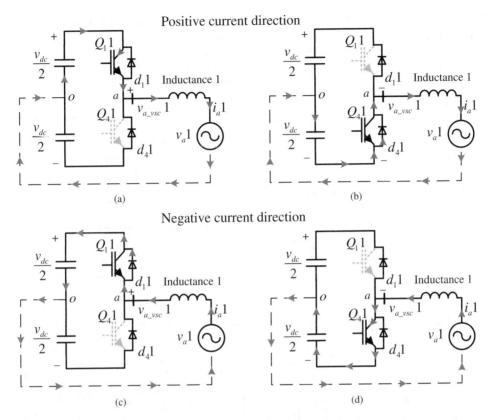

Figure 2.9 Circuit 1 *a* phase current paths.

the converter and the ac source. This is regardless of the current flowing from the converter to the ac circuit (positive direction) or flowing from the ac circuit to the converter (negative direction). Figure 2.9 is used to analyse the voltage and current behaviour of phase *a* in circuit 1 (Figure 2.8) for both positive and negative current direction. For analysis purposes it is assumed there is a common point *o* in the dc circuit that splits the dc voltage v_{dc} into two voltage sources of amplitude $v_{dc}/2$. This point is also connected to the neutral terminal of the ac voltage source. The IGBT firing pulses, and the voltage and current behaviour of this circuit are shown in Figure 2.10.

As shown in Figure 2.10 (a) and (b), the firing pulses $pwm_4 1$ for IGBT4 are inverted from the firing pulses $pwm_1 1$ for IGBT1 to avoid a short circuit between the dc source terminals; also, it can be seen that the polarity of converter voltage $v_{a\_vsc} 1$ depends on the IGBT that is currently turned on (i.e. $v_{a\_vsc} 1 = v_{dc}/2$ when $pwm_1 1 = 1$ and $v_{a\_vsc} 1 = -v_{dc}/2$ when $pwm_1 4 = 1$) but is independent of the current direction. The current $i_a 1$ is constant between the converter and the ac source. If the current $i_a 1$ is positive, it flows through $Q_1 1$ when $pwm_1 1 = 1$ (see Figure 2.9 (a) and $i_{Q_1} 1$ in Figure 2.10 (a)) or through $d_4 1$ when $pwm_4 1 = 1$ (see Figure 2.9 (b) and $i_{d4} 1$ in Figure 2.10 (a)). If the current $i_a 1$ is negative it flows through $d_1 1$ when $pwm_1 1 = 1$ (see Figure 2.9 (c) and $i_{d1} 1$ in Figure 2.10 (b)) or through $Q_4 1$ when $pwm_4 1 = 1$ (see

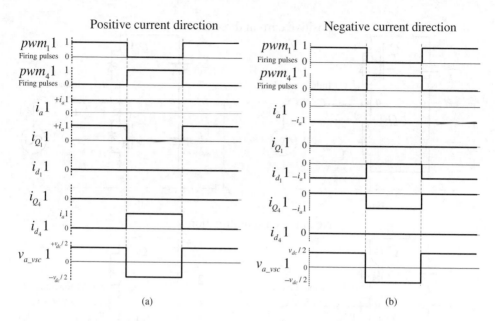

Figure 2.10   Firing pulses, voltage and current behaviour of circuit 1 a phase converter.

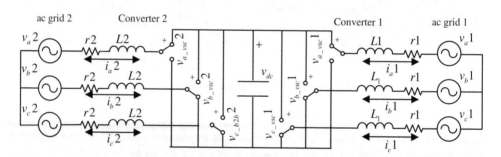

Figure 2.11   Ideal switch representation of a B2B converter.

Figure 2.9 (d) and $i_{Q_4}1$ in Figure 2.10 (b))[1]. The voltages of the converter phase b and c are produced with firing pulses shifted in phase to reproduce the phase shift of the b and c voltages of a three-phase circuit. The current of each phase follows the same flow pattern described above for i_a1.

The efficiency of the B2B converters has been proven to be around 94 to 98% (Blaabjerg *et al.*, 1995). Consequently, it is common practice to represent the IGBTs as ideal switches as shown in Figure 2.11 where the set of two IGBT-diode pairs of each phase branch is

---

[1] Here it is important to notice that IGBTs not necessarily conduct current when they are commanded to turn on, but only when the turn on command is applied and the current has a collector-to-emitter flow direction, otherwise the current flows through the anti-parallel diode of the complementary IGBT of the phase branch.

represented as a single switch which connects either to the positive or negative terminal of the dc capacitor.

The voltage equations for converter 1 in Figure 2.11 are obtained by analysing the circuit formed by the converter and the ac grid 1 (valid for any of the two converters). For a balanced three-phase system the converter voltage equations are given by

$$v_a 1 = r1i_a 1 + L1\frac{di_a 1}{dt} + v_{a\_vsc} = r1i_a 1 + L1\frac{di_a 1}{dt} + v_{dc}\frac{2s_a 1 - s_b 1 - s_c 1}{3}$$

$$v_b 1 = r1i_b 1 + L1\frac{di_b 1}{dt} + v_{b\_vsc} = r1i_b 1 + L1\frac{di_b 1}{dt} + v_{dc}\frac{-s_a 1 + 2s_b 1 - s_c 1}{3} \quad (2.60)$$

$$v_c 1 = r1i_c 1 + L1\frac{di_c 1}{dt} + v_{c\_vsc} = r1i_c 1 + L1\frac{di_c 1}{dt} + v_{dc}\frac{-s_a 1 - s_b 1 + 2s_c 1}{3}$$

where $v_a 1, v_b 1, v_c 1$ are the a, b, c voltages respectively of the ac grid 1, $i_a 1, i_b 1, i_c 1$ are the a, b, c currents respectively that circulate between converter 1 and ac grid 1, $v_{a\_vsc} 1, v_{b\_vsc} 1, v_{c\_vsc} 1$ are the a, b, c voltages respectively in the terminals of converter 1, $r1$ and $L1$ are the resistance and inductance of each phase connected between converter 1 and ac grid 1 and $s_a 1, s_b 1, s_b 1$ indicate the position of the switches in each phase of converter 1, which can be either 0 or 1. When $s_x 1 = 1$ it means that the output x of converter 1 is connected to the positive terminal of the dc capacitor.

2.2.6   Average Modelling of Power Electronic Converters

The switched model of the VSC presented in the previous section describes the steady-state and dynamic behaviour of the converter, including slow transients and high frequency components of voltage and current (due to the switching process). However, for control design and analysis purposes, knowledge of the high-frequency components is often not necessary. This is because the controller loops usually exhibit low-pass filter characteristics and are insensitive to the high-frequency components of the measured and controlled variables. Consequently, it is common practice to use the average values of the variables for control design purposes rather than the instantaneous values (Yazdani et al., 2010a). By doing this, only the fundamental ac dynamics which are important for controller design are considered, and the computation requirements are reduced.

If it is assumed that the IGBTs switch at a frequency at least 10 times the fundamental frequency of the ac grid, then the action of this commutation can be represented with the average of the duty cycle d (i.e. the percentage of time that the IGBT is on state).

To prove this, consider the PMW signal for IGBT 6 $pwm_6(t)$ in Figure 2.12, which shows the 6-pulse pattern of a three-phase pulse-width modulation strategy (for a detailed explanation of the PWM strategy see Appendix A).

The average of the waveform is given by

$$a_6 1 = \frac{1}{T_s}\int_0^{T_s} pwm_6(t)dt \quad (2.61)$$

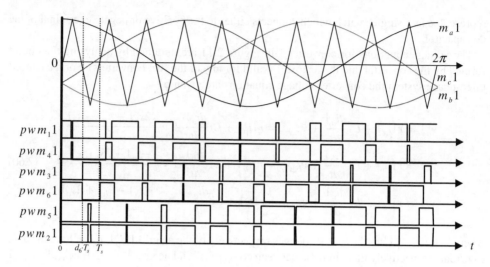

Figure 2.12 PMW signals of converter 1.

where $a_6 1$ is the average value of $pwm_6 1$ and T_s is the period of the carrier (the triangular) waveform. As pwm_6 has a value of 1 during the period $0 < t < d_6 \cdot T_s$ and a value of 0 for the period $d_6 \cdot T_s < t < T_s$ then (2.61) turns into

$$a_6 1 = \frac{1}{T_s} \left(\int_0^{d \cdot T_s} 1(t)dt + \int_{d_6 \cdot T_s}^{T_s} 0(t)dt \right) = \frac{d_6 \cdot T_s}{T_s} = d_6 \qquad (2.62)$$

As shown by (2.62) the average of $pwm_6 1$ is directly dependent on d_6. A similar analysis can be applied to any PMW signal of the converter; thus, by changing the instantaneous voltage equations in (2.60) by their average values, the following equations are obtained:

$$v_a 1 = r1 i_a 1 + L_1 \frac{di_a 1}{dt} + v_{dc} \frac{2d_a - d_b - d_c}{3}$$

$$v_b 1 = r1 i_b 1 + L_1 \frac{di_b 1}{dt} + v_{dc} \frac{-d_a + 2d_b - d_c}{3} \qquad (2.63)$$

$$v_c 1 = r1 i_c 1 + L_1 \frac{di_c 1}{dt} + v_{dc} \frac{-d_a - d_b + 2d_c}{3}$$

where d_a, d_b, d_c are the duty cycles of the IGBTs of the *abc* phases of circuit 1, respectively. Furthermore, it can be shown that the relationship between the duty cycle and modulation signal is given by

$$d = \frac{m + 1}{2} \qquad (2.64)$$

Thus, equations in (2.63) can be expressed in terms of their respective modulation signal as

$$v_a 1 = r1i_a 1 + L_1 \frac{di_a 1}{dt} + v_{dc}\left(\frac{1}{3}m_a 1 - \frac{1}{6}m_b 1 - \frac{1}{6}m_c 1\right)$$

$$v_b 1 = r1i_b 1 + L_1 \frac{di_b 1}{dt} + v_{dc}\left(-\frac{1}{6}m_a 1 + \frac{1}{3}m_b 1 - \frac{1}{6}m_c 1\right) \quad (2.65)$$

$$v_c 1 = r1i_c 1 + L_1 \frac{di_c 1}{dt} + v_{dc}\left(-\frac{1}{6}m_a 1 - \frac{1}{6}m_b 1 + \frac{1}{3}m_c 1\right)$$

and in case the modulation signals are a set of balanced three phase ac voltage commands then it can be shown that (2.65) can further reduced to

$$v_a 1 = r1i_a 1 + L_1 \frac{di_a 1}{dt} + \frac{v_{dc}}{2}m_a 1$$

$$v_b 1 = r1i_b 1 + L_1 \frac{di_b 1}{dt} + \frac{v_{dc}}{2}m_b 1 \quad (2.66)$$

$$v_c 1 = r1i_c 1 + L_1 \frac{di_c 1}{dt} + \frac{v_{dc}}{2}m_c 1$$

Equation (2.66) can also be expressed in term of dq values. If the mathematical transformation T_{dq0} of Eq. (2.25) is applied to (2.66) the following expressions for the voltage and currents for converter 1 in Figure 2.11 are obtained:

$$v_d 1 = r1i_d 1 + L_1 \frac{di_d 1}{dt} - \omega_s L_1 i_q 1 + \frac{v_{dc}}{2}m_d 1$$

$$v_q 1 = r1i_q 1 + L_1 \frac{di_q 1}{dt} + \omega_s L_1 i_d 1 + \frac{v_{dc}}{2}m_q 1 \quad (2.67)$$

Using (2.67) to express the voltages and currents of a VSC is very useful for control design purposes. This is because the active and reactive power between the VSC and the grid can be controlled independently by the d and q components of the current, respectively (see Section 2.3.5). Furthermore, by using dc quantities to represent the voltages and currents of the VSC, simple control structures based on dc command tracking references can be constructed; this prevents the use of controllers with sinusoidal command tracking references, where the compensators and closed loop bandwidth need to increase in complexity and frequency to attain a not-delayed reference signal tracking.

2.2.7   The dc Circuit

The stored energy W in the dc capacitor of the B2B converter is given by the following expression:

$$W = \frac{1}{2}Cv_{dc}^2 \quad (2.68)$$

where C is the capacitance of the circuit. W is giving as a function of the power that flows to/from the dc circuit to the ac grid via any of the two converters:

$$\frac{dW}{dt} = \frac{1}{2}C\frac{dv_{dc}^2}{dt} = -P_1 - P_2 \tag{2.69}$$

where P_1 and P_2 is the power consumed/delivered to the ac grid by converter 1 and converter 2 respectively. Thus, the voltage variation in the dc circuit is given by

$$\frac{dv_{dc}^2}{dt} = \frac{2\left(-P_1 - P_2\right)}{C} \tag{2.70}$$

2.3 Control of the DFIG WT

The key variables to control in a DFIG are the DFIG rotor speed (ω_r), the reactive power of the DFIG (Q_s), the dc voltage of the B2B converter (v_{dc}), and the reactive power of the GSC (Q_{gsc}). These variables are controlled in turn by controlling the direct and quadrature currents circulating in the circuit between the RSC and the rotor windings (i'_{dr}, i'_{qr}), and the ones circulating in the circuit between the GSC and the grid (i_{d_gsc}, i_{q_gsc}). These currents are controlled, in turn, by the direct and quadrature RSC voltages (v'_{dr}, v'_{qr}) and the direct and quadrature GSC voltages (v_{d_gsc}, v_{q_gsc}), respectively. Figure 2.13 shows a block diagram of the control loops for a DFIG wind turbine.

2.3.1 PI Control of Rotor Speed

As seen from Eq. (2.49), the electrical control of the DFIG rotor speed requires a temporal modification of the T_e generated by the IG. Such a modification can be easily achieved if a $dq0$-based control of the DFIG rotor currents is performed since T_e depends almost entirely upon a single current variable. This variable can be controlled independently by the RSC.

Using a reference frame rotating at the synchronous speed to represent the stator voltages in $dq0$ components, and aligning the d component to the grid voltage (at $t = 0$ s, the d axis matches with the rotor phase a) and the q-axis is assumed to be 90° ahead of the d-axis with respect to the rotational direction, as shown in Figure 2.14. Therefore, the following holds true:

$$v_{ds} = v_s \tag{2.71}$$

$$v_{qs} = 0 \tag{2.72}$$

where v_s is the magnitude of the stator voltage.

The stator flux, which lags the stator voltage by 90°, is given in dq components as follow:

$$\psi_{ds} = 0 \tag{2.73}$$

$$\psi_{qs} = -\psi_s \tag{2.74}$$

where λ_s is the magnitude of the stator flux.

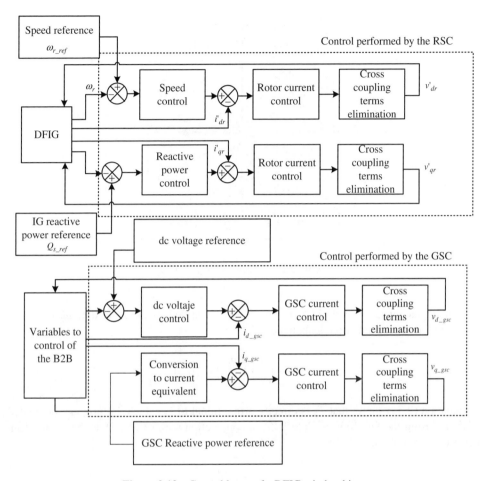

Figure 2.13   Control loops of a DFIG wind turbine.

From Eq. (2.73) it can be seen that

$$\frac{d\psi_{ds}}{dt} = 0 \tag{2.75}$$

Also, since the stator flux is sustained by the stator voltage, which, in steady state, is considered to be constant in amplitude, frequency and phase, it can be assumed that the stator flux varies very little in steady state (Boldea, 2006). In that case:

$$\psi_{qs} = const \rightarrow \frac{d\psi_{qs}}{dt} \approx 0 \tag{2.76}$$

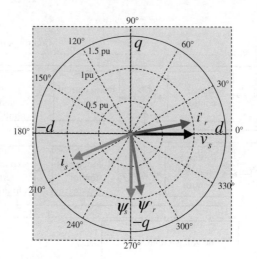

Figure 2.14 Space vectors of an IG.

Now, replacing (2.73) and (2.74) in Eq. (2.39) which defines the T_e in the $dq0$ reference frame, the following equation is obtained:

$$T_e = -\left(\frac{3}{2}\right)\left(\frac{p_g}{2}\right)(i_{ds}\psi_{qs}) \tag{2.77}$$

Equation (2.77) shows that T_e is directly proportional to i_{ds} which, in turn, can be related to i'_{dr} by using the flux linkage Eq. (2.45) in conjunction with (2.73). Thus the relationship between i_{ds} and i'_{dr} turns to be

$$i_{ds} = \frac{-L_m i'_{dr}}{L_{ls} + L_m} \tag{2.78}$$

Finally, replacing (2.78) in (2.77) the electrical torque can be expressed as:

$$T_e = -\left(\frac{3}{2}\right)\left(\frac{p_g}{2}\right)\left(\frac{-L_m\psi_{qs}}{L_{ls} + L_m}\right)i'_{dr} \tag{2.79}$$

Having found the relationship between i'_{dr} and T_e, the PI controller for the rotor speed can be derived as follows.

The mechanical system of a DFIG, considering an aggregated lumped mass, is defined as

$$J\frac{d}{dt}\omega_r = T_e - T_{mech} - B_m\omega_r \tag{2.80}$$

where J is the moment of inertia of the IG and turbine rotor lumped masses in kg·m^2, and B_m is the damping coefficient in N·m/(rad/s).

Considering T_{mech} as a disturbance, (i.e. an external signal affecting the transfer function that cannot be controlled but only compensated for by the controller) not present during the modelling process, then the transfer function $P_\omega(s)$, in terms of the Laplace operator s, from T_e to ω_r is given by

$$P_\omega(s) = \frac{\omega_r(s)}{T_e(s)} = -\frac{1}{Js + B_m} \tag{2.81}$$

from (2.81), it can be seen that the open-loop mechanical system has a stable pole at $-B_m/J$. This pole can be cancelled with the zero provided by the PI rotor speed controller defined as

$$K_\omega(s) = \frac{Kp_\omega s + Ki_\omega}{s} = \frac{Kp_\omega}{s}\left(s + \frac{Ki_\omega}{Kp_\omega}\right) \tag{2.82}$$

where Kp_ω and Ki_ω are the proportional and integral gains of the rotor speed controller. Thus choosing $Ki_\omega/Kp_\omega = B_m/J$ and $Kp_\omega/J = 1/\tau_\omega$ where τ_ω is the time constant of the closed-loop system, the open-loop controller gain is

$$\ell_\omega(s) = K_\omega(s)P_\omega(s) = \left(\frac{Kp_\omega}{Js}\right)\frac{\left(s + \dfrac{Ki_\omega}{Kp_\omega}\right)}{\left(s + \dfrac{B_m}{J}\right)} = \frac{1}{\tau_\omega s} \tag{2.83}$$

and a closed-loop transfer function

$$B_\omega(s) = \frac{\omega_{r\_}ref(s)}{\omega_r} = \frac{K_\omega(s)P_\omega(s)}{1 + K_\omega(s)P_\omega(s)} = \frac{\dfrac{1}{\tau_\omega s}}{1 + \dfrac{1}{\tau_\omega s}} = \frac{1}{\tau_\omega s + 1} \tag{2.84}$$

which is a first-order transfer function with unity gain. The selection of τ_ω is chosen according to the desired speed of response of the closed loop system.

2.3.2 PI Control of DFIG Reactive Power

The reactive power equation of the DFIG is given, in $dq0$, by:

$$Q_s = \frac{3}{2}\left(v_{ds}i_{qs} - v_{qs}i_{ds}\right) \tag{2.85}$$

Substituting (2.71) and (2.72) in (2.85) the Q_s expression reduces to

$$Q_s = \frac{3}{2}v_{ds}i_{qs} \tag{2.86}$$

Equation (2.86) shows Q_s is directly proportional to i_{qs} which, in turn, can be related to i'_{qr} by using the flux linkage Eq. (2.44) together with (2.74). Thus the relationship between i_{qs} and i'_{qr} turns to be

$$i_{qs} = \frac{\psi_{qs} - L_m i'_{qr}}{L_{ls} + L_m}$$

(2.87)

Finally, replacing (2.87) in (2.86), Q_S can be expressed as:

$$Q_s = \frac{3}{2}\left[v_{ds}\left(\frac{\psi_{qs} - L_m i'_{qr}}{L_{ls} + L_m}\right)\right]$$

(2.88)

Now that a relationship between i'_{qr} and Q_s has been established, the controller for the reactive power can be derived.

The relationship from Q_s to i'_{qr} is given by

$$i'_{qr} = \frac{2}{3}Q_s\left(\frac{L_{ls} + L_m}{v_{ds}\psi_{qs} - L_m v_{ds}}\right) \rightarrow \frac{i'_{qr}}{Q_s} = P_{Q_s} = \frac{2}{3}\left(\frac{L_{ls} + L_m}{v_{ds}\psi_{qs} - L_m v_{ds}}\right)$$

(2.89)

Assuming v_{ds} and ψ_{qs} to be constant and able to be measured or estimated then their value can be added to the controller of the reactive power. Thus, the relationship between i'_{qr} and Q_s in (2.89) turns to be directly proportional and is ready to be applied directly as the control function for i'_{qr} according to the desired Q_s. However, in order to compensate for errors in the estimation of ψ_{qs}, an integral term can be added to the Q_s control function. In this way, the controller for Q_s turns out to be:

$$K_Q(s) = \frac{3}{2}\frac{Ki_Q}{s}\left[\frac{v_{ds}(\psi_{qs} - L_m)}{L_{ls} + L_m}\right]$$

(2.90)

where Ki_Q is the integral gain of the reactive power controller. Thus, choosing $Ki_Q = 1/\tau_Q$ where τ_Q is the time constant of the closed-loop system, the open-loop controller gain is

$$\ell_Q(s) = K_Q(s)P_Q = \frac{3}{2}\frac{Ki_Q}{s}\left[\frac{v_{ds}(\psi_{qs} - L_m)}{L_{ls} + L_m}\right]\frac{2}{3}\left[\frac{L_{ls} + L_m}{v_{ds}(\psi_{qs} - L_m)}\right] = \frac{1}{\tau_Q s}$$

(2.91)

and a closed-loop transfer function

$$B_Q(s) = \frac{Q_{s\_}ref(s)}{Q_s} = \frac{K_Q(s)P_Q(s)}{1 + K_Q(s)P_Q(s)} = \frac{\dfrac{1}{\tau_Q s}}{1 + \dfrac{1}{\tau_Q s}} = \frac{1}{\tau_Q s + 1}$$

(2.92)

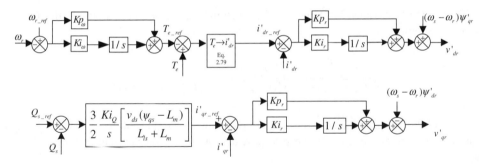

Figure 2.15 Schematic diagram of the RSC controllers.

which is a first-order transfer function with unity gain. The selection of τ_Q is chosen according to the desired speed of response of the closed loop system. It is recommended to set τ_Q 10 times larger than the time response of the i'_{qr} control loop.

2.3.3 PI Control of Rotor Currents

The voltage equations of the rotor circuit in the *dq* reference frame are given in Eqs. (2.42) and (2.43). The transfer functions to control the rotor currents can be reduced to similar expressions if the following three conditions are satisfied:

1. ψ'_{dr} and ψ'_{qr} are expressed in terms of rotor currents in *dq* components.
2. Both *dq* derivative components of the stator flux are zero, as shown in Eqs. (2.75) and (2.76).
3. The cross coupling terms that relate v'_{dr} with ψ'_{qr} and v'_{qr} with ψ'_{dr}, (i.e. $[\omega_s(t) - \omega_r(t)] \psi'_{qr}$ for (2.42) and $[\omega_s(t) - \omega_r(t)] \psi'_{dr}$ for (2.43) respectively), are assumed to be disturbances, not present during the calculation of the rotor current controllers. These disturbances are being numerically compensated for by the control scheme, as shown in Figure 2.15.

Making $L'_r = L'_{lr} + L_m$ and $L_s = L_{ls} + L_m$ for simplification, then the rotor voltage equations can be expressed as

$$v'_{dr} = r'_r i'_{dr} + \frac{d}{dt}\left[\frac{\left(L'_r i'_{dr} L_s - L_m^2 i'_{dr}\right)}{L_s}\right] \tag{2.93}$$

$$v'_{qr} = r'_r i'_{qr} + \frac{d}{dt}\left(\frac{L_m \psi_{qs}}{L_s}\right) + \frac{d}{dt}\left[\frac{\left(L'_r i'_{qr} L_s - L_m^2 i'_{qr}\right)}{L_s}\right] = r'_r i'_{qr} + \frac{d}{dt}\left[\frac{\left(L'_r i'_{qr} L_s - L_m^2 i'_{qr}\right)}{L_s}\right]$$

$$\tag{2.94}$$

Thus, using (2.93) and (2.94), the transfer function from the dq rotor currents to the dq rotor voltages is defined as

$$\frac{i'_{dr}(s)}{v'_{dr}(s)} = \frac{i'_{qr}(s)}{v'_{qr}(s)} = \frac{i'_r(s)}{v'_r(s)} = P_r(s) = \frac{L_s}{s(L'_r L_s - L^2_m) + r'_r L_s} \tag{2.95}$$

From (2.95), it can be seen that the open-loop system has a stable pole at $-r'_r L_s/(L_r L_s - L^2_m)$. This pole can be cancelled with the zero provided by the PI rotor current controller defined as

$$K_r(s) = \frac{Kp_r s + Ki_r}{s} = \frac{Kp_r}{s}\left(s + \frac{Ki_r}{Kp_r}\right) \tag{2.96}$$

where Kp_r and Ki_r are the proportional and integral gains of the rotor current controller. Thus, choosing $Ki_r/Kp_r = r'_r L_s/(L'_r L_s - L^2_m)$ and $Kp_r L_s/(L'_r L_s - L^2_m) = 1/\tau_r$ where τ_r is the time constant of the rotor current closed-loop system, the open-loop controller gain is

$$\ell_r(s) = K_r(s)P_r(s) = \left(\frac{Kp_r L_s}{(L'_r L_s - L^2_m)s}\right)\frac{\left(s + \dfrac{Ki_r}{Kp_r}\right)}{\left(s + \dfrac{r'_r L_s}{(L'_r L_s - L^2_m)}\right)} = \frac{1}{\tau_r s} \tag{2.97}$$

and the closed-loop transfer function:

$$B_r(s) = \frac{i'_r\_ref(s)}{i'_r} = \frac{K_r(s)P_r}{1 + K_r(s)P_r} = \frac{\dfrac{1}{\tau_r s}}{1 + \dfrac{1}{\tau_r s}} = \frac{1}{\tau_r s + 1} \tag{2.98}$$

which is a first-order transfer function with unity gain. The selection of τ_r is chosen according to the desired speed of response of the closed loop system. We must bear in mind that the fastest speed of response of this controller is limited by the speed of response of the RSC, which is usually between 0.5 and 2 ms.

Figure 2.15 shows a schematic diagram of the controllers of the RSC.

2.3.4   PI Control of dc Voltage

The equation describing the dynamics in the dc capacitor voltage is given in (2.70). For a DFIG wind turbine, the active power flowing in the RSC is the rotor active power P_r is defined as:

$$P_r = \frac{3}{2}\left(v'_{dr} i'_{dr} + v'_{qr} i'_{qr}\right) \tag{2.99}$$

And the GSC active power P_{gsc} is defined as:

$$P_{gsc} = \frac{3}{2}\left(v_d i_{d\_gsc} + v_q i_{q\_gsc}\right) \tag{2.100}$$

where v_d and v_q, are the dq components of the grid voltage and i_{d_gsc} and i_{q_gsc} are the d,q components of the GSC currents.

Using a reference frame rotating at the synchronous speed to represent the grid voltages in $dq0$ components, and aligning the d component to phase a of the grid voltage, then (2.100) can be reduced to

$$P_{gsc} = \frac{3}{2} v_d i_{d\_gsc} \tag{2.101}$$

which shows that P_{gsc} can be controlled by means of i_{d_gsc}, since v_d is considered constant (Figure 2.16).

If the power consumed or provided by the RSC is considered a disturbance, not present during the modelling process of the control system, then the transfer function from i_{d_gsc} to v_{dc}, according to (2.70) and (2.101) is defined as

$$\frac{v_{dc}^2(s)}{i_{d\_gsc}(s)} = P_{dc}(s) = -\frac{3v_{d\_gsc}}{Cs} \tag{2.102}$$

As seen in (2.102) $P_{dc}(s)$ has a pole in the origin, which makes it very sensitive to disturbances. This could be a problem when trying to attain a steady v_{dc} control under large disturbances (such an abrupt change of P_r due to a fault) because, even if $P_{dc}(s)$ is controlled by a fast closed-loop system, the disturbance rejection (or damping) is, to a great extent, determined by the dynamics of the plant. This is because the controller does not play any role in determining how the process reacts to a disturbance (on the other hand, when there is a set-point change, the time delay of the controller combines with the physical inertia of the process which damps the process's response to the reference change, making it much less abrupt). A fast set-point tracking controller would require particularly aggressive tuning, but that should not be a problem as long as the controller never needs to reject a disturbance. However, if an unexpected load ever does disturb the process abruptly, a set-point tracking controller will tend to overreact and cause the process variables to oscillate unnecessarily.

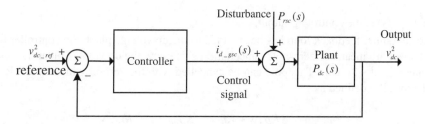

Figure 2.16 Block diagram of the control loop of v_{dc} affected by a disturbance.

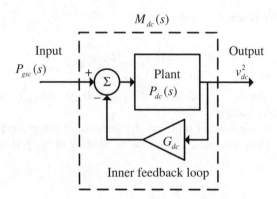

Figure 2.17 The inner feedback loop applied to improve the load disturbance rejection of the dc voltage plant.

In order to improve the disturbance rejection of $P_{dc}(s)$, an additional control loop, used to speed up the natural response of the plant, can be added to the controller design. This control loop is designed to move the pole of the plant away from the origin towards the left-hand side of the complex plane (i.e. into the negative side of the real axis). The configuration of the additional control loop is shown in Figure 2.17. By adding a feedback loop, as shown in Figure 2.17, the transfer function of the improved plant $M_{dc}(s)$ is given by

$$\frac{v_{dc}^2(s)}{P_{gsc}(s)} = M_{dc}(s) = -\frac{P_{dc}(s)}{1 + P_{dc}(s)G_{dc}} = -\frac{3v_d}{Cs + 3v_d G_{dc}} \qquad (2.103)$$

As shown in Eq. (2.103), the artificial pole added by the inner feedback loop has a value of $3v_d G_{dc}$. This value can be chosen according to the desired speed of response of the plant; the larger the value of G_{dc}, the faster the speed of response of the plant is. This translates as a better disturbance rejection (Ottersten, 2003). Some control designers make the plant dynamics as fast as the controller response time, and select G_{dc} accordingly. This procedure is illustrated in Section 2.6.4.

The inner feedback loop can be implemented by making the current control signal of the controller equal to

$$i_{d\_gsc}(s) = i'_{d\_gsc}(s) - G_{dc}v_{dc}^2 \qquad (2.104)$$

where $i'_{d\_gsc}(s)$ is the controller output.

Once the inner feedback loop has been added to the dc voltage plant, the controller tuning can be made by means of zero-pole cancellation since now the dc voltage plant has a stable artificial pole at $-3v_d G_{dc}$. This pole can be cancelled with the zero provided by the PI rotor speed controller defined as

$$K_{dc}(s) = \frac{Kp_{dc}s + Ki_{dc}}{s} = \frac{Kp_{dc}}{s}\left(s + \frac{Ki_{dc}}{Kp_{dc}}\right) \qquad (2.105)$$

where Kp_{dc} and Ki_{dc} are the proportional and integral gains of the dc voltage controller. Thus, choosing $Ki_{dc}/Kp_{dc} = 3v_dG_{dc}/C$ and $3Kp_{dc}v_d/C = 1/\tau_{dc}$ where τ_{dc} is the time constant of the dc voltage closed-loop system, the open-loop controller gain is

$$\ell_{dc}(s) = K_{dc}(s)M_{dc}(s) = \left(\frac{3Kp_{dc}v_d}{Cs}\right)\frac{\left(s + \dfrac{Ki_{dc}}{Kp_{dc}}\right)}{\left(s + \dfrac{3v_dG_{dc}}{C}\right)} = \frac{1}{\tau_{dc}s} \tag{2.106}$$

and the closed-loop transfer function

$$B_{dc}(s) = \frac{v_{dc\_}^2 ref(s)}{v_{dc}^2} = \frac{K_{dc}(s)M_{dc}(s)}{1 + K_{dc}(s)M_{dc}(s)} = \frac{\dfrac{1}{\tau_r s}}{1 + \dfrac{1}{\tau_r s}} = \frac{1}{\tau_r s + 1} \tag{2.107}$$

which is a first-order transfer function with unity gain. The selection of τ_{dc} is chosen according to the desired speed of response of the closed loop system. It is recommended to set τ_{dc} 10 times larger than the time response of the i_{d_gsc} control loop.

2.3.5 PI Control of Grid-side Converter Currents

The equations defining the behaviour of the GSC currents in dq components are

$$v_d = ri_{d\_gsc} + L\frac{d}{dt}i_{d\_gsc} - \omega_s(t)Li_{q\_gsc} + v_{d\_gsc}$$
$$v_q = ri_{q\_gsc} + L\frac{d}{dt}i_{q\_gsc} + \omega_s(t)Li_{d\_gsc} + v_{q\_gsc} \tag{2.108}$$

where r and L are the equivalent resistance and inductance between the GSC and the grid, respectively, v_{d_gsc} and v_{q_gsc} are the dq components of the voltages generated by the GSC. In addition, the reactive power consumed or provided by the GSC, Q_{gsc}, is expressed as:

$$Q_{gsc} = \frac{3}{2}\left(v_d i_{q\_gsc} - v_q i_{d\_gsc}\right) = -\frac{3}{2}v_d i_{q\_gsc} \tag{2.109}$$

The transfer functions used to control the GSC dq currents can be reduced if the cross-coupling terms ($\omega_s Li_{q\_gsc}(t)$, $\omega_s Li_{d\_gsc}(t)$), and the grid voltage components, (v_d, v_q) from (2.108) are considered as disturbances, not present during the calculation of the i_{gsc} control. These disturbances are numerically compensated for by the control scheme, as shown in Figure 2.18. Hence, the GSC current-to-voltage relationships in the dq frame are:

$$\frac{i_{d\_gsc}(s)}{v_{d\_gsc}(s)} = \frac{i_{q\_gsc}(s)}{v_{q\_gsc}(s)} = \frac{i_{gsc}(s)}{v_{gsc}(s)} = P_{gsc}(s) = \frac{1}{Ls + r} \tag{2.110}$$

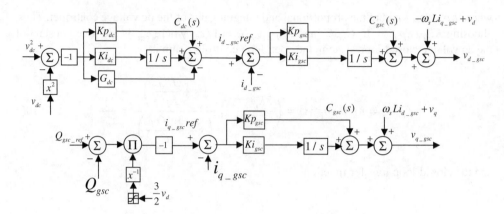

Figure 2.18   Schematic diagram of the GSC controllers.

From (2.111), it can be seen that the open-loop GSC current system has a stable pole at $-r/(L)$. This pole can be cancelled with the zero provided by the PI GSC current controller defined as

$$K_{gsc}(s) = \frac{Kp_{gsc}s + Ki_{gsc}}{s} = \frac{Kp_{gsc}}{s}\left(s + \frac{Ki_{gsc}}{Kp_{gsc}}\right) \tag{2.111}$$

where Kp_{gsc} and Ki_{gsc} are the proportional and integral gains of the rotor speed controller. Thus, choosing $Ki_{gsc}/Kp_{gsc} = r/(L)$ and $Kp_{gsc}/L = 1/\tau_{gsc}$ where τ_{gsc} is the time constant of the closed-loop system, the open-loop controller gain is

$$\ell_{gsc}(s) = K_{gsc}(s)P_{gsc}(s) = \left(\frac{Kp_{gsc}}{Ls}\right)\frac{\left(s + \frac{Ki_\omega}{Kp_\omega}\right)}{\left(s + \frac{r}{L}\right)} = \frac{1}{\tau_{gsc}s} \tag{2.112}$$

and a closed-loop transfer function

$$B_{gsc}(s) = \frac{i_{gsc\_}ref(s)}{i_{gsc}} = \frac{K_{gsc}(s)P_{gsc}(s)}{1 + K_{gsc}(s)P_{gsc}(s)} = \frac{\dfrac{1}{\tau_{gsc}s}}{1 + \dfrac{1}{\tau_{gsc}s}} = \frac{1}{\tau_{gsc}s + 1} \tag{2.113}$$

which is a first-order transfer function with unity gain. The selection of τ_{gsc} is done according to the desired speed of response of the closed-loop system. The speed of response of this controller is limited by the speed of response of the GSC, which is usually between 0.5 and 2 ms (Yazdani *et al.*, 2010b).

As seen in (2.109), the reactive power consumption of the GSC can be controlled by using $i_{q\text{-}gsc}$ in a directly proportional manner since v_d is considered constant. Consequently, there is no need to use a closed-loop controller to set the desired Q_{gsc}, instead it is only required to transform the desired Q_{gsc} to its equivalent $i_{q\text{-}gsc}$, as shown in Figure 2.18.

2.4 DFIG Dynamic Performance Assessment

2.4.1 Three-phase Fault

A DFIG needs to be connected to an excitation source to induce a magnetic field in the stator windings. If that excitation is lost, as happens in a three-phase fault, the magnetic field in the air gap of the machine drops completely. This causes a reduction in the electric torque and a consequent drop in the active power generated. However, during the transient between the steady-state and the condition described above, large currents are induced in the stator and the rotor. The magnitude of these currents depends on the nature of the fault and the value of the operating slip speed.

To derive the magnitude of the stator fault current, an analysis of a three-phase fault close to the generator is presented. The transient currents at the moment of the fault are derived from the dynamic equations that describe the IG. These equations, in terms of space vectors, and making $L'_r = L'_{lr} + L_m$ and $L_s = L_{ls} + L_m$ for simplification, are given by

$$\vec{v}_s = r'_s \vec{i}_r + \frac{d\vec{\psi}_s}{dt} + j\omega_s \vec{\psi}_s \tag{2.114}$$

$$\vec{v}'_r = r_r \vec{i}'_r + \frac{d\vec{\psi}'_r}{dt} + j\left(\omega_s - \omega_r\right)\vec{\psi}'_r \tag{2.115}$$

$$\vec{\psi}_s = L_s \vec{i}_s + L_m \vec{i}'_r \tag{2.116}$$

$$\vec{\psi}'_r = L'_r \vec{i}'_r + L_m \vec{i}_s \tag{2.117}$$

where $\vec{v}_s, \vec{v}'_r, \vec{i}_s, \vec{i}'_r, \vec{\psi}_s$ and $\vec{\psi}'_r$ are the IG stator and rotor voltage, current and flux, respectively expressed in space-vector form (all quantities seen from the stator).

Manipulating Eqs. (2.116) and (2.117) $\vec{i}_s$ and $\vec{i}'_r$ can be written in terms of fluxes as

$$\vec{i}_s = \frac{\vec{\psi}_s}{L_s - \dfrac{L_m^2}{L'_r}} - \left(\frac{L_m}{L'_r}\right)\frac{\vec{\psi}'_r}{L_s - \dfrac{L_m^2}{L'_r}} \tag{2.118}$$

$$\vec{i}'_r = \frac{\vec{\psi}'_r}{L'_r - \dfrac{L_m^2}{L_s}} - \left(\frac{L_m}{L_s}\right)\frac{\vec{\psi}_s}{L'_r - \dfrac{L_m^2}{L_s}} \tag{2.119}$$

Or $\qquad \vec{i}_s = \dfrac{\vec{\psi}_s}{L_{ts}} - \left(k_r\right)\dfrac{\vec{\psi}'_r}{L_{ts}} \tag{2.120}$

$$\vec{i}'_r = \frac{\vec{\psi}'_r}{L_{tr}} - \left(k_s\right)\frac{\vec{\psi}_s}{L_{tr}} \tag{2.121}$$

where $L_{ts} = L_s - \dfrac{L_m^2}{L_r'}$ and $L_{tr} = L_r' - \dfrac{L_m^2}{L_{ls}}$ are the transient inductances of the stator and the

rotor respectively. The stator and rotor coupling factors are $k_s = \dfrac{L_m}{L_s}$ and $k_r = \dfrac{L_m}{L_r'}$ respectively

(Erlich *et al.*, 2007).

In steady-state operation, the IG stator and rotor fluxes can be calculated using the steady-state equations of the machine. The calculation of these steady-state fluxes will help to determine the transient currents during a fault. The transient currents link the two stationary states present during a transition, which, in this case, is the pre-fault stationary state and the post fault-stationary state. To understand how to calculate the transient currents, a review of how ordinary differential equations (ODE) are solved is needed; thus, it is briefly explained here. The solution of an ODE can be divided in two parts: the particular solution and the general solution. The particular solution is obtained fixing a point where the solution of the differential equation must be valid, having a unique constant c and therefore an integral curve that satisfies the equation. In an IG this means solving Eqs. (2.114) and (2.115) for a given voltage and obtaining currents in a steady-estate condition. The general solution is obtained by adding the solution of the homogeneous differential equation to the particular solution. The homogenous equation is calculated making the independent terms and their derivatives equal to zero. The solution of the homogeneous equation gives the transient currents, which separately satisfy the homogenous differential equation; therefore, the transient currents that occur during the fault process can be investigated separately from the stationary currents.

The differential equation to solve is given by (2.114). Changing to a rotor reference frame to simplify calculations, Eq. (2.114) transforms to

$$\vec{v}_s = r_s \vec{i}_s + \frac{d\vec{\psi}_s}{dt} \tag{2.122}$$

Equation (2.122) is solved assuming that the rotor current is zero, the effects of rotor currents are latter considered. Having zero rotor current, Eq. (2.116) is now

$$\vec{\psi}_s = L_{ts} \vec{i}_s \tag{2.123}$$

$$\vec{i}_s = \frac{\vec{\psi}_s}{L_{ts}} \tag{2.124}$$

The homogeneous differential equation is obtained substituting Eq. (2.124) in Eq. (2.122) and making the independent terms equal to zero (i.e. $\vec{v}_s = 0$), which is

$$\frac{d\vec{\psi}_s}{dt} = -\frac{r_s}{L_{ts}} \vec{\psi}_s \tag{2.125}$$

The solution of (2.125) is

$$\vec{\psi}_{s,0} = \vec{\psi}_{s\_initial} e^{-t r_s / L_{ts}} \tag{2.126}$$

where t is the time. Using the procedure above, it can be shown that the homogenous solution of the rotor flux, assuming zero stator current is

$$\overline{\psi}'_{r,0} = \overline{\psi}'_{r\_initial} e^{-t r'_r/L_{tr}} \tag{2.127}$$

where $\overline{\psi}_{s\_initial}$ and $\overline{\psi}'_{r\_initial}$ are the initial value of $\overline{\psi}_s$ and $\overline{\psi}'_r$ respectively, obtained from the steady-state equation of the stator and rotor voltages. The steady-state equations of the stator and rotor voltages in term of phasors rotating at synchronous speed are given by:

$$v_{as} e^{j\omega_s t} = r_s i_{bs} e^{j\omega_s t} + j\omega_s \psi_{as} e^{j\omega_s t} \tag{2.128}$$

$$v_{bs} e^{j\omega_s t - 3\pi/2} = r_s i_{bs} e^{j\omega_s t - 3\pi/2} + j\omega_s \psi_{bs} e^{j\omega_s t - 3\pi/2} \tag{2.129}$$

$$v_{cs} e^{j\omega_s t + 3\pi/2} = r_s i_{cs} e^{j\omega_s t + 3\pi/2} + j\omega_s \psi_{cs} e^{j\omega_s t + 3\pi/2} \tag{2.130}$$

$$v'_{ar} e^{j s_\omega \omega_s t} = r'_r i'_{ar} e^{j s_\omega \omega_s t} + j s_\omega \omega_s \psi'_{ar} e^{j s_\omega \omega_s t} \tag{2.131}$$

$$v'_{br} e^{j s_\omega \omega_s t - 3\pi/2} = r'_r i'_{br} e^{j s_\omega \omega_s t - 3\pi/2} + j s_\omega \omega_s \psi'_{br} e^{j s_\omega \omega_s t - 3\pi/2} \tag{2.132}$$

$$v'_{cr} e^{j s_\omega \omega_s t + 3\pi/2} = r'_r i'_{cr} e^{j s_\omega \omega_s t + 3\pi/2} + j s_\omega \omega_s \psi'_{cr} e^{j s_\omega \omega_s t + 3\pi/2} \tag{2.133}$$

where $s_\omega = (\omega_s - \omega_r)/\omega_s$ is the slip of the IG.

The transient current for i_{sa} can be derived knowing that the following procedure is also valid for the rest of the stator phase currents.

Neglecting the stator and rotor losses (i.e. $r_s = 0$ and $r'_r = 0$), then ψ_{as} and ψ'_{ar} in steady state can be expressed, as

$$\psi_{as} e^{j\omega_s t} = \frac{v_{as} e^{j\omega_s t}}{j\omega_s} = \psi_{as\_initial} \tag{2.134}$$

$$\psi'_{ar} e^{j s_\omega \omega_s t} = \frac{v'_{ar} e^{j s_\omega \omega_s t}}{j s_\omega \omega_s} = \psi'_{ar\_initial} \tag{2.135}$$

Now that the steady-state fluxes of phase a (for the stator and rotor) have been found, these can be used to calculate the transient currents during a fault. This is because even though the current in each winding varies much more sharply during a fault, all these together still contribute to hold constant flux linkages at the moment of the short circuit (according to the theorem of constant flux linkage). Besides, as the stator and rotor are considered short circuited during the fault, the fluxes in both the windings do not change. Clearly, this flux will not hold for too long; instead, it will decay with a stator time constant.

Equation (2.135) shows that ψ'_{ar} is a function of the rotor voltage. This is true in steady state; however, when a three-phase fault occurs, ψ'_{ar} is affected by the frozen (i.e. no longer rotating) flux of the stator. At the moment the short circuit occurs in the stator, the transient i_{as} developed in the winding will act to demagnetise the rotor field. Since the flux linkage with the rotor must remain constant, a transient, i'_{ar}, will be induced on the rotor winding a trying to oppose the demagnetising effect of i_{as}. The transient, i'_{ar}, not being supported by an applied voltage in the rotor (the crowbar protection is active and the RSC is disconnected from the rotor circuit), decays at a rate determined by the rotor's time constant; meanwhile, it adds up to the steady-state rotor current. The transient i'_{ar} creates a transient ψ'_{ar} that can be calculated with Eqs. (2.116), (2.117) and (2.120). To perform the calculation of the transient, ψ'_{ar}, first it is assumed that there is no steady-state rotor current. The effects of the rotor flux created by the steady-state rotor current on the full expression of the transient stator current are introduced later; this is possible due to the principle of superposition.

Thus the ψ'_{ar} created by i_{as} only is

$$\psi'_{ar}(i_{as})e^{j\omega_s t} = i_{as}e^{j\omega_s t}L_m = \frac{L_m}{L_s}\frac{v_{as}e^{j\omega_s t}}{j\omega_s} = k_s\frac{v_{as}e^{j\omega_s t}}{j\omega_s} \tag{2.136}$$

Now, consider the ψ'_{ar} created by the rotor voltage in steady state, which can be calculated using (2.135)

$$\psi'_{ar}e^{js_\omega\omega_s t} = \frac{v'_{ar}e^{js_\omega\omega_s t}}{js_\omega\omega_s} \tag{2.137}$$

The component of ψ'_{ar} created by i_{as} adds up to the component of ψ'_{ar} created by i'_{ar}. ψ'_{ar} can be calculated at the moment of the fault using (2.136), (2.137) and then used to complete Eq. (2.121) to obtain the full expression of the transient i_{as}. Thus, using (2.120), the expression of the transient peak in the stator current in phase a at the moment of the fault $i_{as,0}$ is defined as:

$$i_{as,0} = \frac{\sqrt{2}\psi_{as,0}}{L_{ts}} - (k_r)\frac{\sqrt{2}\psi'_{ar,0}}{L_{ts}} \tag{2.138}$$

where $\psi_{as,0}$ and $\psi'_{ar,0}$ are the fluxes in phase a of the stator and rotor at the moment of the fault, defined as:

$$\psi_{as,0} = \frac{v_{as}}{j\omega_s} \tag{2.139}$$

$$\psi'_{ar,0} = k_s\frac{v_{as}e^{j\omega_s t}}{j\omega_s} + \frac{v'_{ar}e^{js_\omega\omega_s t}}{js_\omega\omega_s}e^{j(1-s_\omega)\omega_s t} \tag{2.140}$$

The first term in the right-hand side of Eq. (2.140) corresponds to the rotor flux created by the stator current, and the second one corresponds to the rotor flux created by the steady-state rotor current sustained by the voltage fed by the RSC. The RSC voltage is applied at an angular frequency of $s_\omega\omega_s$ (i.e. the angular speed of the induced rotor current in phase a). The rotor flux created by the rotor current rotates at a $(1-s_\omega)\omega_s$ speed. This allows the rotor flux created by the rotor current to be 'seen' by the stator as moving at ω_s regardless of the mechanical angular speed of the rotor shaft. This is a characteristic of induction machines.

Now applying a three-phase short circuit at time $t = 0$, then the following stator transient current can be derived using (2.126), (2.127), (2.138), (2.139) and (2.140):

$$i_{as,0} = \frac{\sqrt{2}v_{as}}{j\omega_s L_{ts}} e^{-tr_s/L_{ts}} - (k_r) \left[k_s \frac{\sqrt{2}v_{as}e^{j\omega_s t}}{j\omega_s L_{ts}} + \frac{\sqrt{2}v'_{ar}e^{js_\omega \omega_s t}}{js_\omega \omega_s L_{ts}} e^{j(1-s_\omega)\omega_s t} \right] e^{-tr'_r/L_{tr}} \quad (2.141)$$

Equation (2.142) shows the behaviour of the stator and rotor fluxes at the instant of the fault. The first term in the right-hand side is the frozen stator flux, no longer rotating, but instead creating (just briefly) a dc current, which decreases with a time constant given by

$$\frac{r_s}{L_{ts}} = (T_s)^{-1} \quad (2.142)$$

where T_s is the stator time constant. The second term is the stator current component created by the transient rotor flux; this current also decreases with a time constant given by

$$\frac{r'_r}{L_{tr}} = (T_r)^{-1} \quad (2.143)$$

where T_r is the rotor time constant.

Factoring the common terms in Eq. (2.141) and using the stator and rotor coupling factors previously defined, the stator transient current can be expressed as

$$i_{as,0} = \frac{\sqrt{2}}{jX_{ts}} \left[v_{as} e^{-t/T_s} - \left[(v_{as}k_r k_s + v'_{ar}s_\omega^{-1}k_r)e^{j\omega_s t} \right] e^{-t/T_r} \right] \quad (2.144)$$

which gives a good approximation of the fault current during a three-phase short circuit. From (2.145) it can be seen that the initial fault current magnitude is mostly determined by the magnitude of the stator transient impedance.

2.4.2 Symmetrical Voltage Dips

In the case of the voltage dip, ψ_s becomes non-stationary and the derivation of the transient i_s requires a new formula. If v_s is expressed in pre-fault and post-fault values at $t = 0$, then the stator voltage in phase a v_{as} can be described as

$$v_{as} = \begin{cases} v_{as}e^{j\omega_s t} & t < 0 \\ v_{as}(1 - u)e^{j\omega_s t} & t > 0 \end{cases} \quad (2.145)$$

where u is the magnitude of the voltage dip. Using (2.145) ψ_{as} is then described as

$$
\psi_{as} =
\begin{cases}
\dfrac{v_{as} e^{j\omega_s t}}{j\omega_s} & t < 0 \\[3mm]
\dfrac{v_{as}(1-u)e^{j\omega_s t}}{j\omega_s} & t \geq 0
\end{cases}
\tag{2.146}
$$

To determine the transient current i_{as}, the solution of the homogenous equations obtained in (2.126) and (2.127) are used again. In order to calculate $\psi_{as\_initial}$ for a given voltage dip, first consider that the stator flux must be identical immediately before and after the voltage dip. This is

$$
\psi_s\left(t_0^-\right) = \psi_s\left(t_0^+\right)
\tag{2.147}
$$

Using (2.147) and (2.148), then for ψ_{as} the following expression is obtained

$$
\psi_{as}\left(t_0^-\right) = \psi_{as}\left(t_0^+\right)
$$
$$
\frac{v_{as}}{j\omega_s} e^{j\omega_s t} = \frac{v_{as}(1-u)}{j\omega_s} e^{j\omega_s t} + \overrightarrow{\psi}_{as\_trans}
\tag{2.148}
$$

where $\overrightarrow{\psi}_{as\_trans}$ is the transient stator flux in phase a that links $\psi_{as}\left(t_0^-\right)$ with $\psi_{as}\left(t_0^+\right)$ providing the flux needed to fulfil (2.147). Thus $\overrightarrow{\psi}_{as\_trans}$ has an initial value of

$$
\psi_{as\_initial} = \frac{uv_{as}}{j\omega_s}
\tag{2.149}
$$

Therefore, the homogeneous solution of $\overrightarrow{\psi}_{as\_trans}$, according to (2.126) and (2.142) is

$$
\overrightarrow{\psi}_{as\_trans} = \frac{uv_{as}}{j\omega_s} e^{-t/T_s}
\tag{2.150}
$$

Using (2.148) and (2.150) the time behaviour of ψ_{as} during and after the fault occurrence can be described by:

$$
\psi_{as}(t) = \frac{v_{as}(1-u)e^{j\omega_s t}}{j\omega_s} + \frac{uv_{as}}{j\omega_s} e^{-\frac{t}{T_s}}
\tag{2.151}
$$

Using (2.151) the transient current i_{sa} produced by a voltage dip can be calculated using the same procedure explained before for the three-phase fault case. The i_{as} induced by ψ_{as} only can be calculated as

$$i_{as}e^{j\omega_s t} = \frac{v_{as}(1-u)e^{j\omega_s t} + uv_{as}e^{-t/T_s}}{j\omega_s L_{ts}} \qquad (2.152)$$

Thus, the rotor flux produced by the stator current only can be expressed as

$$\psi'_{ar}(i_{as})e^{j\omega_s t} = i_{as}e^{j\omega_s t}L_m = \frac{L_m}{L_{ls}}\frac{v_{as}(1-u)e^{j\omega_s t} + uv_{as}e^{-t/T_s}}{j\omega_s} = k_s\frac{v_{as}(1-u)e^{j\omega_s t} + uv_{as}e^{-t/T_s}}{j\omega_s} \qquad (2.153)$$

Substituting (2.151) in (2.120) and adding the ψ'_{ar} produced by v'_{ar} the transient i_{as} for a voltage dip can be expressed as

$$i_{as,0} = \frac{v_{as}(1-u)e^{j\omega_s t}}{j\omega_s L_{ts}} + \frac{uv_{as}}{j\omega_s L_{ts}}e^{-t/T_s} - (k_r)$$
$$\times \left[k_s\frac{\left[v_{as}(1-u)e^{j\omega_s t} + uv_{as}e^{t/T_s}\right]e^{-t/T_r}}{j\omega_s L_{ts}} + \frac{v'_{ar}}{s_\omega}e^{j\omega_s t} \right] \qquad (2.154)$$

Equation (2.154) states that when a voltage dip occurs, a dc current with magnitude $uv_{as}/j\omega_s L'_s$ (the second term in the right-hand side of the equation) will appear, and will decay with time constant of e^{-t/T_s} until reaching the new steady-state condition created by the new stator voltage (the first term of the equation). The rotor flux will add some dc current as well with a magnitude proportional to the magnitude of the stator voltage dip.

2.4.3 Asymmetrical Faults

The effect of asymmetrical faults in phase voltages can be studied by means of symmetrical components theory. According to this theory, a three-phase system can be expressed as the sum of positive, negative and zero sequence components (Anderson, 1995). Therefore, the stator voltage space vector can be expressed as

$$\vec{v}_s = \vec{v_1}e^{j\omega_s t} + \vec{v_2}e^{-j\omega_s t} + \vec{v_0} \qquad (2.155)$$

where $\vec{v_1}$, $\vec{v_2}$ and $\vec{v_0}$ are the positive, negative and zero sequence components of $\vec{v}_s$, respectively. Just as in the case of voltage dips, unbalanced faults generate a transient ψ_s (the solution of the homogeneous equation) at the start of the fault. The transient ψ_s guarantees that the total ψ_s remains the same before and just after the fault inception. Thus, (2.126) is also used to calculate the transient fluxes of unbalanced faults; however, the initial value of the flux $\vec{\psi}_{s\_initial}$

does not only depend on the type of fault (single-phase, phase-to-phase, etc.) but also on the instant at which the fault occurs.

2.4.4 Single-Phase-to-Ground Fault

A single-phase-to-ground fault causing a voltage drop in the stator phase a can be represented as (Lopez $et\ al.$, 2008):

$$v_{as} = v_s(1-u)$$
$$v_{bs} = v_s\vec{a^2} \qquad (2.156)$$
$$v_{cs} = v_s\vec{a}$$

where $\vec{a} = e^{j(2\pi/3)}$. The sequence components for the stator voltages of (2.156) are

$$\begin{bmatrix} \vec{v_1} \\ \vec{v_2} \\ \vec{v_0} \end{bmatrix} = \frac{1}{3} \begin{bmatrix} 1 & \vec{a} & \vec{a^2} \\ 1 & \vec{a^2} & \vec{a} \\ 1 & 1 & 1 \end{bmatrix} \begin{bmatrix} v_{as}(1-u) \\ v_{bs}\vec{a^2} \\ v_{cs}\vec{a} \end{bmatrix} = v_s \begin{bmatrix} 1-u/3 \\ -u/3 \\ -u/3 \end{bmatrix} \qquad (2.157)$$

Assuming symmetric impedances in the IG, then the positive sequence $\vec{v_1}$ creates a flux that rotates at synchronous speed, whereas the negative sequence $\vec{v_2}$ creates a flux that rotates in the opposite direction. The zero sequence $\vec{v_0}$ does not create any flux. Thus, the steady-state $\vec{\psi}_s$ is only composed of the fluxes created by $\vec{v_1}$ and $\vec{v_2}$, referred to as $\vec{\psi}_{s1}$ and $\vec{\psi}_{s2}$, respectively. These fluxes can be calculated, if the stator losses are neglected, as:

$$\vec{\psi}_{s1} = \frac{\vec{v_1}e^{j\omega_s t}}{j\omega_s} \qquad (2.158)$$

$$\vec{\psi}_{s2} = \frac{\vec{v_2}e^{-j\omega_s t}}{-j\omega_s} \qquad (2.159)$$

$\vec{\psi}_{s1}$ rotates anticlockwise and $\vec{\psi}_{s2}$ rotates clockwise. If the voltage dip starts at $t = 0$ the transient $\vec{\psi}_s$ is zero because $\vec{\psi}_{s1}$ and $\vec{\psi}_{s2}$ are aligned and its sum is equal to the $\vec{\psi}_s$ previous to the fault. This can be shown by solving (2.147) for $t = 0$:

$$\vec{\psi}_s(t_0^-) = \frac{\vec{v_s}e^{(0)}}{j\omega_s} = \frac{\vec{v_s}}{j\omega_s}$$
$$\vec{\psi}_s(t_0^-) = \vec{\psi}_s(t_0^+) = \vec{\psi}_{s1}(t_0^+) + \vec{\psi}_{s2}(t_0^+)$$
$$\vec{\psi}_s(t_0^+) = \frac{\vec{v_1}e^{(0)}}{j\omega_s} + \frac{\vec{v_2}e^{-(0)}}{-j\omega_s} = \frac{v_s(1-u/3)}{j\omega_s} + \frac{v_s(-u/3)}{-j\omega_s} \qquad (2.160)$$
$$\vec{\psi}_s(t_0^+) = \frac{v_s}{j\omega_s}$$

As demonstrated by (2.160) the magnitude of the $\vec{\psi}_s$ prior to the fault is the same just after the fault, meaning that no transient flux $\vec{\psi}_{s\_trans}$ was generated in order to fulfil the law of flux conservation. However, in case the fault happens at a time $t = T/4$, where T is the period of one cycle of the voltage sine waveform, then the initial value of the $\vec{\psi}_{s\_trans}$ is at its maximum since $\vec{\psi}_{s1}$ and $\vec{\psi}_{s2}$ are totally opposed at that specific instant of time. This can be proved by solving (2.147) for $t = T/4$:

$$\vec{\psi}_s(t_{1/4}^-) = \frac{\vec{v}_s e^{\left(j\frac{\pi}{2}\right)}}{j\omega_s} = \frac{\vec{v}_s}{\omega_s}$$

$$\vec{\psi}_s\left(t_{1/4}^-\right) = \vec{\psi}_s\left(t_{1/4}^+\right) = \vec{\psi}_{s1}\left(t_{1/4}^+\right) + \vec{\psi}_{s2}\left(t_{1/4}^+\right)$$

$$\vec{\psi}_s\left(t_{1/4}^+\right) = \frac{\vec{v}_1 e^{\left(j\frac{\pi}{2}\right)}}{j\omega_s} + \frac{\vec{v}_2 e^{-\left(j\frac{\pi}{2}\right)}}{-j\omega_s} = \frac{v(1 - u/3)(j)}{j\omega_s} + \frac{v(-u/3)(-j)}{-j\omega_s}$$

$$\vec{\psi}_s(t_{1/4}^+) = \frac{v_s(1 - 2/3u)}{\omega_s}$$

(2.161)

The derivation of the stator fault currents associated with the transient flux can be conducted by following the procedure presented in Eqs. (2.126)–(2.144).

2.4.5 Phase-to-Phase Fault

In the case of a phase-to-phase fault, the $\vec{\psi}_{s\_trans}$ is also a function of the instant at which the fault occurs. The voltages for a phase-to-phase fault between phases b and c can be expressed as (Lopez *et al.*, 2008):

$$v_{as} = v_s$$

$$v_{bs} = v_s \left(\vec{a}^2 + j\frac{\sqrt{3}}{2}u\right)$$

$$v_{cs} = v_s \left(\vec{a} - j\frac{\sqrt{3}}{2}u\right)$$

(2.162)

The sequence components for the stator voltages of (2.162) are

$$\begin{bmatrix} \vec{v}_1 \\ \vec{v}_2 \\ \vec{v}_0 \end{bmatrix} = \frac{1}{3} \begin{bmatrix} 1 & \vec{a} & \vec{a}^2 \\ 1 & \vec{a}^2 & \vec{a} \\ 1 & 1 & 1 \end{bmatrix} \begin{bmatrix} v_s \\ v_s\left(\vec{a}^2 + j\frac{\sqrt{3}}{2}u\right) \\ v_s\left(\vec{a} - j\frac{\sqrt{3}}{2}u\right) \end{bmatrix} = v_s \begin{bmatrix} 1 - u/2 \\ u/2 \\ u/2 \end{bmatrix}$$

(2.163)

For a phase-to-phase fault the maximum $\overrightarrow{\psi}_{s\_trans}$, contrary to the single-phase fault, happens at $t = 0$ which can be shown by solving (2.147) for $t = 0$:

$$\overrightarrow{\psi}_s\left(t_0^-\right) = \frac{\overrightarrow{v}_s e^{(0)}}{j\omega_s} = \frac{\overrightarrow{v}_s}{j\omega_s}$$

$$\overrightarrow{\psi}_s\left(t_0^-\right) = \overrightarrow{\psi}_s\left(t_0^+\right) = \overrightarrow{\psi}_{s1}\left(t_0^+\right) + \overrightarrow{\psi}_{s2}\left(t_0^+\right)$$

$$\overrightarrow{\psi}_s\left(t_0^+\right) = \frac{\overrightarrow{v}_1 e^{(0)}}{j\omega_s} + \frac{\overrightarrow{v}_2 e^{-(0)}}{-j\omega_s} = \frac{v_s\left(1 - u/2\right)}{j\omega_s} + \frac{v_s\left(u/2\right)}{-j\omega_s}$$

$$\overrightarrow{\psi}_s\left(t_0^+\right) = \frac{v_s}{j\omega_s}\left(1 - u\right)$$

(2.164)

In contrast, $\overrightarrow{\psi}_{s\_trans}$ is zero if the phase-to-phase fault happens at a time $t = T/4$, which can be shown by solving (2.147) for $t = 0$:

$$\overrightarrow{\psi}_s\left(t_{1/4}^-\right) = \frac{\overrightarrow{v}_s e^{\left(j\frac{\pi}{2}\right)}}{j\omega_s} = \frac{\overrightarrow{v}_s}{\omega_s}$$

$$\overrightarrow{\psi}_s\left(t_0^-\right) = \overrightarrow{\psi}_s\left(t_0^+\right) = \overrightarrow{\psi}_{s1}\left(t_0^+\right) + \overrightarrow{\psi}_{s2}\left(t_0^+\right)$$

$$\overrightarrow{\psi}_s\left(t_0^+\right) = \frac{\overrightarrow{v}_1 e^{\left(j\frac{\pi}{2}\right)}}{j\omega_s} + \frac{\overrightarrow{v}_2 e^{-\left(j\frac{\pi}{2}\right)}}{-j\omega_s} = \frac{v_s\left(1 - u/2\right)j}{j\omega_s} + \frac{v_s\left(u/2\right)\left(-j\right)}{-j\omega_s}$$

$$\overrightarrow{\psi}_s\left(t_0^+\right) = \frac{v_s}{\omega_s}$$

(2.165)

The derivation of the stator fault currents associated with the transient flux can be done by following the procedure presented in Eqs. (2.126)–(2.144).

2.4.6 Torque Behaviour under Symmetrical Faults

The T_e generated by the IG decreases accordingly to the magnitude of the voltage dip caused by a fault in the ac network. The relationship between voltage and T_e in steady state can be obtained expanding the steady state equation of T_e given by

$$T_e = \frac{3}{\omega_s} \frac{\left(i_r'\right)^2 r_r'}{s_\omega}$$

(2.166)

Writing i_s in terms of v_s then (2.166) transforms to

$$T_e = \frac{3}{\omega_s} \frac{r'_r}{s_\omega} \left[\frac{|v_s|^2}{\left(r'_s + \frac{r'_r}{s_\omega}\right)^2 + (X_s + X'_r)^2} \right] \qquad (2.167)$$

where X_s is the stator reactance and X'_r is the rotor reactance seen from the stator circuit. Equation (2.167) shows that the T_e generated by the IG at any rotor speed is proportional to the square of the supply voltage. This means that a reduction in v_s conveys a further reduction of T_e.

The mechanical torque T_{mech} produced can be considered constant during the fault period. Therefore, the dynamic behaviour of the rotor speed can be described by the equation of motion of the IM, which in the case of a single-mass system is defined as

$$J\frac{d}{dt}\omega_r = T_{mech} - T_e - B_m\omega_r \qquad (2.168)$$

$$\omega_r = \frac{1}{J}\int (T_{mech} - T_e - B_m\omega_r)dt + \omega_{r0} \qquad (2.169)$$

Equation (2.171) shows that if case T_e is less than T_{mech}, then ω_r increments according to a slope proportional to the difference $T_{mech} - T_e$. During such a time, ω_r increases according to the ω_r vs. T_e curve shown in Figure 2.19, also increasing T_e up to the critically stable point. If the critically stable point is not reached, the IG continues working in a new steady state, where both T_{mech} and T_e are again equal. However, if ω_r increases beyond the critically stable point the machine will run away and T_e will collapse (over-speed protection is then activated).

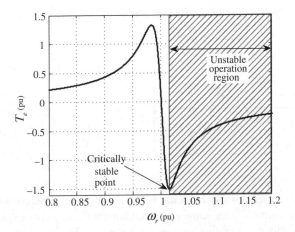

Figure 2.19   Characteristic ω_r/T_e curve for a 2MW IG.

2.4.7 Torque Behaviour under Asymmetrical Faults

When unbalanced faults occur, positive and negative sequence components of voltage, flux and current appear in the stator and rotor circuits of the IG. As mentioned in previous sections, positive sequence stator voltages produce an air-gap flux which rotates at synchronous speed anticlockwise, making the rotor speed spin in 'forward' direction with a slip speed s_ω^+. This is given by the following expression:

$$s_\omega^+ = (\omega_s - \omega_r)/\omega_s \tag{2.170}$$

For positive sequence voltage, v_1, the expression of the positive sequence torque T_e^+ is the same as that in (2.167); however, the negative sequence voltage, v_2, produces an air-gap flux which rotates synchronously in 'reverse' (clockwise) direction. In this case, the negative sequence slip speed, s_ω^-, is given as follows:

$$s_\omega^- = (-\omega_s - \omega_r)/ - \omega_s$$
$$s_\omega^- = (2 - s_\omega^+) \tag{2.171}$$

Substituting the negative sequence voltage and slip speed in Eq. (2.167) and remembering that for negative sequence voltages the synchronous speed is $(-\omega_s)$, the expression for the negative sequence torque T_e^- is given by:

$$T_e^- = \frac{3}{-\omega_s} \frac{r'_r}{2 - s_\omega^+} \left[\frac{|v_2|^2}{\left(r_s + \dfrac{r'_r}{2 - s_\omega^+}\right)^2 + \left(X_s + X'_r\right)^2} \right] \tag{2.172}$$

Thus, the total T_e production of the IG is given by the following expression:

$$T_e = T_e^+ + T_e^-$$

$$T_e = \frac{3}{\omega_s} \left[\frac{r'_r}{s_\omega} \left[\frac{|v_1|^2}{\left(r_s + \dfrac{r'_r}{s_\omega}\right)^2 + \left(X_s + X'_r\right)^2} \right] - \frac{r'_r}{2 - s_\omega^+} \left[\frac{|v_2|^2}{\left(r_s + \dfrac{r'_r}{2 - s_\omega^+}\right)^2 + \left(X_s + X'_r\right)^2} \right] \right] \tag{2.173}$$

The interaction between positive sequence flux and positive sequence rotor currents produce a positive sequence torque T_e^+; the same can be said about the negative sequence flux and rotor currents, which produce a negative sequence torque T_e^-. The interaction between positive sequence flux and negative sequence currents and vice-versa also produces torques; however, these torques are pulsating in nature. More specifically, the rotating field product of the negative sequence voltage and currents move in space at $\omega_s(1 - 2s_\omega)$. This field induces currents in the stator that, for a slip speed range of $-1 \leq s_\omega \leq 0.5$, produce a positive torque in the stator but a negative torque in the rotor; thus, for $-1 \leq s_\omega \leq 0.5$, the torque acting in the rotor is the difference between T_e^+ and T_e^-. Similarly, it can be explained that for $0.1 \leq s_\omega \leq 1$ the two torques add. The pulsating torque causes vibrations, which reduce the lifetime of the IG.

2.4.8 Effects of Faults in the Reactive Power Consumption of the IG

The RSC of the DFIG wind turbine can modify the electromagnetic flux of the IG either to provide reactive power to the IG or to demand reactive power from the IG. However, when the wind turbine is subject to a fault the RSC is blocked and disconnected from the rotor circuit, causing the DFIG to behave as a squirrel-cage IG with added rotor impedance.

The Q_s of the IG depends on the total impedance of the machine, which, in turn, depends on s_ω. This can be shown by analysing the equivalent impedance of the IG circuit which, in steady-state, is given by

$$Z_{im} = r_s + jX_{ls} + \frac{jX_m\left(\dfrac{r_r'}{s} + jX_{lr}'\right)}{\dfrac{r_r'}{s_\omega} + j(X_{lr}' + X_m)} \tag{2.174}$$

where Z_{im} is the equivalent IG impedance, X_{lr}' leakage reactance seen from the stator circuit, X_m is the mutual reactance of the IG. The IG current and its Q_s consumption are given, in phasor representation as follows:

$$i_s e^{j\omega_s t} = \frac{V_s e^{j\omega_s t}}{Z_{im}} \tag{2.175}$$

$$Q_s = 3 \operatorname{Im}\left\{ V_s e^{j\omega_s t} i_s e^{j\omega_s t} \right\} \tag{2.176}$$

As seen in (2.176), Q_s is proportional to the magnitude of the stator reactive current and the voltage level at the machine terminals.

Figure 2.20 shows the effects of a voltage drop in T_e and Q_s consumption at the terminals of an IG.

As shown in Figure 2.20, even a slight increase or decrease in ω_r means high Q_s consumption and a dangerous approximation to the critically stable point of the IG. This point does not change with the fault magnitude. Without the right control, a successful fault ride-through (FRT) is very unlikely. The use of the crowbar protection can enhance the FRT capabilities of the machine, up to some extent. The effects of the crowbar protection as well as the advantages and disadvantages for a successful FRT are presented in the following sections.

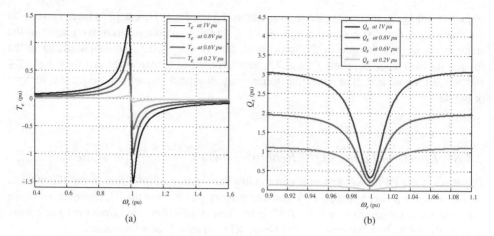

Figure 2.20 T_e and Q_s consumption for different magnitudes of voltage dips.

2.5   Fault Ride-Through Capabilities and Grid Code Compliance

2.5.1   Advantages and Disadvantages of the Crowbar Protection

The crowbar protection disconnects the RSC from the rotor circuits and short-circuits the rotor phases using an external crowbar resistance r_{cb}. The disconnection of the RSC for a period of time means the total loss of control over ω_r and the reactive power consumption of the IG Q_s, causing the IG to behave like a squirrel-cage type with a much lower speed variation range and much higher Q_s consumption. However, the use of r_{cb}, modifies the $\omega_r - T_e$ characteristic of the IG as well as the ω_r vs. reactive power consumption of the IG Q_s. Figure 2.21 (a) and (b) show the behaviour of T_e and Q_s respectively for different values of r_{cb}. It can be seen

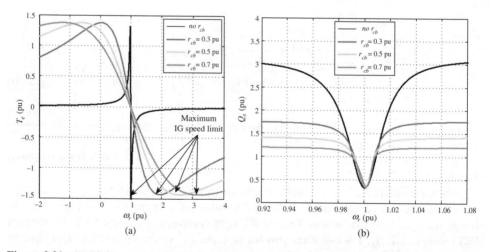

Figure 2.21   (a) IG T_e vs. ω_r curves for different values of r_{cb} (b) IG Q_s vs. ω_t curves for different values of r_{cb}.

that a larger r_{cb} means a higher maximum IG speed limit and a lower Q_s consumption of the uncontrolled IG (Hansen *et al.*, 2007). Thus, due to the use of r_{cb}, it is possible to allow a transient over-speed of the IG without reaching the maximum speed limit or consuming excessive Q_s. This allows the recovery of control over ω_r and Q_s after the crowbar protection is removed. r_{cb} also allows the rotor transient currents to decrease with a faster time constant, allowing a faster re-connection of the RSC to the rotor circuit.

When the crowbar protection is triggered, the transient currents induced in the rotor windings decrease faster with a modified time constant T_{r_cb} defined by

$$\frac{r'_r + r_{cb}}{L_{tr}} = \left(T_{r\_cb}\right)^{-1} \tag{2.177}$$

Because of this, a faster reconnection and in consequence a faster regain of control of the machine can be achieved.

The use of added rotor impedance also changes the speed-torque characteristics of the IM, changing its maximum speed equation to

$$\pm s\omega_{max} = \frac{\pm(r'_r + r_{cb})}{\left(r'_r + r_{cb}\right)^2 + \left(X_s + X'_r\right)^2} \tag{2.178}$$

where s_{max} is the maximum speed of the IG.

It can be seen in (2.178) that there is a proportional relationship between the maximum rotor speed and the magnitude of r_{cb}. Under fault conditions the rotor speed of a DFIG increases because of the sudden reduction of T_e. However, if the crowbar protection is triggered, the extra over speed limit of the IG provided by the protection avoids the machine entering an unstable operation zone. This allows the DFIG to remain connected to the faulted network during the fault period.

Also, the addition of r_{cb} to the rotor circuit decreases the amount of reactive power consumption of the IM at higher rotor speeds. This also allows the IM to remain connected to the faulted network.

The main disadvantage of using the crowbar protection concept is the unavoidable loss of control of the DFIG variables for a period of time. This includes the loss of control over Q_s. Another disadvantage of the crowbar protection has to do with its negative effects on the DFIG controllers, which require extra robustness to remain stable during the fault and the recurrent triggering of the crowbar protection (Campos-Gaona *et al.*, 2013).

2.5.2 Effects of DFIG Variables over Its Fault Ride-Through Capabilities

Even though the crowbar protection enables the IM to remain connected to a faulted network for an extended period of time, variables such as the rotor speed and the wind speed during and after the fault happening can affect the fault ride-through capabilities of the machine. This is because during the fault period the rotor speed will increase in accordance with T_{mech} (which is dependent of the prevailing wind speed) and, depending on the pre-fault rotor speed value, may surpass the maximum power handling capability of the B2B converter. This can be evidenced when reviewing the equation of the rotor power in terms of the air gap power of the IM, which is

$$P_r = s_\omega P_{ag} \tag{2.179}$$

P_{ag} is the total power across the air gap. P_r is related to T_e in per unit (pu), by

$$P_{em} = T_e \omega_r = P_{ag} - P_r \tag{2.180}$$

Substituting P_{ag} and s_ω into (2.180) P_r can be expressed as

$$P_r = T_e \left(\omega_s - \omega_r \right) \tag{2.181}$$

From (2.181), if $T_e = 1$ pu, (i.e. the wind turbine is working at full power) then P_r becomes directly proportional to ω_r. For example, a DFIG B2B converter rated a 0.3 pu would allow a $\pm$ 0.3 pu speed deviation from the synchronous speed under full T_e production. Thus, if the post-fault ω_r is beyond the maximum speed deviation allowed by the power rating of the B2B converter, the DFIG will not be able to regain control over the IM variables. Instead, the overloading of the RSC converter may cause the triggering of the crowbar protection.

In this way, the behaviour of ω_r circumscribes a wide range of variables that affect the DFIG during a fault happening, the more evident are listed below:

- The fault magnitude (by means of the magnitude of reduction of T_e and the consequent effect in ω_r).
- Fault time period (the longer the fault, the more the over speeding of ω_r).
- Fault type (because of its effect on T_e behaviour).
- The wind speed (because it affects the pre and post fault value of T_e).
- The value of J and the speed of the pitch angle controller (because their effect on the rate of change of ω_r).

As mentioned in Section 2.4.6, during the fault period ω_r accelerates accordingly to the value of $T_e - T_{mech}$. In general terms, a larger pre-fault value of ω_r means a larger value of the post fault ω_r. Thus a DFIG working at super synchronous speeds prior to the fault is less likely to ride through a fault than a DFIG working at sub synchronous speeds.

2.6 Enhanced Control Strategies to Improve DFIG Fault Ride-Through Capabilities

2.6.1 The Two Degrees of Freedom Internal Model Control (IMC)

The IMC technique relies on the 'internal model' principle, which includes a model of the plant to be controlled in the control structure (Morari and Zafiriou, 1989). Figure 2.22 shows the structure of the IMC controller in terms of the Laplace operator s.

An analysis of the IMC controller structure in Figure 2.22 shows that if the model of the plant being controlled, $P'(s) \equiv P(s)$, is an exact representation of the real plant and no disturbance

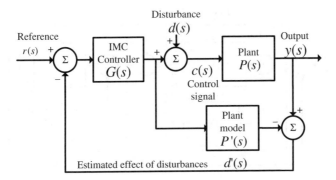

Figure 2.22   The IMC controller structure.

is present, then $d'(s) = 0$, and so the closed-loop relationship becomes equal to the open-loop one. In this condition, an IMC controller of the form $G(s) = P'^{-1}(s)$ implies the 'ideal control'. However, such ideal control cannot be implemented in the case that the model of the plant is proper (requiring the use of an improper controller), as it will require the use of pure differentiators (i.e. a real-time unfiltered differentiation of a continuous-time signal, which cannot be implemented by any physical device, such as a digital or analogue computer) in the controller structure (Harris and Tyreus, 1987). To make the control possible, the IMC structure introduces a low-pass filter $L(s)$ in cascade with $P'^{-1}(s)$, which gives $G(s) = L(s)P'^{-1}(s)$. $L(s)$ is designed to add poles to $G(s)$ and is chosen such that the closed-loop system retains its asymptotic tracking properties (i.e. zero offset at steady state for asymptotical constant inputs and step disturbances). It is usually of the following type:

$$L(s) = \left(\alpha \cdot [s + \alpha]^{-1} \right)^{n} \tag{2.182}$$

where the filter order, n, is chosen according to the order of $P(s)$, and where α can be regarded as the bandwidth of the filter, for a first-order filter. Consequently, when considering an exact representation of the plant, the controller action $A(s)$ is

$$A(s) = G(s)P(s) = L(s)\,P'^{-1}(s)\,P(s) = L(s) \tag{2.183}$$

Using this approach, the controller parameters are linked in a unique, straightforward manner to the model parameters; α is now conveniently the only parameter to be tuned to influence the speed of response of the closed-loop system. In addition, the IMC controller also has fast and accurate set-point tracking characteristics in the open-loop configuration while keeping the benefits of a feedback system (Harris and Tyreus, 1987). It is evident that the use of a filter to detune the controller imposes the trade-off of sacrificing performance to attain robustness; however, such a trade-off is inherent to any control system. Nevertheless, if $P'(s)$ is a good representation of $P(s)$, then a high speed of response can be demanded while still keeping robust stability; this topic is analysed in Section 2.6.6.

The IMC control scheme can be further improved by including an inner feedback loop to $P(s)$ as previously done in Section 2.3.4. This element provides an additional degree of freedom

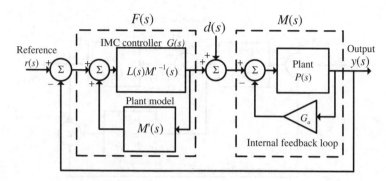

Figure 2.23 Two-degrees-of-freedom IMC controller configured as a classical controller.

(i.e. an additional control loop for disturbances, in addition to the primarily set-point tracking loop) to speed up the load disturbance rejection of the plant, which is still determined by $P(s)$ even with the use of the IMC controller. Figure 2.23 illustrates the 'two-degrees-of-freedom IMC' control scheme used to control the B2B converter. The addition of the inner feedback of gain G_a, as shown in Figure 2.23, changes $P(s)$ to:

$$M(s) = P(s) \cdot \left[1 + P(s)G_a\right]^{-1} = \left[P^{-1}(s) + G_a\right]^{-1} \qquad (2.184)$$

As seen in (2.184), the new transfer function of the plant, $M(s)$, is augmented with an inner-feedback loop gain G_a, which proves especially useful for poorly-damped systems such as the dc circuit of the DFIG.

All variables controlled by the B2B converter are represented by a first-order transfer function, implying a first-order filter for each IMC in the B2B converter. In this way $F(s)$ Figure 2.23 is:

$$F(s) = \frac{L(s)M'^{-1}(s)}{1 - L(s)M'^{-1}(s)M'(s)} = \frac{\alpha}{s}M'^{-1}(s) \qquad (2.185)$$

where $M'(s)$ is the model of $M(s)$.

The B2B converter controls all the variables with a scheme of the type given by (2.185).

The additional degree-of-freedom contributed by the inclusion of the inner feedback loop is set in each case to match the dynamics of the plant with those of the controller. This allows the load-disturbance rejection of the plant to be as fast as the controller is. Through this process, the pole created by G_a is set to match the pole of the IMC controller in the transfer function from the disturbance $d(s)$ to the output signal of the plant $y(s)$:

$$\frac{y(s)}{d(s)} = \frac{M(s)}{1 + F(s)M(s)} = \left(\frac{s}{s + \alpha}\right)\frac{1}{P^{-1}(s) + G_a} \qquad (2.186)$$

G_a can be calculated in order to reduce (2.186) to an expression of the following type:

$$\frac{y(s)}{d(s)} = \left[\left(\frac{s}{s + \alpha} \right) \frac{K}{s + \alpha} \right] = K \left[\frac{s}{(s + \alpha)^2} \right] \tag{2.187}$$

where K is a constant. As can be seen in (2.187), $d(s)$ and the control loop are damped with a similar time constant. On the other hand, α is chosen to obtain the rise time t_r needed for $y(s)$. The relationship between t_r and α for under-damped single-pole systems can be approximated by

$$\alpha \approx 0.35/t_r \text{ (Hz)} \quad \text{or} \quad \alpha \approx 2.2/t_r \text{ (rad)} \tag{2.188}$$

2.6.2 IMC Controller of the Rotor Speed

If an inner feedback loop of gain G_ω is added to the rotor speed plant, a transfer function of the following type is obtained:

$$M_\omega(s) = \frac{P_\omega(s)}{1 + P_\omega(s)G_\omega} = -\frac{1}{G_\omega + Js + B_m} \tag{2.189}$$

The inner feedback loop is implemented to the plant by means of making the input signal to the plant equal to

$$T_e(s) = T_e'(s) - \omega_r(s)G_\omega \tag{2.190}$$

where T_e' is the output of the IMC controller $F_\omega(s)$, which, according to Eq. (2.185) (and considering T_{mech} as a disturbance, not present during the calculation of the speed control), turns to be

$$F_\omega(s) = \frac{\alpha_\omega}{s} \left(-\frac{1}{G_\omega + Js + B_m} \right)^{-1} = -\alpha_\omega J - \frac{\alpha_\omega(B_m - G_\omega)}{s} \tag{2.191}$$

where α_ω is the bandwidth of the closed loop system of ω_r.

From (2.191) the controller gains are obtained as,

$$\begin{align} Kp_\omega &= -\alpha_\omega J \\ Ki_\omega &= \alpha_\omega G_\omega - \alpha_\omega B_m \end{align} \tag{2.192}$$

G_ω can be selected to match the pole of $M_\omega(s)$ with the pole of $F_\omega(s)$ on the transfer function from the disturbance $d_\omega(s)$ to the output of the plant $\omega(s)$, which is described in (2.186). Thus, selecting G_ω to be expressed as follows:

$$G_\omega = B_m - \alpha_\omega J \tag{2.193}$$

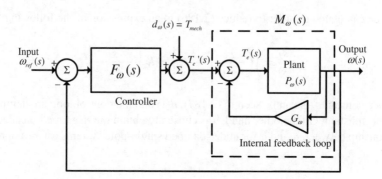

Figure 2.24   DFIG speed IMC control loop.

it can be shown that the relationship $d_\omega(s)/\omega(s)$ is given by

$$\frac{\omega(s)}{d_\omega(s)} = -\frac{s}{J\left(\alpha_\omega + s\right)^2}$$

(2.194)

As seen in (2.194), the disturbances coming from $d_\omega(s)$ are rejected by the plant and the controller with the same time constant, which is in turn dependent on α_ω. Figure 2.24 shows the IMC speed control loop of the DFIG.

2.6.3   IMC Controller of the Rotor Currents

If an inner feedback loop of gain G_r is added to the rotor current plant, a transfer function of the following type is obtained:

$$M_{ir}(s) = \frac{P_{ir}(s)}{1 + P_{ir}(s)G_r} = \frac{L_s}{r'_r L_s + s L'_r L_s - s L_m^2 + L_s G_r}$$

(2.195)

The inner feedback loop is implemented to the plant by means of making the input signal to the plant equal to

$$v'_r(s) = v''_r(s) - i'_r(s)G_r$$

(2.196)

where $v''_r(s)$ is the output of the IMC controller $F_{ir}(s)$, which, by following Eq. (2.185), turns to be

$$F_{ir}(s) = \frac{\alpha_r}{s}\left(\frac{L_s}{r'_r L_s + s L'_r L_s - s L_m^2 + L_s G_r}\right)^{-1} = \frac{\alpha_r\left(L'_r L_s - L_m^2\right)}{L_s} + \frac{\alpha_r\left(r'_r L_s + L_s G_r\right)}{s L_s}$$

(2.197)

where α_r is the bandwidth of the closed loop system of i'_r.

From (2.197) the controller gains are expressed as:

$$Kp_{ir} = \frac{\alpha_r \left(L_r' L_s - L_m^2\right)}{L_s}$$

$$Ki_{ir} = \frac{\alpha_r \left(r_r' L_s + L_s G_r\right)}{L_s} \tag{2.198}$$

G_r can be selected to match the pole of $M_{ir}(s)$ with the pole of $F_{ir}(s)$ on the transfer function from the disturbance $d_r(s)$ to the output of the plant $i_r'(s)$, which is described in (2.186). Thus, selecting G_ω to have the following value:

$$G_r = \left(-r_r' L_s + \alpha_r \left(L_r' L_s - L_m^2\right)\right) / L_s \tag{2.199}$$

It can be shown that the relationship $i_r'(s)/d_{ir}(s)$ turns to

$$\frac{i_r'(s)}{d_{ir}(s)} = -\frac{L_s s}{\left(\alpha_r + s\right)^2 \left(-L_s L_r' + L_m^2\right)} \tag{2.200}$$

As seen in (2.200) the disturbances coming from $d_\omega(s)$ are rejected by the plant and the controller with the same time constant, which is, in turn, dependent on α_r. Figure 2.25 shows the DFIG rotor current IMC control loop.

2.6.4 IMC Controller of the dc Voltage

The IMC controller $F_{dc}(s)$ for the dc voltage plant described in Eq. (2.102), which following Eq. (2.185), is given as

$$F_{dc}(s) = \frac{\alpha_{dc}}{s} M_{dc}^{-1}(s) = \frac{\alpha_{dc}}{s} \left(-\frac{3v_d}{Cs - 3v_d G_{dc}}\right)^{-1} = -\frac{\alpha_{dc} C}{3v_d} + \frac{\alpha_{dc} G_{dc}}{s} \tag{2.201}$$

where α_{dc} is the bandwidth of the closed loop system of v_{dc}.

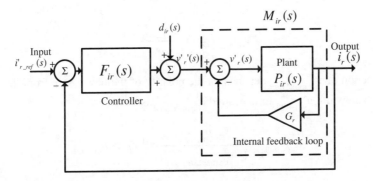

Figure 2.25 DFIG rotor current IMC control loop.

From (2.201) the controller gains are expressed by

$$Kp_{dc} = -\frac{\alpha_{dc}C}{3v_d}$$
$$Ki_{dc} = \alpha_{dc}G_{dc}$$

(2.202)

G_{dc} can be selected to match the pole of $M_{dc}(s)$ with the pole of $F_{dc}(s)$ on the transfer function from the disturbance P_r to the output of the plant v_{dc}^2, which is described in (2.186). Thus, selecting G_{dc} to have the value of

$$G_{dc} = -\frac{1}{3}\frac{\alpha_{dc}C}{v_d}$$

(2.203)

it can be shown that the relationship v_{dc}^2/P_r turns to

$$\frac{v_{dc}^2(s)}{P_r} = \frac{2s}{(Cs - 3v_dG_{dc})(s + \alpha_{dc})} = \frac{2s}{C(s + \alpha_{dc})^2}$$

(2.204)

Here α_{dc} can be calculated by attaining the minimum error for v_{dc} in case of a power surge in the circuits connected to a B2B converter. In the case of the DFIG, the v_{dc} behaviour under power surges ($v_{dc\,max\_step}^2$) can be assessed by applying a step of magnitude $|P_{r\_max}|$, (the maximum power that the RSC can deliver), to the transfer function from the disturbance P_r to the output $v_{dc}^2(s)$ of the v_{dc} plant, in the time domain (Ottersten, 2003):

$$v_{dc\,max\_step}^2(t) = L^{-1}\left\{\frac{2s|P_{r\_max}|}{C(s + \alpha_{dc})^2 s}\right\} = \frac{2|P_{r\_max}|}{C}te^{-\alpha_{dc}t}$$

(2.205)

To find the maximum error, the derivative of $v_{dc\,max\_step}^2(t)$ is computed by

$$\frac{dv_{dc\,max\_step}^2}{dt} = \frac{2P_{r\_max}}{C}(1 - \alpha_{dc}t)e^{-\alpha_w t}$$

(2.206)

The local maximum is at $t = 1/\alpha_{dc}$. Substituting such value in (2.205) the maximum error v_{dc}^2 is

$$v_{dc\,max\_step}^2 = \frac{2P_{r\_max}}{C\alpha_{dc}}e^{-1}$$

(2.207)

The direction of $v_{dc\,max\_step}^2$ depends on the polarity of the power step from the rotor. The minimum bandwidth α_{dc_min} for a desired $v_{dc\,max\_step}^2$ is

$$\alpha_{dc\_min} = \frac{\pm 2P_{r\_max}}{(\pm v_{dc\,max\_step}^2)(C)}e^{-1} \quad \alpha_{dc} \geq \alpha_{dc\_min}$$

(2.208)

Figure 2.26 shows the dc voltage IMC control loop.

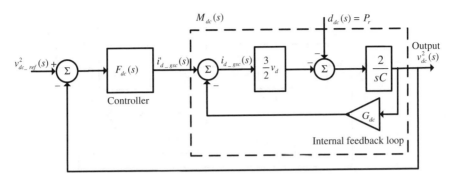

Figure 2.26   dc voltage IMC control loop.

2.6.5   IMC Controller of the Grid-Side Converter Currents

If an inner feedback loop of gain G_{gsc} is added to the GSC current plant, we have a transfer function of the following type:

$$M_{gsc}(s) = \frac{1}{G_{gsc} + Ls + r} \qquad (2.209)$$

The inner feedback loop is implemented for the plant by means of making the input signal to the plant equal to

$$v_{gsc}(s) = v'_{gsc}(s) - i_{gsc}(s)G_{gsc} \qquad (2.210)$$

where $v'_{gsc}(s)$ is the output of the IMC controller $F_{gsc}(s)$, which, according to Eq. (2.185) turns to be

$$F_{gsc}(s) = \alpha_{gsc}L + \frac{\left(r + G_{gsc}\right)\alpha_{gsc}}{s} \qquad (2.211)$$

where α_{gsc} is the bandwidth of the closed loop system of i_{gsc}.
    From (2.211) the controller gains are expressed as

$$\begin{aligned} Kp_{gsc} &= \alpha_{gsc}L \\ Ki_{gsc} &= \left(r + G_{gsc}\right)\alpha_{gsc} \end{aligned} \qquad (2.212)$$

G_{gsc} can be selected to match the pole of $M_{gsc}(s)$ with the pole of $F_{gsc}(s)$ on the transfer function from the disturbance $d_{gsc}(s)$ to the output of the plant $i_{gsc}(s)$, which is described in (2.186). Thus, selecting G_{gsc} to have the value of

$$G_{gsc} = \alpha_{gsc}L - r \qquad (2.213)$$

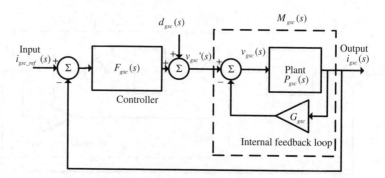

Figure 2.27   GSC currents IMC control loop.

It can be shown that the relationship $i_{gsc}(s)/d_{gsc}(s)$ is

$$\frac{i_{gsc}(s)}{d_{gsc}(s)} = \frac{s}{L\left(\alpha_{gsc} + s\right)^2}$$

(2.214)

As seen in (2.214), the disturbances coming from $d_{gsc}(s)$ are rejected by the plant and the controller with the same time constant, which in turn depends on α_{gsc}. Figure 2.27 shows the GSC currents IMC control loop.

2.6.6   DFIG IMC Controllers Tuning for Attaining Robust Control

A large bandwidth for all IMC control loops implies a lower attenuation of the reference signal, a more effective disturbance rejection and a faster response. Analysing (2.183) and (2.185), and assuming an exact representation of the plant, it can be observed that the closed-loop poles of any of the DFIG IMC control loops are in the left-half plane for any $\alpha > 0$; that is:

$$\frac{F(s)M(s)}{1 + F(s)M(s)} = \frac{\alpha/s}{1 + \alpha/s} = \frac{\alpha}{s + \alpha}$$

(2.215)

Consequently, the internal stability issue becomes trivial and the bandwidth selection is only limited by the maximum speed-of-response of the B2B converter. Yet, to attain robust stability and a good degree of performance, the selection of α must follow the requirements of the robust stability theorem but shaped by an optimal control criterion. This is especially true in the inevitable case of an internal model mismatch; in the case of a DFIG, this is more apparent in the value of L_m and to a lesser degree in L_{ls} and L_{lr}.

2.6.7   The Robust Stability Theorem

The robust stability theorem is derived from the Nyquist stability criterion and considers all the plants P in a family of plants Π. It states that for any uncertainty, l_m, in the plant's model,

P' (e.g. bounds of parameters in the linear model, bounds on nonlinearities, frequency domain bounds etc.); that is:

$$\Pi = \left\{ P(s) : \left[|P(s) - P'(s)| \, / P'(s) \right] \leq l_m(\omega) \right\} \tag{2.216}$$

with Π having the same number of right-half poles and a particular controller $G(s)$ stabilising $P'(s)$, then the system is robustly stable with the controller $G(s)$, if and only if, the complementary sensitive function η satisfies the following bound (Morari and Zafiriou, 1989, Lee *et al.*, 1995):

$$\left| \eta l_m(\omega) \right| < 1 \quad \forall \omega \tag{2.217}$$

where η, which relates the reference signal $r(s)$ to the output $y(s)$ (i.e. the performance of the controller), is defined for the IMC structure Figure 2.22 as $\eta = G(s)P'(s)$ in the case that the plant model is exact.

As explained in Section 2.6.1, an IMC controller $G(s)$ has to be detuned by $L(s)$; therefore, the bound in (2.218) can be defined for $L(s)$, substituting s by $i\omega$, as

$$|L(i\omega)| < \left[\left| G(i\omega)P'(i\omega)l_m(\omega) \right| \right]^{-1} \quad \forall \omega \tag{2.218}$$

In order to satisfy the bound in (2.218) $|L(i\omega)|$ can be designed arbitrarily small. However, such a condition may imply a very poor controller performance. Consequently, $L(i\omega)$ is shaped using a performance objective. For IMC controllers, the H_∞ performance has been proposed by some authors because the mathematical formulation of the optimisation problem allows the stability and the robust performance of the controller to be expressed in terms of the same optimisation equation. When applying the H_∞ performance objective to the IMC controller, the robust performance condition is found to be

$$\begin{aligned} &\left| G(i\omega)P(i\omega)L(i\omega)l_m(\omega) \right| + \\ &\left| [1 + G(i\omega)P(i\omega)L(i\omega)]^{-1} \cdot \varpi \right| < 1 \quad \forall \omega \end{aligned} \tag{2.219}$$

where ϖ is the normalised input to the control system (a specific input or a set of bounded inputs).

From (2.219) it can be seen that when $|L(i\omega)|$ is decreased (i.e. small α), the second term of (2.219) increases, and depending on the value of ϖ, the bound given by the H_∞ performance objective can be exceeded. Further analysis of (2.218) and (2.219) shows that a small $|l_m(\omega)|$ allows the use of a larger $|L(i\omega)|$ (i.e. higher α) without exceeding the bounds for robust stability and nominal performance. In the case $P(s) = P'(s)$, then $l_m(\omega) = 0$ and both (2.218) and (2.219) bound requirements are satisfied for any $\alpha > 0$ selection. Nevertheless, $l_m(\omega)$ always increases for any real system on large frequencies because of phase uncertainty. Therefore, whether the frequency range over control is possible will always be limited by the model's constraints.

References

Ackermann, T. (2012) *Wind Power in Power Systems*, Wiley.

Akhmatov, V. (2003) Analysis of dynamic behaviour of electric power systems with large amount of wind power. PhD thesis, Technical University of Denmark.

Anderson, P.M. (1995) *Analysis of faulted power systems*, IEEE Press.

Blaabjerg, F., Jaeger, U. and Munk-Nielsen, S. 1995. Power losses in PWM-VSI inverter using NPT or PT IGBT devices. *Power Electronics, IEEE transactions on*, **10**, 358–367.

Boldea, I. (2006) 2. *Variable Speed Generators*, Taylor & Francis.

Boldea, I. and Nasar, S.A. (2001) *The Induction Machine Handbook*, Taylor & Francis.

Campos-Gaona, D., Moreno-Goytia, E.L. and Anaya-Lara, O. (2013) Fault ride-through improvement of DFIG-WT by integrating a two-degrees-of-freedom internal model control. *Industrial Electronics, IEEE Transactions on*, **60**, 1133–1145.

Erlich, I., Wrede, H. and Feltes, C. (2007) Dynamic Behavior of DFIG-Based Wind Turbines during Grid Faults. Power Conversion Conference - Nagoya, 2007. PCC '07, 2–5 April 2007, pp. 1195–1200.

Hansen, A.D., Iov, F., Sørensen, P. *et al.* (2007) Dynamic wind turbine models in power system simulation tool DIgSILENT. Risø National Laboratory, Technical University of Denmark.

Harris, T.J. and Tyreus, B.D. (1987) Internal model control. 4. PID controller design. Comments. *Industrial & Engineering Chemistry Research*, **26**, 2161–2162.

Hearn, E.J. (1997) *Mechanics of Materials 2: The Mechanics of Elastic and Plastic Deformation of Solids and Structural Materials*, Elsevier Science.

Lee, T.H., Low, T.S., Al-Mamun, A. and Tan, C.H. (1995) Internal model control (IMC) approach for designing disk drive servo-controller. *Industrial Electronics, IEEE Transactions on*, **42**, 248–256.

Lopez, J., Gubia, E. and Sanchis, P. (2008) Wind turbines based on doubly fed induction generator under asymmetrical voltage dips. *Energy Conversion, IEEE Transactions on* **23**, 321–330.

Manwell, J.F., Mcgowan, J.G. and Rogers, A.L. (2010) *Wind Energy Explained: Theory, Design and Application*, Wiley.

Morari, M. and Zafiriou, E. (1989) *Robust Process Control*, Prentice-Hall, New Jersey.

Nelson, V.C. (2009) *Wind Energy: Renewable Energy and the Environment*, Taylor & Francis.

Novotny, D.W. and Lipo, T.A. (1996) *Vector Control and Dynamics of AC Drives*, Clarendon Press.

Ong, C.-M. (1998) 6.8 Simulation of an induction machine on the stationary reference frame, in *Dynamic Simulations of Electric Machinery: Using MATLAB/SIMULINK*, Prentice Hall PTR.

Ottersten, R. (2003) On the Control od Back-to-Back Converters and Sensorless Induction Machine Drives. PhD Thesis, Chalmers University of Technology.

Pannell, G., Atkinson, D.J. and Zahawi, B. (2010) Minimum-threshold crowbar for a fault-ride-through grid-code-compliant DFIG wind turbine. *Energy Conversion, IEEE Transactions on*, **25**, 750–759.

Paul Gardner, A.G., Lars, F.H., Peter, J. *et al.* (2009) Wind Energy-The facts Part I Tecnology. European Wind Energy Association.

Petersson, A. (2005) Analysis, Modeling and Control of Doubly-Fed Induction Generators for Wind Turbines. PhD Thesis, Chalmers University Of Technology.

Shima, Y., Takahashi, R., Murata, T. *et al.* (2008) Transient stability simulation of wind generator expressed by two-mass model. *Electrical Engineering in Japan*, Vol. 162, pp. 27–37.

Yazdani, A. and Iravani, R. (2010a) 2.5 Converter averaged model, in *Voltage-Souced Converters in Power Systems* (ed. Wiley), Wiley, New Jersey.

Yazdani, A. and Iravani, R. (2010b) 3. Control of half-bridge converter, in *Voltage-Souced Converters in Power Systems* (ed. Wiley), Wiley, New Jersey.

3

Fully-Rated Converter Wind Turbine (FRC-WT)

3.1 Synchronous Machine Fundamentals

3.1.1 Synchronous Generator Construction

A synchronous generator consists of two elements: the field and the armature. The field is located on the rotor and the armature on the stator (Figure 3.1). The armature has concentrated three-phase windings, whilst the field winding carries direct current and produces a magnetic field that rotates with the rotor and induces alternating voltages in the armature windings (Kundur, 1994; Anaya-Lara *et al.*, 2009).

There are two basic rotor structures: salient and cylindrical (Figure 3.2) (Ong, 1998). A salient-pole rotor is mostly used in low-speed generators where the diameter-to-length ratio of the rotor can be made larger to accommodate the large number of poles required to produce the rated frequency. A rotor with salient poles and concentrated windings is better suited mechanically to this situation. Salient-pole synchronous machines are often used in hydro generators to match the low operating speed of the hydraulic turbines. Such rotors often have damper windings in the form of copper or brass rods embedded in the pole face. These bars are connected to end rings to form short-circuited windings similar to those of a squirrel-cage induction machine. They are intended to damp out speed oscillations.

Steam and gas turbines operate instead at high speeds and their generators have round (or cylindrical) rotors made up of solid steel forgings. They have two or four poles, formed by distributed windings placed in slots milled in the solid rotor and held in place by steel wedges. They often do not have special damper windings, but the solid steel rotor offers paths for eddy currents, which have comparable effects to those of damper windings. Under steady-state conditions, the only rotor current that exists is the direct current in the field winding. However, under dynamic conditions eddy currents are induced on the rotor surface and slot wall, and in damper windings, to produce additional damping.

Offshore Wind Energy Generation: Control, Protection, and Integration to Electrical Systems, First Edition.
Olimpo Anaya-Lara, David Campos-Gaona, Edgar Moreno-Goytia and Grain Adam.
© 2014 John Wiley & Sons, Ltd. Published 2014 by John Wiley & Sons, Ltd.
Companion Website: www.wiley.com/go/offshore_wind_energy_generation

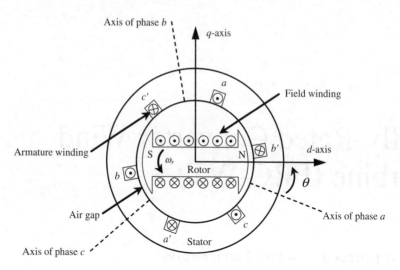

Figure 3.1 Schematic of the cross section of a three-phase synchronous machine with one pair of field poles.

3.1.2 The Air-Gap Magnetic Field of the Synchronous Generator

The concentrated stator windings of three phases a, b and c are represented by three equivalent windings a-a', b-b' and c-c' (Figure 3.1). When the rotor is driven by a prime mover, the magnetic field produced by the field winding rotates in space at synchronous speed ω_s. This magnetic field cuts the stator conductors which induces three voltages, displaced by 120° (in time), in the three windings a-a', b-b' and c-c'. If these windings are connected to three identical loads, the resulting three phase currents are also displaced by 120°, as shown in Figure 3.3. These currents will, in turn, each produce a magnetic field. The resultant magnetic

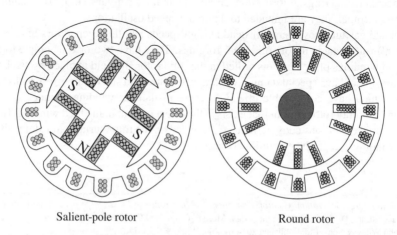

Salient-pole rotor   Round rotor

Figure 3.2 Cross-sections of salient and cylindrical four-pole synchronous generators (Ong, 1998).

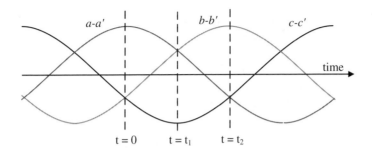

Figure 3.3   Three phase currents in the stator windings *a-a'*, *b-b'* and *c-c'*.

field in the air-gap is the combination of the stator magnetic field, ϕ_s, produced by the stator currents, and the rotor magnetic field, ϕ_r, produced by the field winding. For production of a steady torque, the magnetic fields of stator and rotor must operate at the same speed, that is at the synchronous speed given by (Kundur, 1994):

$$n = \frac{120f}{p_f}$$
(3.1)

where n is the speed in rev/min, f is the frequency in Hz, and p_f is the number of field poles. The number of field poles is determined by the mechanical speed of the rotor and electric frequency of stator currents.

As shown in Figure 3.3, when $t = 0$ the current in phase a is at its positive maximum (I_m) and the currents in phases b and c are at their negative half maxima $(-I_m/2)$. If the effective number of turns of each phase of the stator windings is N_s, then the current in phase a produces a component of the stator magnetic field, ϕ_a, where the magnitude is proportional to the number of ampere-turns, $N_s I_m$, along the axis of phase a (Hindmarsh and Renfrew, 1996). Similarly, the currents in phase b and c produce two components of the stator magnetic field, ϕ_b and ϕ_c, whose magnitudes are proportional to the number of amperes-turns, $N_s I_m/2$ along the axes of phases b and c, respectively. These three magnetic fields and the resultant stator magnetic field at $t = 0$ are shown in Figure 3.4a. At time $t = t_1$ the current in phases a and b produce two magnetic fields whose magnitudes are proportional to the number of ampere-turns, $N_s I_m/2$ along the axes of phases a and b, respectively, and the current in phase c produces a magnetic field whose magnitude is proportional to the number of ampere-turns, $N_s I_m$, along the axis of phase c. The resultant stator magnetic field at $t = t_1$ is then shifted by $\pi/3$ as shown in Figure 3.4b. Similarly, at time $t = t_2$, the stator magnetic field further shifts by $\pi/3$ as shown in Figure 3.4c.

From Figure 3.4, it is clear that in each of the two time intervals $t_1 - 0 = \pi/3\omega_s$ and $t_2 - t_1 = \pi/3\omega_s$, the stator magnetic field has rotated by $\pi/3$. In other words, the field has rotated at the synchronous speed, ω_s. The peak value of the stator magnetic field is proportional to $3N_s I_m/2$.

A component of the stator magnetic field, ϕ_s, links with the component of the rotor magnetic field, ϕ_r, at the air-gap. The resultant magnetic field in the air-gap is then given by the vector

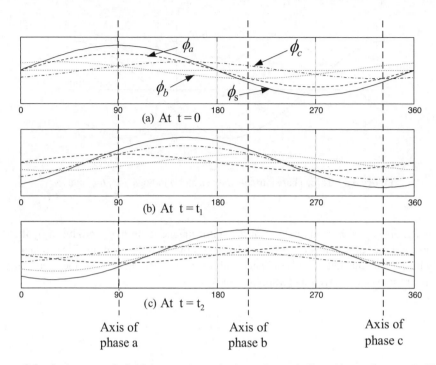

Figure 3.4 Stator magnetic field due to currents in three-phase windings (Anaya-Lara *et al.*, 2009).

sum of these magnetic fields. The components, which are not contributing to the air-gap magnetic field, are called leakage fluxes.

3.1.2.1 Coil Representation of the Synchronous Generator

Consider three coils, a, b and c, each carrying direct currents $I_m cos 0 = I_m$, $I_m cos(0 - 2\pi/3) = -I_m/2$ and $I_m cos(0 - 4\pi/3) = -I_m/2$, respectively. They rotate at a speed ω_s as shown in Figure 3.5a. The resultant magnetic field produced by three-phase windings will be proportional to $3N_s I_m/2$ in the direction of the axis of coil a. As time elapses, the magnitude of this magnetic field remains the same but rotates at synchronous speed, ω_s. Therefore, this three-coil structure fed with direct current and rotating at synchronous speed can be used as an analogue for the stator of a synchronous generator.

To define the two-phase system, two orthogonal coils are selected, one placed on the d axis, which is chosen to align with the rotor field winding position, and the other on the q axis, that leads the d axis by 90° (Figure 3.5b).

Resolving the resultant magnetic field produced by the three-phase windings, ϕ_s aligned with phase a, into the direction of d and q, Eqs. (3.2) and (3.3) are obtained:

$$\phi_d = \phi_s \cos \theta \tag{3.2}$$

$$\phi_q = -\phi_s \sin \theta \tag{3.3}$$

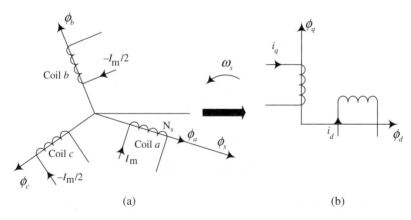

Figure 3.5 Three-phase to two-phase transformation: (a) three-coil representation; (b) two-coil representation. The three coils are placed such that each coils makes 0 rad with the corresponding axes at $t = 0$, that is, coil a is placed on the axis of phase a, coil b on the axis of phase b and coil c on the axis of phase c.

If the number of turns in the two-phase windings is N', the corresponding current relationships can be derived from Eqs. (3.2) and (3.3) as

$$N'i_d = \frac{3}{2}N_s I_m \cos \theta \qquad (3.4)$$

$$N'i_q = -\frac{3}{2}N_s I_m \sin \theta \qquad (3.5)$$

Various authors select the ratio N_s/N' either as $2/3$ or as $\sqrt{2/3}$ (Fitzgerald *et al.*, 1992; Kundur, 1994; Krause *et al.*, 2002). When N_s/N' is selected as $\sqrt{2/3}$ the power calculated in the *dq* coordinate system is the same as that in the *abc* system and therefore called the *power-invariant dq* transformation. If $N_s/N' = 2/3$ is used instead, then the *dq* transformation is said to be *amplitude-invariant*. In this chapter N_s/N' is selected as $\sqrt{2/3}$.

For a viewer on a platform which is rotating at synchronous speed (the synchronous rotating reference frame), the fluxes in the synchronous generator can be described by three stationary coils, two representing the stator field and one representing the rotor field (Figure 3.6). The stator coils d and q carry direct currents of $\sqrt{3/2}I_m \cos \theta$ and $-\sqrt{3/2}I_m \sin \theta$ respectively and the rotor coil carries the dc field current.

3.1.2.2 Mutually Coupled Stationary Coils

Consider two mutually coupled stationary coils as shown in Figure 3.7. The flux associated with coil 1 may be expressed as (Krause *et al.*, 2002)

$$\phi_1 = \phi_{l1} + \phi_{m1} + \phi_{m2} \qquad (3.6)$$

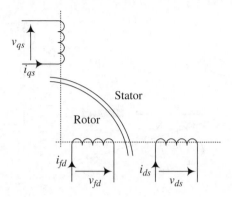

Figure 3.6 Two-coil representation of the synchronous generator.

where ϕ_{l1} is the leakage flux due to coil 1, ϕ_{m1} is the flux between coils 1 and 2 due to the current in coil 1 and ϕ_{m2} is the flux between coils 1 and 2 due to the current in coil 2.

The voltage equation for coil 1 can be expressed as

$$v_1 = r_1 i_1 + N_1 \frac{d\phi_1}{dt} = r_1 i_1 + \frac{d\psi_1}{dt} \tag{3.7}$$

where r_1 is the resistance of coil 1, N_1 is the number of turns in coil 1 and ψ_1 is the flux linkage with coil 1 (where $\psi_1 = N_1\phi_1$).

For modelling purposes, it is convenient to express flux linkage in terms of inductance and currents. From Eqs. (3.6) and (3.7), the flux linking with coil 1 can be written as (Krause *et al.*, 2002)

$$\psi_1 = L_{l1} i_1 + L_m i_1 + L_m i_2 \tag{3.8}$$

where L_{l1} is the leakage inductance of coil 1 and L_m is the mutual inductance between coils 1 and 2. The term $L_{l1} + L_m$, which is associated with coil 1 is generally referred to as the self-inductance and L_m is referred to as the mutual inductance.

The self- and mutual inductances, which govern the voltage equations of the synchronous generator, vary with angle θ, which in turn varies with time. However, in the synchronously rotating reference frame, both stator and rotor fluxes are seen as stationary. Hence, the flux linkage and thus inductances are constant.

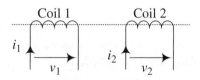

Figure 3.7 Two mutually coupled stationary coils.

3.2 Synchronous Generator Modelling in the *dq* Frame

The generator stator and rotor equations in the *dq* reference frame, where the *d* axis is orientated with the field flux vector and the *q* axis is assumed to be 90° ahead of the *d* axis in the direction of rotation, are given in Eqs. (3.9)–(3.14) (Kundur, 1994; Krause *et al.*, 2002; Anaya-Lara *et al.*, 2009). When deriving these equations it was assumed that the three-phase currents are balanced.

Stator voltage equations:

$$v_{ds} = -r_s i_{ds} - \omega_s \psi_{qs} + \frac{d\psi_{ds}}{dt} \tag{3.9}$$

$$v_{qs} = -r_s i_{qs} + \omega_s \psi_{ds} + \frac{d\psi_{qs}}{dt} \tag{3.10}$$

Stator flux equations:

$$\psi_{ds} = -L_{ls} i_{ds} + L_{md}(-i_{ds} + i_f) \tag{3.11}$$

$$\psi_{qs} = -L_{ls} i_{qs} + L_{mq}(-i_{qs}) \tag{3.12}$$

Rotor voltage equation:

$$v_f = r_f i_f + \frac{d\psi_f}{dt} \tag{3.13}$$

Rotor flux equation:

$$\psi_f = L_{lf} i_f + L_{md}(-i_{ds} + i_f) \tag{3.14}$$

The voltage equations Eqs. (3.9) and (3.10) are very similar to the voltage equation Eq. (3.7) derived for two stationary coils. However, an additional term of speed is present in these equations. This term results from the transformation into the synchronous reference frame and is referred to as 'speed voltage' (due to flux changes in space) (Kundur, 1994; Krause *et al.*, 2002). The 'speed voltage' term does not appear in the rotor voltage equation as the field coil is stationary in the synchronous reference frame.

In order to obtain a per unit (pu) representation of the voltage equations, consider Eq. (3.9). Dividing Eq. (3.9) by the base value of impedance, Z_b, given by $Z_b = V_b/I_b = \omega_b L_b$:

$$\frac{v_{ds}}{Z_b} = -\frac{r_s}{Z_b} i_{ds} - \frac{\omega_s \psi_{qs}}{Z_b} + \frac{1}{Z_b} \frac{d\psi_{ds}}{dt}$$

$$\frac{v_{ds}}{V_b} = -\frac{r_s}{Z_b} \frac{i_{ds}}{I_b} - \frac{\omega_s}{\omega_b} \frac{\psi_{qs}}{I_b L_b} + \frac{1}{\omega_b} \frac{d(\psi_{ds}/I_b L_b)}{dt} \tag{3.15}$$

As the base value of the flux linkage is given by $\psi_b = I_b L_b$, Eq. (3.15) can be represented by

$$\bar{v}_{ds} = -\bar{r}_s \bar{i}_{ds} - \bar{\omega}_s \bar{\psi}_{qs} + \frac{1}{\omega_b} \frac{d\bar{\psi}_{ds}}{dt} \tag{3.16}$$

with pu quantities represented by an upper bar. In Eq. (3.16), all quantities are in pu except time, which is in seconds and base angular frequency, which is in radians per second.

Similarly, Eqs. (3.17)–(3.22) represent the synchronous generator equations in the dq domain and in per unit.

Stator voltage equations:

$$\bar{v}_{ds} = -\bar{r}_s \bar{i}_{ds} - \bar{\omega}_s \bar{\psi}_{qs} + \frac{1}{\omega_b} \frac{d\bar{\psi}_{ds}}{dt} \tag{3.17}$$

$$\bar{v}_{qs} = -\bar{r}_s \bar{i}_{qs} + \bar{\omega}_s \bar{\psi}_{ds} + \frac{1}{\omega_b} \frac{d\bar{\psi}_{qs}}{dt} \tag{3.18}$$

Stator flux equations:

$$\bar{\psi}_{ds} = -\bar{L}_{ls} \bar{i}_{ds} + \bar{L}_{md} \left(-\bar{i}_{ds} + \bar{i}_f \right) \tag{3.19}$$

$$\bar{\psi}_{qs} = -\bar{L}_{ls} \bar{i}_{qs} + \bar{L}_{mq} \left(\bar{i}_{qs} \right) \tag{3.20}$$

Rotor voltage equation:

$$\bar{v}_f = \bar{r}_f \bar{i}_f + \frac{1}{\omega_b} \frac{d\bar{\psi}_f}{dt} \tag{3.21}$$

Rotor flux equation:

$$\bar{\psi}_f = \bar{L}_{lf} \bar{i}_f + \bar{L}_{md} \left(-\bar{i}_{ds} + \bar{i}_f \right) \tag{3.22}$$

When the synchronous generator carries unbalanced currents, the zero sequence current component, $\bar{i}_{0s}$, should also be considered. Under such conditions, in addition to Eqs. (3.17)–(3.22), the following equations should also be considered:

$$\bar{v}_{0s} = -\bar{r}_s \bar{i}_{0s} + \frac{1}{\omega_b} \frac{d\bar{\psi}_{0s}}{dt} \tag{3.23}$$

$$\bar{\psi}_{0s} = -\bar{L}_{ls} \bar{i}_{0s} \tag{3.24}$$

The generator electromagnetic torque is given by the cross-product of the stator flux and stator current:

$$\bar{T}_e = \bar{\psi}_{ds} \cdot \bar{i}_{qs} - \bar{\psi}_{qs} \cdot \bar{i}_{ds} \tag{3.25}$$

3.2.1 Steady-State Operation

Under steady-state conditions, the d/dt terms in Eqs. (3.17), (3.18) and (3.21) are equal to zero. With $\overline{L}_d = \overline{L}_{ls} + \overline{L}_{md}$, $\overline{L}_q = \overline{L}_{ls} + \overline{L}_{mq}$ and $\overline{L}_f = \overline{L}_{lf} + \overline{L}_{md}$, Eqs. (3.17)–(3.22) can be reduced as follow:

Stator voltage equations:

$$\overline{v}_{ds} = -\overline{r}_s\overline{i}_{ds} - \overline{\omega}_s\overline{\psi}_{qs} \tag{3.26}$$

$$\overline{v}_{qs} = -\overline{r}_s\overline{i}_{qs} + \overline{\omega}_s\overline{\psi}_{ds} \tag{3.27}$$

Stator flux equations:

$$\overline{\psi}_{ds} = -\overline{L}_{ds}\overline{i}_{ds} + \overline{L}_{md}\overline{i}_f \tag{3.28}$$

$$\overline{\psi}_{qs} = -\overline{L}_{qs}\overline{i}_{qs} \tag{3.29}$$

Rotor voltage equation:

$$\overline{v}_f = \overline{r}_f\overline{i}_f \tag{3.30}$$

Rotor flux equation:

$$\overline{\psi}_f = \overline{L}_{lf}\overline{i}_f - \overline{L}_{md}\overline{i}_{ds} \tag{3.31}$$

Substituting for flux terms in Eqs. (3.26) and (3.27) from Eqs. (3.28) and (3.29) the following two equations can be obtained:

$$\overline{v}_{ds} = -\overline{r}_s\overline{i}_{ds} + \overline{\omega}_s\overline{L}_{qs}\overline{i}_{qs} = -\overline{r}_s\overline{i}_{ds} + \overline{X}_{qs}\overline{i}_{qs} \tag{3.32}$$

$$\overline{v}_{qs} = -\overline{r}_s\overline{i}_{qs} - \overline{\omega}_s\overline{L}_{ds}\overline{i}_{ds} + \overline{\omega}_s\overline{L}_{md}\overline{i}_f = -\overline{r}_s\overline{i}_{qs} - \overline{X}_{ds}\overline{i}_{ds} + \overline{\omega}_s\overline{L}_{md}\overline{i}_f \tag{3.33}$$

where $\overline{X}_{qs} = \overline{\omega}_s\overline{L}_{qs}$ and $\overline{X}_{ds} = \overline{\omega}_s\overline{L}_{ds}$.

From Eq. (3.30), $\overline{i}_f$ in Eq. (3.33) can be replaced by $\overline{v}_f/\overline{r}_f$ and with the definition of $\overline{E}_{fd} = \overline{\omega}_s\overline{L}_{md}\overline{v}_f/\overline{r}_f$, then

$$\overline{v}_{qs} = -\overline{r}_s\overline{i}_{qs} - \overline{X}_{ds}\overline{i}_{ds} + \overline{E}_{fd} \tag{3.34}$$

The armature terminal voltage is expressed as $\overline{E}_t = \overline{v}_{ds} + j\overline{v}_{qs}$ and from Eqs. (3.32) and (3.34), the following steady-state equation of the synchronous machine can be obtained:

$$\overline{E}_t = \overline{v}_{ds} + j\overline{v}_{qs} = -\overline{r}_s\left(\overline{i}_{ds} + j\overline{i}_{qs}\right) + \left(\overline{X}_{qs}\overline{i}_{qs} - j\overline{X}_{ds}\overline{i}_{ds}\right) + j\overline{E}_{fd} \tag{3.35}$$

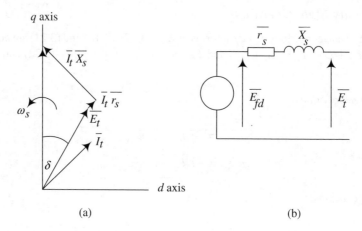

Figure 3.8   Phasor diagram and the equivalent circuit for steady-state operation.

If saliency is neglected, $\overline{X}_{qs} = \overline{X}_{ds} = \overline{X}_s$ and, $\overline{i}_t = \overline{i}_{ds} + j\overline{i}_{qs}$ Eq. (3.35) can be reduced to

$$
\begin{aligned}
\overline{E}_t &= -\overline{r}_s \overline{i}_t + \overline{X}_s \left(\overline{i}_{qs} - j\overline{i}_{ds} \right) + j\overline{E}_{fd} \\
&= -\overline{r}_s \overline{i}_t + \overline{X}_s \left(-j^2 \overline{i}_{qs} - j\overline{i}_{ds} \right) + j\overline{E}_{fd} \qquad\qquad (3.36) \\
&= -\left(\overline{r}_s + j\overline{X}_s \right) \overline{i}_t + j\overline{E}_{fd}
\end{aligned}
$$

Eq. (3.36) defines the steady-state equation of the synchronous machine and can be represented by the phasor diagram shown in Figure 3.8a and the equivalent circuit shown in Figure 3.8b.

3.2.2   Synchronous Generator with Damper Windings

In both, salient-pole and cylindrical-pole generators, solid copper bars run through the rotor to provide additional paths for circulating damping currents. The currents in the damper windings interact with the air-gap flux and produce a torque which provides damping of rotor oscillations following a transient disturbance. The currents in the damper windings can be resolved into two components. The circulating damping current under a pole forms the d axis damping current; whereas the circulating damping current between two pole faces forms the q axis damping current. In the generator model shown in Figure 3.9, these currents were assumed to flow in sets of closed circuits: one set whose flux is in line with that of the field along the d axis and the other set whose flux is along the q axis. In the simplified model representing the synchronous generator, only one damper winding along the q axis is used, but often two damper windings, kq_1 and kq_2 are represented (Figure 3.9). Although the same basic representation can be used for both salient-pole and cylindrical-pole generators, the circuit parameters representing the damper windings are widely different.

The synchronous generator equations in the dq domain including damper windings are as follows.

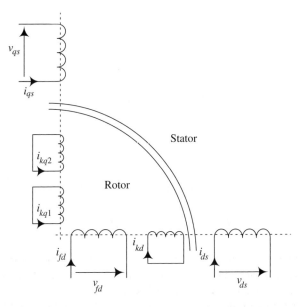

Figure 3.9   Stator and rotor circuits of a synchronous generator.

Stator voltage equations:

$$\overline{v}_{ds} = -\overline{r}_s\overline{i}_{ds} - \overline{\omega}_s\overline{\psi}_{qs} + \frac{1}{\omega_b}\frac{d\overline{\psi}_{ds}}{dt} \tag{3.37}$$

$$\overline{v}_{qs} = -\overline{r}_s\overline{i}_{qs} + \overline{\omega}_s\overline{\psi}_{ds} + \frac{1}{\omega_b}\frac{d\overline{\psi}_{qs}}{dt} \tag{3.38}$$

Stator flux equations:

$$\overline{\psi}_{ds} = -\overline{L}_{ls}\overline{i}_{ds} + \overline{\psi}_{md} \tag{3.39}$$

$$\overline{\psi}_{qs} = -\overline{L}_{ls}\overline{i}_{qs} + \overline{\psi}_{mq} \tag{3.40}$$

Rotor voltage equations:

$$\overline{v}_{fd} = \overline{r}_{fd}\overline{i}_{fd} + \frac{1}{\omega_b}\frac{d\overline{\psi}_{fd}}{dt} \tag{3.41}$$

$$\overline{v}_{kd} = \overline{r}_{kd}\overline{i}_{kd} + \frac{1}{\omega_b}\frac{d\overline{\psi}_{kd}}{dt} \tag{3.42}$$

$$\overline{v}_{kq1} = \overline{r}_{kq1}\overline{i}_{kq1} + \frac{1}{\omega_b}\frac{d\overline{\psi}_{kq1}}{dt} \tag{3.43}$$

$$\overline{v}_{kq2} = \overline{r}_{kq2}\overline{i}_{kq2} + \frac{1}{\omega_b}\frac{d\overline{\psi}_{kq2}}{dt} \tag{3.44}$$

Rotor flux equations:

$$\overline{\psi}_{fd} = \overline{L}_{lfd}\overline{i}_{fd} + \overline{\psi}_{md} \tag{3.45}$$

$$\overline{\psi}_{kd} = \overline{L}_{lkd}\overline{i}_{kd} + \overline{\psi}_{md} \tag{3.46}$$

$$\overline{\psi}_{kq1} = \overline{L}_{lkq1}\overline{i}_{kq1} + \overline{\psi}_{mq} \tag{3.47}$$

$$\overline{\psi}_{kq2} = \overline{L}_{lkq2}\overline{i}_{kq2} + \overline{\psi}_{mq} \tag{3.48}$$

where

$$\overline{\psi}_{md} = \overline{L}_{md}\left(-\overline{i}_{ds} + \overline{i}_{fd} + \overline{i}_{kd}\right)$$

$$\overline{\psi}_{mq} = \overline{L}_{mq}\left(-\overline{i}_{qs} + \overline{i}_{kq1} + \overline{i}_{kq2}\right)$$

The stator voltage equations Eqs. (3.37) and (3.38) and the rotor voltage equations Eqs. (3.41)–(3.44) are written in terms of currents and flux linkages. The flux linkages and the currents are related and both cannot be independent.

The currents in terms of flux linkages are obtained from Eqs. (3.39) and (3.40), and Eqs. (3.45)–(3.48) are given as follows:

$$\overline{i}_{ds} = -\frac{1}{\overline{L}_{ls}}\left(\overline{\psi}_{ds} - \overline{\psi}_{md}\right) \tag{3.49}$$

$$\overline{i}_{qs} = -\frac{1}{\overline{L}_{ls}}\left(\overline{\psi}_{qs} - \overline{\psi}_{mq}\right) \tag{3.50}$$

$$\overline{i}_{fd} = \frac{1}{\overline{L}_{lfd}}\left(\overline{\psi}_{fd} - \overline{\psi}_{md}\right) \tag{3.51}$$

$$\overline{i}_{kd} = \frac{1}{\overline{L}_{lkd}}\left(\overline{\psi}_{kd} - \overline{\psi}_{md}\right) \tag{3.52}$$

$$\overline{i}_{kq1} = \frac{1}{\overline{L}_{lkq1}}\left(\overline{\psi}_{kq1} - \overline{\psi}_{mq}\right) \tag{3.53}$$

$$\overline{i}_{kq2} = \frac{1}{\overline{L}_{lkq2}}\left(\overline{\psi}_{kq2} - \overline{\psi}_{mq}\right) \tag{3.54}$$

The non-reduced order model of the synchronous generator includes stator transients and rotor transients as well as the damper windings. The following differential equations are directly derived from Eqs. (3.37), (3.38), (3.50) and (3.51):

$$\frac{d\overline{\psi}_{ds}}{dt} = \omega_b\left[\overline{v}_{ds} + \overline{\omega}_s\overline{\psi}_{qs} + \frac{\overline{r}_s}{\overline{L}_{ls}}\left(\overline{\psi}_{md} - \overline{\psi}_{ds}\right)\right] \tag{3.55}$$

$$\frac{d\overline{\psi}_{qs}}{dt} = \omega_b\left[\overline{v}_{qs} - \overline{\omega}_s\overline{\psi}_{ds} + \frac{\overline{r}_s}{\overline{L}_{ls}}\left(\overline{\psi}_{mq} - \overline{\psi}_{qs}\right)\right] \tag{3.56}$$

The rotor dynamic equations, with two damper windings in the q axis and one in the d axis, are as follows [from Eqs. (3.41)–(3.44) and (3.51)–(3.54)]:

$$\frac{d\overline{\psi}_{fd}}{dt} = \omega_b \left[\frac{\overline{r}_{fd}}{\overline{L}_{md}} \overline{e}_{xfd} + \frac{\overline{r}_{fd}}{\overline{L}_{lfd}} \left(\overline{\psi}_{md} - \overline{\psi}_{fd} \right) \right] \tag{3.57}$$

$$\frac{d\overline{\psi}_{kd}}{dt} = \omega_b \left[\overline{v}_{kd} + \frac{\overline{r}_{kd}}{\overline{L}_{lkd}} \left(\overline{\psi}_{md} - \overline{\psi}_{kd} \right) \right] \tag{3.58}$$

$$\frac{d\overline{\psi}_{kq1}}{dt} = \omega_b \left[\overline{v}_{kq1} + \frac{\overline{r}_{kq1}}{\overline{L}_{lkq1}} \left(\overline{\psi}_{mq} - \overline{\psi}_{kq1} \right) \right] \tag{3.59}$$

$$\frac{d\overline{\psi}_{kq2}}{dt} = \omega_b \left[\overline{v}_{kq2} + \frac{\overline{r}_{kq2}}{\overline{L}_{lkq2}} \left(\overline{\psi}_{mq} - \overline{\psi}_{kq2} \right) \right] \tag{3.60}$$

The excitation dynamics of the generator are given by Eq. (3.57), where $\overline{e}_{xfd} = \overline{L}_{md} \cdot \overline{i}_{rfd}$ represents the excitation voltage of the generator at base speed ω_b.

If the zero sequence currents are present in the stator, then the following equation should also be considered:

$$\frac{d\overline{\psi}_{0s}}{dt} = \omega_b \left(\overline{v}_{0s} - \frac{\overline{r}_s}{\overline{L}_{ls}} \overline{\psi}_{0s} \right) \tag{3.61}$$

A reduced-order model may be obtained by neglecting the stator transients in Eqs. (3.55), (3.56) and (3.61) as follows:

$$\overline{\psi}_{ds} = \frac{1}{\omega_s} \left[\overline{v}_{qs} + \frac{\overline{r}_s}{\overline{L}_{ls}} \left(\overline{\psi}_{mq} - \overline{\psi}_{qs} \right) \right] \tag{3.62}$$

$$\overline{\psi}_{qs} = -\frac{1}{\omega_s} \left[\overline{v}_{ds} + \frac{\overline{r}_s}{\overline{L}_{ls}} \left(\overline{\psi}_{md} - \overline{\psi}_{ds} \right) \right] \tag{3.63}$$

$$\overline{\psi}_{0s} = \frac{\overline{L}_{ls}}{\overline{r}_s} \overline{v}_{0s} \tag{3.64}$$

3.3 Control of Large Synchronous Generators

A large power system consists of a number of generators and loads connected through transmission and distribution circuits. Loads connected to the power system have different characteristics and continuously vary in time. In order to operate the power system within the limits required (voltage and frequency), and in order to maintain the stability of the system in case

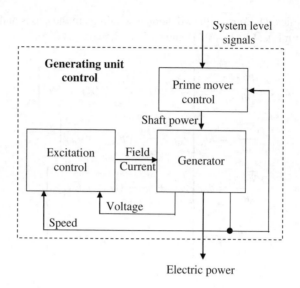

Figure 3.10   Synchronous generator control (Kundur, 1994; Anaya-Lara *et al.*, 2009).

of a disturbance, large generators are controlled individually and collectively. The different
controls associated with a synchronous generator are shown in Figure 3.10 (Kundur, 1994).
These functional blocks perform two basic control actions: reactive power/voltage control and
active power/frequency control (Anaya-Lara *et al.*, 2009).

3.3.1   Excitation Control

As conditions vary on the power system, the active and reactive power demand varies. Under
heavy-load conditions, both the transmission system and the loads absorb reactive power
and the synchronous generators need to inject reactive power into the network. Under light-
load conditions the capacitive behaviour of the transmission lines can become dominant and
under such conditions it is desirable for synchronous generators to absorb reactive power. The
variations in reactive power demand on a synchronous generator can be accommodated by
adjusting its excitation voltage. The excitation system performs the basic function of automatic
voltage regulation. It also performs the protective functions required to operate the machine
and other equipment within their capabilities. A block diagram of an excitation control system
is shown in Figure 3.11.

- **Regulator**

    A synchronous generator employs an automatic voltage regulator (AVR) to maintain
    the generator stator terminal voltage close to a predefined value. If the generator terminal
    voltage falls due to increased reactive power demand, the change in voltage is detected and
    a signal is fed into the exciter to produce an increase in excitation voltage. The generator
    reactive power output is thereby increased and the terminal voltage is returned close to its
    initial value.

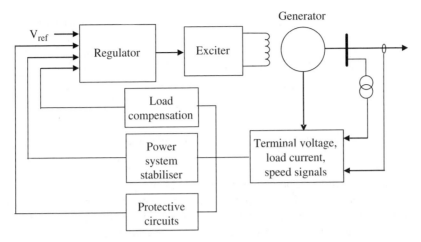

Figure 3.11   Block diagram of an excitation control system.

- **Exciter**

    The purpose of an exciter is to supply an adjustable direct current to the generator field winding. The exciter may be a dc generator on small set sizes. On larger sets, commutation problems prohibit the use of dc generators and ac generators are employed, supplying the field via a rectifier. Static excitation systems are also widely used. These comprise a controlled rectifier usually powered from the generator terminals and permit fast response excitation control. In all the mentioned cases the dc supply is connected to the synchronous generator field winding via slip rings.

- **Load Compensation**

    The AVR normally controls the generator stator terminal voltage. Building an additional loop to the AVR control allows the voltage at a remote point on the network to be controlled. The load compensator has adjustable resistance and reactance that simulates the impedance between the generator terminals and the point at which the voltage is being effectively controlled. Using this impedance and the measured current the voltage drop is computed and added to the terminal voltage.

- **Power System Stabiliser**

    The basic function of the power system stabiliser (PSS) is to add damping to the generator rotor oscillations by controlling its excitation. The commonly used auxiliary stabilising signals to control the excitation are shaft speed, terminal frequency and power.

3.3.2   Prime Mover Control

The governing systems of the generator prime movers provide the means of adjusting the power outputs of the generators of the network to match the power demand of the network load. If, for example, the network load increases then this imposes increased torques on the generators and this causes them to decelerate. The resulting fall in speed is detected by the governor of each regulating prime mover and used to increase its power output. The change in power produced in an individual generator is determined by the droop setting of its governor. A 4% droop setting indicates that the regulation is such that a 4% change in speed would result in a 100%

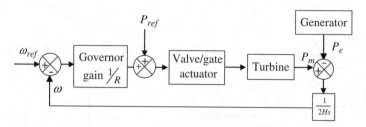

Figure 3.12   Speed governor system (Wood, 1996; Anaya-Lara *et al.*, 2009).

change in the generator power output. In steady state, all the generators of the network operate at the same frequency and this frequency determines the operating speeds of the individual generator prime movers. Hence, following a network load increase, the network frequency will fall until the sum of the power output changes it produces in the regulating generators matches the change in the network load. The basic elements of a governor power control loop are shown in the block diagram of Figure 3.12 and the droop characteristic is shown graphically in Figure 3.13. By changing the load reference set point, P_{ref}, the generators governor characteristics (Figure 3.13) can be set to give the reference frequency, f_0 (50 or 60 Hz), at any desired unit output. In other words it shifts the characteristic vertically.

3.4   Fully-Rated Converter Wind Turbines

In recent years there has been an increased interest in wind turbines equipped with fully-rated converters as shown in Figure 3.14 (Fox *et al.*, 2007; Anaya-Lara *et al.*, 2009). This type of wind turbine can or cannot have a gearbox and a wide range of electrical generator types such as asynchronous, conventional synchronous and permanent magnet can be employed. As all the power from the wind turbine is transferred through the power converter the specific characteristics and dynamics of the electrical generator are effectively isolated from the power network. Hence, the electrical frequency of the generator may vary as the wind speed changes, while the network frequency remains unchanged, enabling variable-speed operation. The rating of the power converter in this wind turbine corresponds to the rated power of the generator.

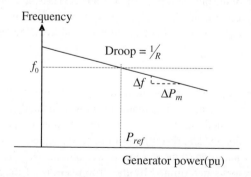

Figure 3.13   Droop characteristic (Anaya-Lara *et al.*, 2009).

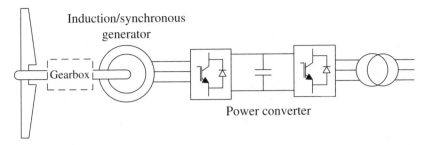

Figure 3.14 Typical configuration of a fully-rated converter connected wind turbine (Fox *et al.*, 2007; Anaya-Lara *et al.*, 2009).

The power converter can be arranged in various ways. Whilst the Generator-Side Converter (GSC) can be a diode-based rectifier or a PWM Voltage Source Converter, the Network-Side Converter (NSC) is typically a PWM Voltage Source Converter. The strategy to control the operation of the generator and power flows to the network depend very much on the type of power converter arrangement employed.

3.5   FRC-WT with Synchronous Generator

In an FRC wind turbine based on synchronous generators, the generator can be electrically excited or it can have a permanent magnet rotor. In the direct-drive arrangement, the turbine and generator rotors are mounted on the same shaft without a gearbox and the generator is specially designed for low-speed operation with a large number of poles. The synchronous generators of direct-drive turbines tend to be very large due to the large number of poles. However, if the turbine includes a gearbox then a smaller generator with lower number of poles can be employed (Akhmatov *et al.*, 2003).

Today, almost all wind turbines use standard generators (4 pole) for speeds between 750 and 1800 rpm. The turbine speed is much lower than the generator speed, typically between 20 and 60 rpm. Therefore, in a conventional wind turbine a gearbox is used between the turbine and the generator. An alternative is to use a generator for very low speeds. The generator can then be directly connected to the turbine shaft. The drive trains of a conventional wind turbine and one with a direct-driven generator are shown schematically in Figure 3.15 (Grauers, 1996).

Direct-driven generators are favoured for some applications due to reduction in losses in the drive train and less noise (Grauers, 1996). A significant difference between conventional and direct-driven wind turbine generators is that the low speed of the direct-driven generator makes a very high-rated torque necessary. This is an important difference, since the size and the losses of a low-speed generator depend on the rated torque rather than on the rated power. A direct-driven generator for a 500-kW, 30-rpm wind turbine has the same rated torque as a 50-MW, 3000-rpm steam-turbine generator. Because of the high-rated torque, direct-driven generators are usually heavier and less efficient than conventional generators. To increase the efficiency and reduce the weight of the active parts, direct-driven generators are usually designed with a large diameter.

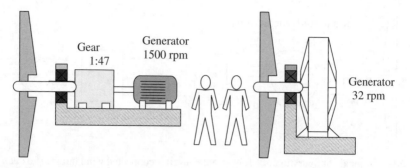

Figure 3.15 Drive trains of a conventional wind turbine (left) and one with a direct-drive generator (right) (Grauers, 1996).

3.5.1 Permanent Magnets Synchronous Generator

Just as with any type of synchronous machine, the PMSG produces electric energy by rotating a magnetic field around a set of windings. Here, an induced EMF (electromotive force) is generated each time the field 'cuts' across the winding conductors. The rotating magnetic field of the PMSG is created by a rotor composed from a core (usually iron) and permanent magnets either glued to or buried inside the rotor core. Figure 3.16 shows some typical cross sections of PMSGs (Hanselman, 2006).

As seen in Figure 3.16 the cross section shape of the rotor can assume different forms depending on the application where the PMSG is used. Rotors of the type (a) and (b) are the most commonly used in motor applications. Rotor type (c) is preferred for high speed operation since the rectangular magnets are entirely enclosed in the rotor structure, meaning this, a better mechanical protection against centrifugal forces. This configuration is also used to add a reluctance component to the produced torque. Rotor type (d) promotes flux concentration because the magnet surface area is greater than the rotor surface area; this is useful for gaining better performance from ferrite magnetic material. This configuration is also used to add a reluctance component to the produced torque. Rotor type (e) refers to the transverse flux

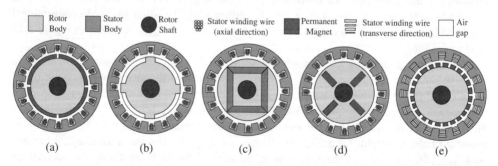

Figure 3.16 Cross section of different types of PMSG. (a) Surface-mounted permanent magnets (b) inset surface-mounted permanent magnets (c) interior permanent magnet (d) buried (spoke) permanet magnets (e) transverse flux permanent magnet.

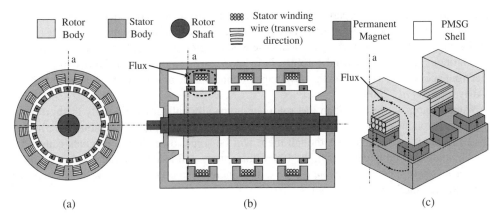

Figure 3.17 (a) radial cross section of a transverse flux permanent magnet generator (b) transversal cross section of a transverse flux permanent magnet generator (c) a-phase section of the transverse flux permanent magnet generator.

permanent magnet topology, which is gaining preference for gearless (direct drive) wind energy systems, where low rotational speed and high torque density are required. Here the generator is built in single phase configuration with ring shaped stator coils (transverse direction respecting the rotor axis) and an array of surface PMs on the rotor.

Figure 3.17 shows the cross and transversal section of a transverse flux permanent magnet generator as well as a section of the stator core. As seen in Figure 3.17 (b) and (c) the ring shaped stator coil is embraced by the stator's C-shaped iron cores that create a variable reluctance structure with two poles. These poles are exposed, on each end of the core, to the array of alternating polarity permanent magnets displaced on the rotor surface thereby producing an alternating flux in the stator iron.

The transverse flux permanent magnet generator is characterized by the fact that all the permanent magnet fluxes of all north poles at one instant, combine in the ring shaped stator, and then, after the rotor travels one permanent magnet pole angle, all south poles add their flux into the coil. Thus, by attaching multiple C-shaped core units in the direction of the movement, a multiple-pole machine can be produced which enables an increased current loading, and as a result, a higher value of torque density. Since the system conformed from the C-shaped cores and the permanent magnets generate a single phase EMF, there is no common rotating field built by the three phase winding. So, to generate a 3-phase EMF, three independent 120 electrical degrees-shifted alternating fields, are created by means of stacking together, in the lateral direction, three single phase C-core array units with mechanically displaced magnets on the rotor surface, placed to provide the required phase-shift.

3.5.1.1 Permanent Magnets Versus Electrically-Excited Synchronous Generators

The synchronous machine has the ability to provide its own excitation on the rotor. Such excitation may be obtained either by means of a current-carrying winding, or by means of permanent magnets (PM). The wound-rotor synchronous machine has a very desirable feature compared to its PM counterpart: adjustable excitation current and, consequently, control of its output voltage independent of load current. This feature explains why most constant-speed, grid-connected hydro and turbo generators use wound-rotor instead of PM-excited rotors. The synchronous generator in wind turbines is in most cases connected to the network via an electronic converter. Therefore, the advantage of controllable no-load voltage is not as critical (Anaya-Lara *et al.*, 2009).

Wound rotors are heavier than PM rotors and typically bulkier (particularly in short pole-pitch synchronous generators). Also, electrically-excited synchronous generators have higher losses in the rotor windings. Although there will be some losses in the magnets caused by the circulation of eddy currents in the PM volume, they will usually be much lower than the copper losses of electrically-excited rotors. This increase in copper losses will also increase when increasing the number of poles.

3.5.2 FRC-WT Based on Permanent Magnet Synchronous Generator

Permanent magnet excitation avoids the field current supply or reactive power compensation facilities needed by wound-rotor synchronous generators and induction generators, and it also removes the need for slip rings (Chen *et al.*, 1998). Figure 3.18 shows the arrangement with an uncontrolled diode-based rectifier as the generator-side converter. A dc-booster is used to stabilise the dc link voltage whereas the network-side converter (PWM-based VSC) controls the operation of the generator. The NSC can be controlled using load-angle techniques or current controllers developed in a voltage-orientated *dq* reference frame (Fox *et al.*, 2007; Anaya-Lara *et al.*, 2009).

The topology with a permanent magnet synchronous generator and back-to-back power converters is shown in Figure 3.19. In this arrangement the generator-side converter controls the operation of the generator, and the network-side converter controls the dc link voltage by exporting active power to the network.

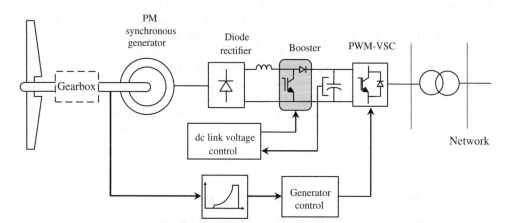

Figure 3.18 Permanent Magnet synchronous generator with diode rectifier (Fox *et al.*, 2007; Anaya-Lara *et al.*, 2009).

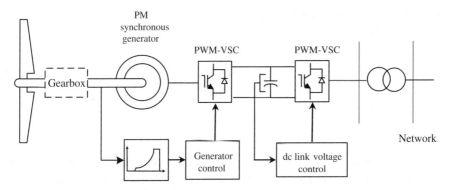

Figure 3.19 Permanent magnets synchronous generator with two back-to-back voltage source converters (Fox *et al.*, 2007; Anaya-Lara *et al.*, 2009).

3.5.3   Generator-Side Converter Control

The generator-side converter controls the operation of the wind turbine and two control techniques are explained namely, load angle and vector control.

3.5.3.1   Load Angle Control Technique

The load angle control strategy employs steady-state power flow equations (Kundur, 1994; Fox *et al.*, 2007; Anaya-Lara *et al.*, 2009) to determine the transfer of active and reactive power between the generator and the dc link. With reference to Figure 3.20 the term E_g represents the magnitude of the generator internal voltage, X_g the synchronous reactance, the term V_t represents the voltage (magnitude) at the converter terminals, and α_g is the phase difference between the voltages E_g and V_t.

The active and reactive power flows in the steady state are defined as:

$$P = \frac{E_g V_t}{X_g} \sin \alpha_g \tag{3.65}$$

$$Q = \frac{E_g^2 - E_g V_t \cos \alpha_g}{X_g} \tag{3.66}$$

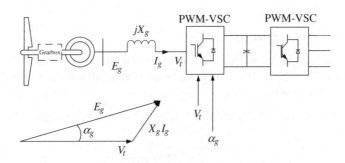

Figure 3.20 Load angle control of a synchronous generator wind turbine.

As the load angle α_g is generally small, $\sin \alpha_g \approx \alpha_g$ and $\cos \alpha_g \approx 1$. Hence, Eqs. (3.65) and (3.66) can be simplified to:

$$P = \frac{E_g V_t}{X_g} \alpha_g \tag{3.67}$$

$$Q = \frac{E_g^2 - E_g V_t}{X_g} \tag{3.68}$$

From Eqs. (3.67) and (3.68) it is seen that the active power transfer depends mainly on the phase angle α_g. The reactive power transfer depends mainly on voltage magnitudes, and it is transmitted from the point with higher voltage magnitude to the point with lower magnitude.

The operation of the generator and the power transferred from the generator to the dc link are controlled by adjusting the magnitude and angle of the voltage at the ac terminals of the generator-side converter. The magnitude, V_t, and angle, α_g, required at the terminals of the generator-side converter are calculated using Eqs. (3.67) and (3.68) as:

$$\alpha_g = \frac{P_{g_{ref}} X_g}{E_g V_t} \tag{3.69}$$

$$V_t = E_g - \frac{Q_{g_{ref}} X_g}{E_g} \tag{3.70}$$

where $P_{g_{ref}}$ is the reference value of the active power that needs to be transferred from the generator to the dc link, and $Q_{g_{ref}}$ is the reference value for the reactive power.

The reference value $P_{g_{ref}}$ is obtained from the maximum power extraction curve shown for a given generator speed, ω_r. As the generator has permanent magnets, it does not require magnetising current through the stator, thus the reactive power reference value can be set to zero, $Q_{g_{ref}} = 0$ (i.e. V_t and E_g are equal in magnitude). The implementation of the load angle control scheme is shown in Figure 3.21.

The major advantage of the load angle control is its simplicity. However, as in this technique the dynamics of the generator are not considered it may not be very effective in controlling the generator during a transient operating condition (Fox *et al.*, 2007; Anaya-Lara *et al.*, 2009).

3.5.3.2 Vector Control Strategy

Vector control techniques are implemented based on the dynamic model of the synchronous generator expressed in the dq frame. The dq frame is defined as the d axis aligned with the magnetic axis of the rotor (field). For the vector control $\bar{i}_{ds_{ref}}$ is set to zero and $\bar{i}_{qs_{ref}}$ is derived from Eq. (3.25). From Eqs. (3.25) and (3.28) with $\bar{i}_{ds} = 0$, the following can be obtained:

$$\bar{T}_e = \bar{\psi}_{ds} \bar{i}_{qs} \tag{3.71}$$

$$\bar{\psi}_{ds} = \bar{L}_{md} \bar{i}_f \tag{3.72}$$

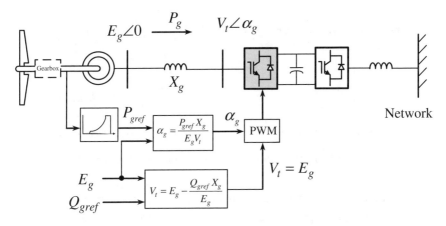

Figure 3.21 Load angle control of the generator-side converter.

Defining $\overline{L}_{md}\overline{i}_f = \overline{\psi}_{fd}$ and substituting for $\overline{\psi}_{ds}$ from Eq. (3.72) into Eq. (3.71):

$$\overline{T}_e = \overline{\psi}_{fd}\overline{i}_{qs} \tag{3.73}$$

From Eq. (3.73) for a given torque reference $\overline{T}_{sp}$:

$$\overline{i}_{qs_{ref}} = \frac{\overline{T}_{sp}}{\overline{\psi}_{fd}} \tag{3.74}$$

Once the reference currents, $\overline{i}_{qs_{ref}}$ and $\overline{i}_{ds_{ref}}$ are determined by the controller the corresponding voltage magnitudes can be calculated from Eqs. (3.32) and (3.34) as,

$$\overline{v}_{ds} = -\overline{r}_s\overline{i}_{ds} + \overline{X}_{qs}\overline{i}_{qs} \tag{3.75}$$

$$\overline{v}_{qs} = -\overline{r}_s\overline{i}_{qs} - \overline{X}_{ds}\overline{i}_{ds} + \overline{E}_{fd} \tag{3.76}$$

A PI controller is used to regulate the error between the reference and actual current values, which relate to the $\overline{r}_s$ term in the right-hand side of Eqs. (3.75) and (3.76). Additional terms are included to eliminate the cross-coupling effect as shown in Figure 3.22.

The current reference $\overline{i}_{ds_{ref}}$ is kept to zero when the generator operates below the base speed, and it is set to a negative value to cancel some of the flux linkage when the generator operates above the base speed. The current reference $\overline{i}_{qs_{ref}}$ is determined from the torque equation. The implementation of the vector control technique is shown in Figure 3.23.

The torque control is exercised in the q axis, and the magnetisation of the generator is controlled in the d axis. The reference value of the stator current in the q axis, $\overline{i}_{qs_{ref}}$, is calculated from Eq. (3.74) and compared with the actual value, $\overline{i}_{qs}$. The error between these two signals is processed by a PI controller whose output is the voltage in the q axis, $\overline{v}_{qs}$, required to control the generator-side converter. To calculate the required voltage in the d axis,

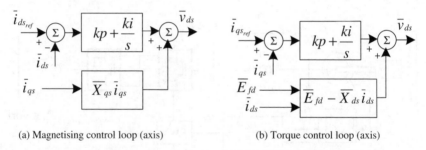

(a) Magnetising control loop (axis)   (b) Torque control loop (axis)

Figure 3.22 Control loops in the vector control strategy.

$\bar{v}_{ds}$, the reference value of the stator current in the d axis, $\bar{i}_{ds_{ref}}$, is compared against the actual current in the d axis, $\bar{i}_{ds}$, and processing the error between these two signals by a PI controller. The reference $\bar{i}_{ds_{ref}}$ may be assumed zero for the permanent magnets synchronous generator (Fox *et al.*, 2007; Anaya-Lara *et al.*, 2009).

3.5.4 Modelling of the dc Link

For simulation purposes the reference value for the active power, $P_{g_{ref}}$, that needs to be transmitted to the grid can be determined by examining the dc-link dynamics with the aid

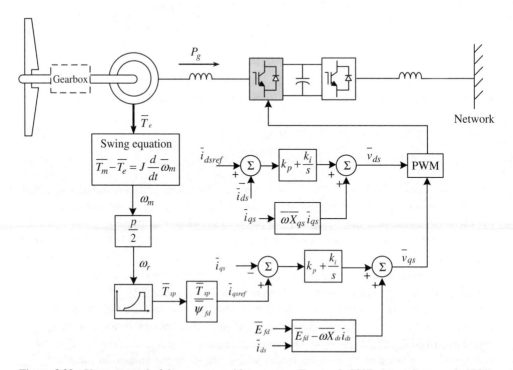

Figure 3.23 Vector control of the generator-side converter (Fox *et al.*, 2007; Anaya-Lara *et al.*, 2009).

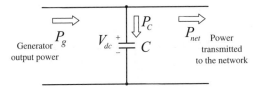

Figure 3.24   Power flow in the dc link.

of Figure 3.24 (Fox *et al.*, 2007; Anaya-Lara *et al.*, 2009). This figure illustrates the power balance at the dc-link, which is expressed as,

$$P_C = P_g - P_{net} \tag{3.77}$$

where P_C is the power that goes through the dc-link capacitor, C, P_g is the active power output of the generator (and transmitted to the dc link), and P_{net} is the active power transmitted from the dc link to the grid.

The power flow through the capacitor is given as,

$$\begin{aligned} P_C &= V_{DC} \cdot I_{DC} \\ &= V_{DC} \cdot C \frac{dV_{dc}}{dt} \end{aligned} \tag{3.78}$$

From Eq. (3.78) the dc-link voltage V_{dc} is determined as follows:

$$\begin{aligned} P_C &= V_{dc} \cdot C \frac{dV_{dc}}{dt} = \frac{C}{2} \cdot 2 \cdot V_{dc} \frac{dV_{dc}}{dt} \\ &= \frac{C}{2} \cdot \frac{dV_{dc}^2}{dt} \end{aligned} \tag{3.79}$$

Rearranging Eq. (3.79) and integrating both sides of the equation,

$$V_{dc}^2 = \frac{2}{C} \int P_C dt \tag{3.80}$$

then

$$V_{dc} = \sqrt{\frac{2}{C} \int P_C dt} \tag{3.81}$$

By substituting P_C in Eq. (3.81) using Eq. (3.77) the dc-link voltage, V_{dc} can be expressed in terms of the generator output power, P_g, and the power transmitted to the grid, P_{net}, as,

$$V_{dc} = \sqrt{\frac{2}{C} \int (P_g - P_{net}) dt} \tag{3.82}$$

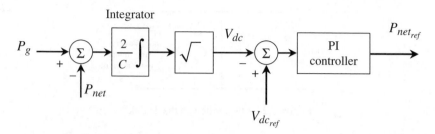

Figure 3.25 Calculation of the active power reference, $P_{net_{ref}}$, (suitable for simulation purposes).

Equation (3.82) calculates the actual value of V_{dc}. The reference value of the active power, $P_{net_{ref}}$, to be transmitted to the network is calculated by comparing the actual dc link voltage, V_{dc}, against the desired dc-link voltage reference, $V_{dc_{ref}}$. The error between these two signals is processed by a PI controller whose output provides the reference active power $P_{net_{ref}}$, as shown in Figure 3.25. It should be noted that in a physical implementation the actual value of the dc-link voltage, V_{dc}, is obtained from measurements via a transducer.

3.5.5 Network-Side Converter Control

The objective of the network-side converter controller is to maintain the dc-link voltage at the reference value by exporting active power to the network. In addition the controller is designed to enable the exchange of reactive power between the converter and the network as required by the application specifications.

3.5.5.1 Load-Angle Control Technique

A methodology to control the network-side converter is also the load-angle control technique, where the network-side converter is the sending source, $V_{VSC}\angle\delta$, and the network is the receiving source, $V_{net}\angle 0$. As the network voltage is known it is selected as the reference, hence the phase angle δ is positive. The inductor coupling these two sources is the reactance X_{net}.

To implement the load-angle controller the reference value of the reactive power, $Q_{net_{ref}}$, may be set to zero for unity power factor operation. Hence, the magnitude, V_{VSC}, and angle, δ, required at the terminal of the network-side converter are calculated as:

$$\delta = \frac{P_{net_{ref}}X_{net}}{V_{VSC}V_{net}} \tag{3.83}$$

$$V_{VSC} = V_{net} + \frac{Q_{net_{ref}}X_{net}}{V_{VSC}}; \quad Q_{net_{ref}} = 0 \tag{3.84}$$

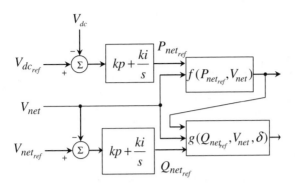

Figure 3.26   Control of active and reactive power by load angle and magnitude control.

From Eqs. (3.83) and (3.84) the network-side converter voltage magnitude V_{VSC} and angle δ can be obtained as:

$$V_{VSC} = g(Q_{net_{ref}}, V_{net}, \delta) = \frac{V_{VSC}\cos\delta + \sqrt{(V_{net}\cos\delta)^2 + 4\cdot(Q_{net_{ref}}/3)\cdot X_{net}}}{2} \tag{3.85}$$

$$\delta = f(P_{net_{ref}}, V_{net}) = \sin^{-1}\left(\frac{P_{net_{ref}}X_{net}}{3V_{VSC}V_{net}}\right) \tag{3.86}$$

A second-order quadratic equation given in (3.85) needs to be solved to determine the value of V_{VSC}, where only one solution is appropriate. Figure 3.26 shows the control block diagram of the load-angle control methodology. The dc-link voltage reference, $V_{dc_{ref}}$, is compared with the actual (or measured) dc voltage, V_{dc}, and the error regulated by a PI controller. The PI controller output $P_{net_{ref}}$ and the reactive power $Q_{net_{ref}}$ are used to find the network-side converter voltage magnitude and angle.

3.5.5.2   Vector Control Strategy

The block diagram of the vector control of the network-side converter is shown in Figure 3.27. The dc-link voltage is maintained by controlling the q-axis current and the ac voltage at the network terminal is controlled in the d-axis. The reference currents are initially determined in the dq-frame of the voltage V_{net}, where the voltage vector is aligned with the q-axis. Then the reference currents are transformed to the network reference frame and compared with the actual currents. Current error signals are regulated by PI controllers and then decoupling components are added to eliminate the coupling effect between the two axes. Finally, dq components of the voltage V_{net} are added to find the required voltage components at the terminals of the network-side converter in the network reference frame.

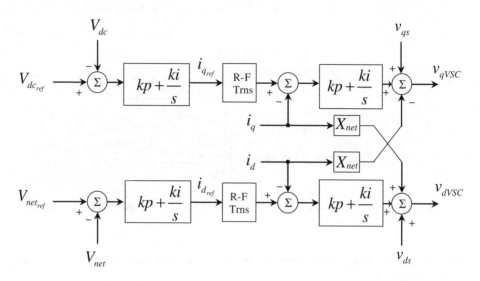

Figure 3.27 Network-side converter control in the *dq*-frame.

3.6 FRC-WT with Squirrel-Cage Induction Generator

Figure 3.28 shows the block diagram of a fully-rated converter wind turbine with a squirrel-cage induction generator (Caliao, 2008) (in the following all electrical quantities are in pu unless otherwise indicated). The generator-side converter is controlled using rotor flux-oriented control. The network-side converter is controlled using load-angle control or vector control (Vas, 1990; Krause *et al.*, 2002; Caliao, 2008; Anaya-Lara *et al.*, 2009).

3.6.1 Control of the FRC-IG Wind Turbine

The equations of the induction machine (in *dq* coordinates and in pu), used to design the controller of the grid-side converter are as follows (Kundur, 1994; Ackermann, 2005; Anaya-Lara *et al.*, 2009):

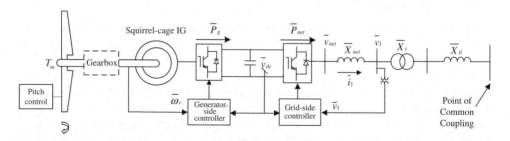

Figure 3.28 Block diagram of fully-rated converter wind turbine based on squirrel-cage induction generator (Caliao 2008).

Voltage equations:

$$\bar{v}_{ds} = -\bar{r}_s \bar{i}_{ds} - \bar{\omega}_s \bar{\psi}_{qs} + \frac{1}{\omega_b} \frac{d\bar{\psi}_{ds}}{dt} \tag{3.87}$$

$$\bar{v}_{qs} = -\bar{r}_s \bar{i}_{qs} + \bar{\omega}_s \bar{\psi}_{ds} + \frac{1}{\omega_b} \frac{d\bar{\psi}_{qs}}{dt} \tag{3.88}$$

$$\bar{v}_{dr} = \bar{r}_r \bar{i}_{dr} - s\bar{\omega}_s \bar{\psi}_{qr} + \frac{1}{\omega_b} \frac{d\bar{\psi}_{dr}}{dt} \tag{3.89}$$

$$\bar{v}_{qr} = \bar{r}_r \bar{i}_{qr} + s\bar{\omega}_s \bar{\psi}_{dr} + \frac{1}{\omega_b} \frac{d\bar{\psi}_{qr}}{dt} \tag{3.90}$$

Flux equations:

$$\bar{\psi}_{ds} = -\bar{L}_{ss} \bar{i}_{ds} + \bar{L}_m \bar{i}_{dr} \tag{3.91}$$

$$\bar{\psi}_{qs} = -\bar{L}_{ss} \bar{i}_{qs} + \bar{L}_m \bar{i}_{qr} \tag{3.92}$$

$$\bar{\psi}_{dr} = \bar{L}_{rr} \bar{i}_{dr} - \bar{L}_m \bar{i}_{ds} \tag{3.93}$$

$$\bar{\psi}_{qr} = \bar{L}_{rr} \bar{i}_{qr} - \bar{L}_m \bar{i}_{qs} \tag{3.94}$$

where, $\bar{L}_{ss} = \bar{L}_s + \bar{L}_m$ and $\bar{L}_{rr} = \bar{L}_r + \bar{L}_m$

dq voltages behind a transient reactance:

$$\bar{e}_d = -\frac{\bar{\omega}_s \bar{L}_m}{\bar{L}_{rr}} \bar{\psi}_{qr} \tag{3.95}$$

$$\bar{e}_q = \frac{\bar{\omega}_s \bar{L}_m}{\bar{L}_{rr}} \bar{\psi}_{dr} \tag{3.96}$$

Torque equation:

$$\bar{T}_e = \frac{\bar{e}_d \bar{i}_{ds} + \bar{e}_q \bar{i}_{qs}}{\bar{\omega}_s} \tag{3.97}$$

Rotor mechanics equation:

$$2H \frac{d\bar{\omega}_r}{dt} = \bar{T}_m - \bar{T}_e \tag{3.98}$$

The rotor flux-orientated control was used in the generator-side converter controller. Figure 3.29 shows the vector diagram representing the operating conditions of an induction generator in a reference frame fixed to the rotor flux (thus $\bar{\psi}_{qr} = 0$). As shown in Figure 3.29,

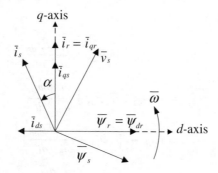

Figure 3.29 Vector diagram representation of the operating conditions of an induction generator in a reference frame fixed to the rotor flux.

the rotor flux is aligned with the d axis, which rotates at the synchronous speed $\overline{\omega}$ (Vas, 1990; Krause *et al.*, 2002).

From Eq. (3.94), $\overline{\psi}_{qr} = \overline{L}_{rr}\overline{i}_{qr} - \overline{L}_m\overline{i}_{qs} = 0$; therefore, the rotor current $\overline{i}_{qr}$ is

$$\overline{i}_{qr} = \frac{\overline{L}_m}{\overline{L}_{rr}}\overline{i}_{qs} \tag{3.99}$$

Using the expressions for the dq voltages behind a transient reactance $\overline{e}_d$ and $\overline{e}_q$ [from Eqs. (3.95) and (3.96)], the electromagnetic torque $\overline{T}_e$ [given in Eq. (3.97)] is calculated as

$$\overline{T}_e = \frac{\overline{L}_m}{\overline{L}_{rr}}\left(-\overline{\psi}_{qr}\overline{i}_{ds} + \overline{\psi}_{dr}\overline{i}_{qs}\right) = \frac{\overline{L}_m}{\overline{L}_{rr}}\overline{\psi}_{dr}\overline{i}_{qs} \tag{3.100}$$

For an FRC-IG, since the rotor is short circuited, $\overline{v}_{qr} = \overline{v}_{dr} = 0$. Further, if $\overline{\psi}_{qr} = 0$, then $d\overline{\psi}_{qr}/dt = 0$ and if $\overline{\psi}_{dr}$ is a constant then $d\overline{\psi}_{dr}/dt = 0$. Therefore, the rotor voltage equations [Eqs. (3.89) and (3.90)] can be simplified to

$$\overline{v}_{dr} = \overline{r}_r\overline{i}_{dr} = 0 \tag{3.101}$$

$$\overline{v}_{qr} = \overline{r}_r\overline{i}_{qr} + s\overline{\omega}_s\overline{\psi}_{dr} = 0 \tag{3.102}$$

From Eq. (3.102), the slip speed can be obtained as

$$s\overline{\omega}_s = \frac{\overline{r}_r\overline{i}_{qr}}{\overline{\psi}_{dr}} \tag{3.103}$$

Substituting Eq. (3.101) into Eq. (3.93), $\overline{\psi}_{dr}$ is obtained as

$$\overline{\psi}_{dr} = -\overline{L}_m\overline{i}_{ds} \tag{3.104}$$

Substituting Eq. (3.104) into Eqs. (3.100) and (3.103), the electromagnetic torque and slip speed can be rewritten as

$$\overline{T}_e = -\frac{\overline{L}_m^2}{\overline{L}_{rr}}\overline{i}_{ds}\overline{i}_{qs} \tag{3.105}$$

$$s\overline{\omega}_s = \frac{\overline{r}_r}{\overline{L}_m}\frac{\overline{i}_{qr}}{\overline{i}_{ds}} \tag{3.106}$$

Substituting Eq. (3.99) into Eq. (3.106):

$$s\overline{\omega}_s = \frac{\overline{r}_r}{\overline{L}_m}\frac{\overline{i}_{qs}}{\overline{i}_{ds}} \tag{3.107}$$

With $\overline{i}_{dr} = 0$, Eq. (3.91) reduces to:

$$\overline{\psi}_{ds} = -\overline{L}_{ss}\overline{i}_{ds} \tag{3.108}$$

Substituting Eq. (3.99) into Eq. (3.92), the following equation was obtained:

$$\overline{\psi}_{qs} = -\overline{L}_{ss}\overline{i}_{qs} + \overline{L}_m\frac{\overline{L}_m}{\overline{L}_{rr}}\overline{i}_{qs} = -\left(\overline{L}_{ss} - \frac{\overline{L}_m^2}{\overline{L}_{rr}}\right)\overline{i}_{qs} = -\overline{L}'\overline{i}_{qs} \tag{3.109}$$

Where $\overline{L}' = \overline{L}_{ss} - (\overline{L}_m^2/\overline{L}_{rr})$. Substituting $\overline{\psi}_{ds}$ and $\overline{\psi}_{qs}$ from Eqs. (3.108) and (3.109) into Eqs. (3.87) and (3.88), the stator voltages in the steady state are given by

$$\overline{v}_{ds} = -\overline{r}_s\overline{i}_{ds} + \overline{\omega}_s\overline{L}'\overline{i}_{qs} \tag{3.110}$$

$$\overline{v}_{qs} = -\overline{r}_s\overline{i}_{qs} - \overline{\omega}_s\overline{L}_{ss}\overline{i}_{ds} \tag{3.111}$$

The stator voltage $\overline{v}_{ds}$ includes the voltage $\overline{\omega}_s\overline{L}'\overline{i}_{qs}$ and $\overline{v}_{qs}$ includes the voltage $-\overline{\omega}_s\overline{L}_{ss}\overline{i}_{ds}$. These terms give the cross-coupling of the dq axes voltages with dq axes currents. It follows that the d-axis stator voltage is also affected by the q-axis stator current, and the q-axis stator voltage is also affected by the d-axis stator current. To eliminate the coupling effect, $\overline{\omega}_s\overline{L}'\overline{i}_{qs}$ and $-\overline{\omega}_s\overline{L}_{ss}\overline{i}_{ds}$ are added in the control system. Then $\overline{i}_{ds}$ is controlled through $\overline{v}_{ds}$, and $\overline{i}_{qs}$ is controlled through $\overline{v}_{qs}$ independently. The flux and torque control loops of the generator-side converter controller are shown in Figure 3.30 (Caliao, 2008; Anaya-Lara et al., 2009).

In the flux control loop, the reference d-axis stator current i_{ds}^{-ref} sets the air-gap flux level. The reference d–axis current is compared with its actual value and the error signal is regulated by the PI controller. The PI controller output and the decoupling term are added to obtain the d-axis stator voltage. The reference q-axis stator current is obtained using the generator torque-speed curve and Eq. (3.105). This is compared with its actual value and the error signal

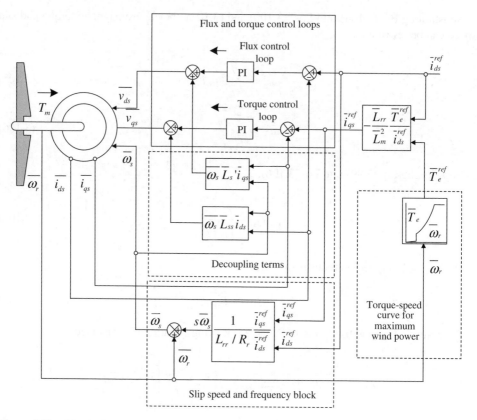

Figure 3.30 Block diagram of rotor flux orientated control of the grid-side converter (Caliao, 2008; Anaya-Lara *et al.*, 2009).

is regulated by the PI controller. The output of the controller is added to the decoupling term to determine the *q*-axis stator voltage.

As shown in Figure 3.30, the generator terminal frequency was controlled by adding the rotor speed and slip speed given in Eq. (3.107).

The ratio of the stator reactance to the stator resistance for a large induction machine is much higher than that of a small induction machine (Krause *et al.*, 2002). For example, this ratio is about 20 for a 2 MW induction generator that is employed for a wind turbine (in contrast to a ratio of 2 for a 3 HP induction machine). Therefore, the PI controllers defined by Eqs. (3.110) and (3.111) can be simplified to (Anaya-Lara *et al.*, 2009):

$$\overline{v}_{ds} = \overline{\omega}_s \overline{L}' \overline{i}_{qs} \tag{3.112}$$

$$\overline{v}_{qs} = -\overline{\omega}_s \overline{L}_{ss} \overline{i}_{ds} \tag{3.113}$$

Hence, the two PI controllers can be simplified as shown in Figure 3.31 without the decoupling terms.

The network-side controller is controlled as described in Section 3.5.5.

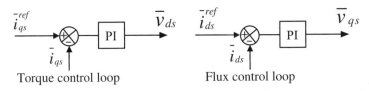

Torque control loop   Flux control loop

Figure 3.31 Simplified generator controller.

3.7 FRC-WT Power System Damper

Large scale integration of wind generation will influence the levels of damping of the network. Inadequate network damping causes instabilities, the most frequent of which are the electromechanical modes of oscillations. These oscillations are in the range of 0.1–2 Hz and occur in the rotor of the synchronous machines.

Studies show that wind turbines can contribute to damping power system oscillations. This capability has been demonstrated for both fixed speed induction generator (FSIG) and doubly fed induction generator (DFIG) wind turbines. A damping facility for FRC-WT is incorporated into the grid side converter controller. The responsibilities of the grid side converter controller then include providing support to damp power system oscillations, maintaining the dc-link voltage and providing voltage and frequency control for grid code compliance. When providing the grid support, the FCWT grid side controller maintains the decoupling between the grid side and generator side controllers.

A FRC-WT with power oscillation damping controller was connected into a single bus equivalent of the large power system. Small signal analysis and time domain simulations were used to design and evaluate the performance of the controllers. The burden on the dc link capacitance when providing the damping of power system oscillations was also evaluated.

3.7.1 Power System Oscillations Damping Controller

Figure 3.32 shows the fully-rated converter wind farm (FRC-WF) connected into a single bus equivalent of a large network. The large power system was modelled by its lumped inertia. A network oscillation was initiated by generating a voltage sag at the point of connection. The network disturbance causes changes in the electrical power and frequency of the network.

The change in the network frequency serves as an input signal to the FRC-WT power system oscillations damping controller (PDC). Figure 3.33 shows the power system oscillations damping controller. The change in the network frequency passes through a wash-out filter to eliminate control contributions under steady-state conditions. After this, it passes through a compensator that provides the gain and phase shift for a positive contribution to network damping.

The washout filter time constant (Tw) was chosen to be 20 seconds. The phase compensator time constants were T1 = 1 and T2 = 0.9. The value of the gain KPDC was 1 or 2. These values were chosen by observing how the dominant eigenvalues of the network shifted in the complex plane. The values were selected so as to damp a local mode of oscillation of approximately 1 Hz, which is caused by the swinging of the FRC-WF against the large network.

Figure 3.34 shows the change in active power ΔPn of the network during a disturbance. The network is in a stable condition when ΔPn is located to the left of the vertical axis. This

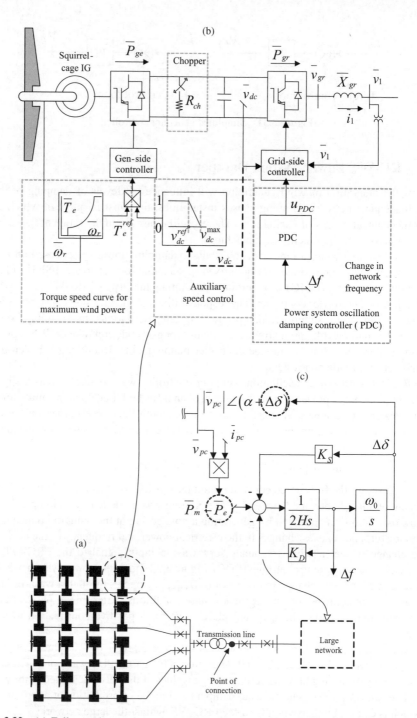

Figure 3.32 (a) Fully-rated converter wind farm (FRC-WF) connected to a large power system; (b) Schematic diagram of the fully-rated converter wind turbine (FRC-WT) with power system oscillation damping controller (PDC); (c) Large power system model.

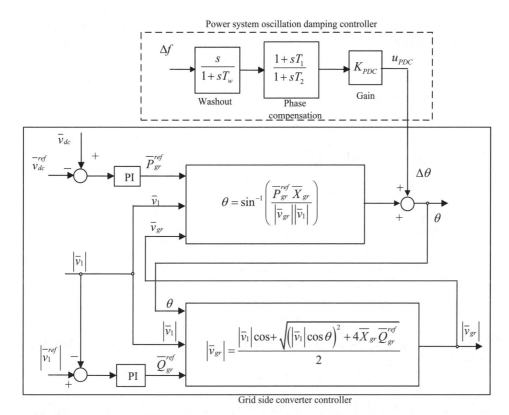

Figure 3.33 Schematic diagram of the grid side converter controller with power system oscillation damping controller (PDC).

means that the damping component of the active power of the network is directed to the left of the vertical axis. The system will then exhibit positive damping. After a small disturbance the system will have damped oscillation.

The network is unstable when ΔPn is located to the right of the vertical axis. The damping component of the active power is then directed to the right of the vertical axis. The system will then exhibit negative damping.

The natural damping characteristic of the network can be improved by adding a signal that can contribute positively to the damping component of the active power. The power oscillation damping controller will then act on the damping component to prevent ΔPn from shifting to the right-hand side of the vertical axis. In this case the damping controller will control the angle theta (θ) in the grid side converter controller so that ΔPn will stay on the left-hand side of the vertical axis in Figure 3.34.

3.7.2 Influence of Wind Generation on Network Damping

The influence of wind generation on the network damping was investigated by using FSIG and FRC-WT technologies. FSIG-based generation was first connected to a large power system.

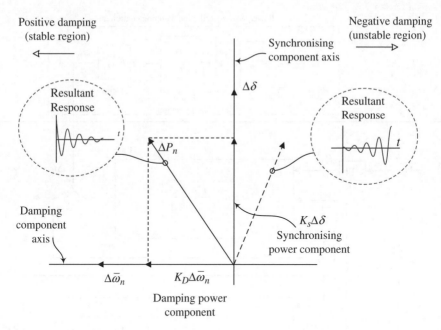

Figure 3.34 Components of the active power of the network during a disturbance (Caliao, 2008; Caliao *et al.*, 2010).

The FSIG-based generation capacity was increased from 10% to 15% and then to 20% of the power system capacity. The FSIG-based generation was then replaced by an FRC-WT-based generation. The FRC-WT-based generation capacity was increased from 10% to 15% and then to 20% of the power system capacity. Initially, the FRC-WT do not have damping facility. The power system was rated at 21 000 MVA. The influence of FSIG and FRC-WT generation on network damping can be assessed by observing the dominant eigenvalues and damping ratio of the network.

Figure 3.35 shows the dominant eigenvalues of the network with different capacity of the FSIG and FRC-WT generation. It is shown that the FSIG wind farm was oscillating at about 1.0 Hz with respect to the large power system. The FRC-WT wind farm was also oscillating at about 1.0 Hz with respect to the large power system. It is seen that the eigenvalues moved away from the imaginary line when the capacity of the FSIG wind farm was increased, whereas they move towards the imaginary line in the case of increased FRC-WT based wind generation. At 20% capacity the eigenvalues are very close to the imaginary line. Figure 3.36 shows the damping ratio when the amount of wind generation is increased.

3.7.3 Influence of FRC-WT Damping Controller on Network Damping

The damping controller was added to the FRC-WT. Figure 3.37 shows the dominant oscillatory mode in the network of approximately 1 Hz (6.6 rad/s), whereby the wind farm is oscillating

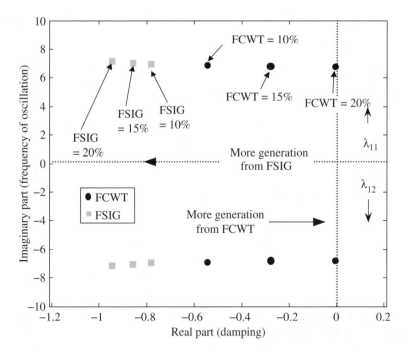

Figure 3.35 Comparison of the impact of FRC-WT and FSIG generation capacities on the network dominant eigenvalues (Caliao, 2008; Caliao *et al.*, 2010).

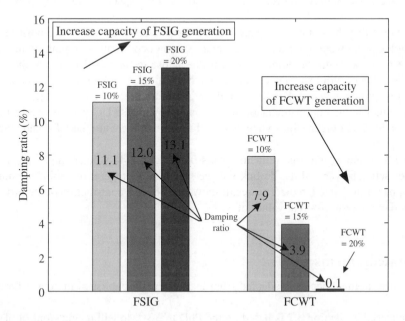

Figure 3.36 Network damping ratio with increasing capacity of wind generation (Caliao, 2008; Caliao *et al.*, 2010).

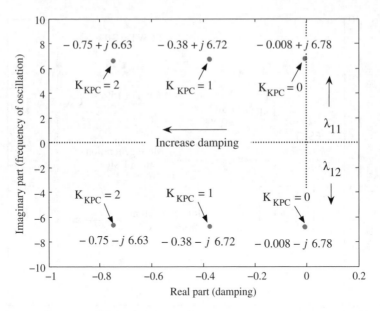

Figure 3.37 Influence of KPDC on network dominant eigenvalues ($\lambda 11$ and $\lambda 12$) (KPDC = 0, no damping) (Caliao, 2008; Caliao *et al.*, 2010).

against the main system (with FRC-WT generation at 20% capacity of the large power system capacity).

It is shown that the dominant eigenvalues of the network are close to the positive side of the complex plane when the FRC-WT generators do not contribute damping (with KPDC = 0) to the network. With the dominant eigenvalues close to the positive side of the complex plane, the network will experience oscillations that will take a long time to damp out. When damping is introduced (with KPDC = 1), the eigenvalues are shifted to the left of the imaginary line indicating improvement on the damping. When KPDC was increased to 2, the network's dominant eigenvalues were observed to shift further into the left-hand side of the complex plane.

Figure 3.38 shows the time domain results of the grid side voltage angle, the grid side converter active power and the dc-link voltage when the voltage at the point of connection was dropped from 1.0 to 0.5 pu. This result shows the positive contribution of the FRC-WT to network damping for different values of KPDC.

Acknowledgements

The material in this chapter is adapted from that originally published in Anaya-Lara *et al.*, 2009.

The material in Section 3.7 is based on the PhD thesis 'Modelling and Control of a Fully Rated Converter Wind Turbine', University of Manchester, 2008, by Nolan D. Caliao and used with his permission.

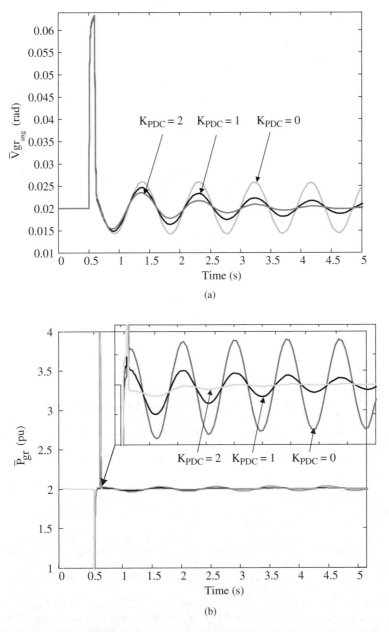

Figure 3.38 Time domain results of the influence of KPDC on eigenvalues (λ11 and λ12), changes in the grid side converter voltage angle, grid side active power and dc-link voltage (KPDC = 0, no damping) (Caliao, 2008; Caliao *et al.*, 2010). (a) Network-side converter voltage angle; (b) Network-side converter active power; (c) dc link voltage. (*Continued*)

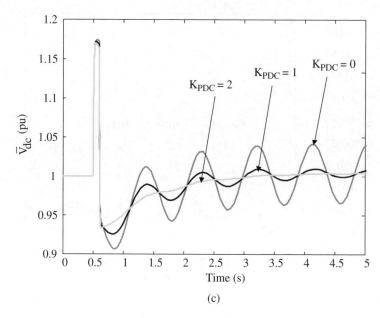

Figure 3.38 (*Continued*)

References

Ackermann, T. (ed.) (2005) *Wind Power in Power Systems*, Wiley & Sons Ltd.

Akhmatov, V., Nielsen, A.H. and Pedersen, J.K. (2003) Variable-speed wind turbines with multi-pole synchronous permanent magnet generators. Part I: Modelling in dynamic simulation tools. *Wind Engineering*, **27**, 531–548.

Anaya-Lara, O., Jenkins, N., Ekanayake, J. *et al.* (2009) *Wind Generation Systems – Modelling and Control*, Wiley & Sons Ltd, ISBN 0-470-71433-6.

Caliao, N. (2008) Modelling and Control of a Fully-Rated Converter Wind Turbine. PhD Thesis, University of Manchester.

Caliao, N., Ramtharan, G., Ekanayake, J. and Jenkins, N. (2010) Power oscillation damping controller for fully rated converter wind turbines, UPEC.

Chen, Z. and Spooner, E. (1998) Grid interface options for variable-speed permanent-magnet generators. *IEE Proceedings – Electric Power Applications*, **145**(4), 273–283.

Fitzgerald, A.E., Kingsley, C. Jr. and Umans, S.D. (1992) *Electrical Machinery*, McGraw-Hill Series.

Fox, B., Flynn, D., Bryans, L. *et al.* (2007) Wind Power Integration: Connection and System Operational Aspects. *IET Power and Energy Series* **50**, ISBN-10: 0863414494.

Grauers, A. (1996) Design of Direct Driven Permanent Magnet Generators for Wind Turbines. MSc Thesis, Chalmers University of Technology, Rep. No. 292 L.

Hanselman, D. C. (2006) *Brushless Permanent Magnet Motor Design*, Magna Physics Publishing.

Hindmarsh, J. and Renfrew, A. (1996) *Electrical Machines and Drive Systems*, Butterworth-Heinemann.

Krause, P.C., Wasynczuk, O. and Sudhoff, S.D. (2002) *Analysis of Electrical Machinery and Drive Systems*, 2nd edn, IEEE Press, Wiley & Sons, Inc., Piscataway, NJ.

Kundur, P. (1994) *Power System Stability and Control*, McGraw-Hill.

Ong, C.-M. (1998) *Dynamic Simulation of Electric Machinery*, Prentice-Hall, ISBN 0-13-723785-5.

Vas, O. (1990) *Vector Control of AC Machines*, Oxford University Press, New York.

Wood, A.J. and Wollenberg, B.F. (1996) *Power Generation, Operation, and Control*, 2nd edn, Wiley & Sons Ltd.

4

Offshore Wind Farm Electrical Systems

4.1 Typical Components

Figure 4.1 shows a schematic with the typical components in an offshore wind farm installation. These may be grouped into wind turbine level, wind farm level and offshore transmission level (power systems) (Anaya-Lara *et al.*, 2013).

Figure 4.2 sketches the control boundaries between these subsystems (Anaya-Lara *et al.*, 2013). In general terms, the control objectives at the different levels are as follows:

Wind turbine level:
- Maximise power capture.
- Optimise power quality.
- Provide support to power system operation.
- Mitigate turbine loads.

Wind farm level:
- Optimise power quality.
- Minimise wake losses and electrical losses in cables.

Offshore transmission:
- Provide support to power system operation.
- Comply with Grid Code requirements.

4.2 Wind Turbines for Offshore – General Aspects

To maximise the potential of working offshore larger turbines tend to be used, having longer blades mounted on taller towers than onshore designs (giving hub heights of up to 130 metres).

Offshore Wind Energy Generation: Control, Protection, and Integration to Electrical Systems, First Edition.
Olimpo Anaya-Lara, David Campos-Gaona, Edgar Moreno-Goytia and Grain Adam.
© 2014 John Wiley & Sons, Ltd. Published 2014 by John Wiley & Sons, Ltd.
Companion Website: www.wiley.com/go/offshore_wind_energy_generation

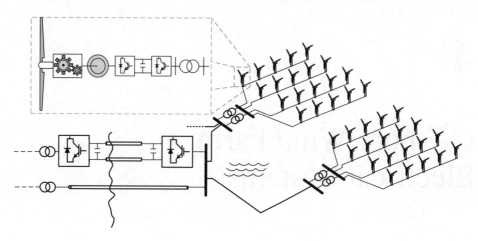

Figure 4.1   Offshore wind energy system.

There is no consensus on how large offshore wind turbines will become, although most agree that no physical limit prevents building turbines larger than 10 MW. A critical issue in developing very large machines is that the physical scaling laws do not allow some components to be increased in size without a change in the fundamental technology. New size-enabling technologies will be required to extend the design space for offshore wind turbines beyond the current

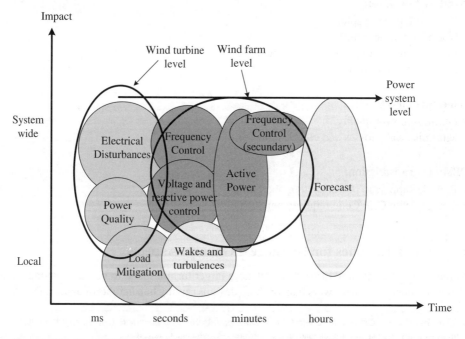

Figure 4.2   Control boundaries in an offshore wind energy system (Anaya-Lara *et al.*, 2013).

5–7 MW size. Some of these technologies may include a variety of stiffer, lightweight composite materials and new composite manufacturing methods; lightweight, low-maintenance drive-trains; lightweight, high-speed downwind rotors; direct-drive generators; and large gearbox and bearing technologies that can tolerate slower rotational speeds and larger scales (Musial *et al.*, 2010). In addition, other considerations with offshore wind turbines are (E.ON Climate and Renewables, 2012):

- Offshore wind turbines must be highly robust and reliable to avoid costly standstills, since they cannot be accessed at some times because of wind, waves and other weather conditions.
- There may be damage in addition to normal 'wear and tear' from sources such as:
  o Corrosion due to aggressive salty environment.
  o High wear due to heavy mechanical loads and higher utilisation.
  o Failures affecting gearbox, generator, transformer, blades and transmission cables.
- Size and capacity of offshore wind turbines have increased considerably. At the moment, wind turbines with a capacity of up to 7 MW are being tested.
- While at onshore sites the size of wind turbines is often limited by restrictions on height and rotor diameter, offshore wind turbines do not encounter these limits in the open sea.
- Wind turbine components need to be pre-assembled onshore to allow quick installation offshore. The most important pre-assembly works are the stacking of tower segments.
- From arrival at site the installation of a wind turbine takes a minimum of 24 hours.

4.3   Electrical Collectors

As the power capacity of offshore wind farms increases, the adequacy of the wind farm electrical system becomes critical. The efficiency, cost, reliability and performance of the overall wind farm will depend, to a great extent, on the electrical system design (Quinonez-Varela *et al.*, 2007). The overall function of the electrical collector system is to collect power from individual wind turbines and maximise the overall energy generation. An electrical collector can be designed using different layouts depending on the wind farm size and the desired level of collector reliability. There are various arrangements for wind farm collector systems employed in existing offshore wind farms, whilst others are in a conceptual stage. Four basic designs shown in Figure 4.3 are discussed below.

a. *Radial design*
   The most straightforward arrangement of a wind farm collector system is a radial design (Figure 4.3a), in which a number of wind turbines are connected to a single cable feeder within a string. The maximum number of wind turbines on each string feeder is determined by the capacity of the generators and the maximum rating of the subsea cable in the string. This design offers the benefits of being simple to control and also inexpensive because the total cable length is smaller with tapering of cable capacity away from the hub being possible. The major drawback of this design is its poor reliability as in the case of a cable or switchgear fault at the hub end of the radial string, it has the potential to prevent all downstream turbines from exporting power.

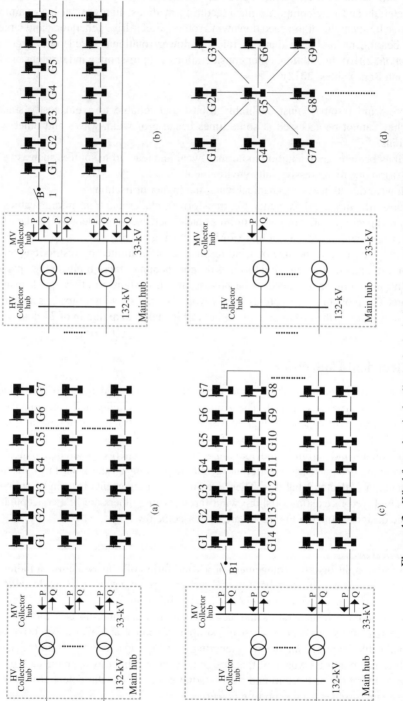

Figure 4.3 Wind farm electrical collector basic designs (Quinonez-Varela *et al.*, 2007).

b. *Single-sided ring design*

With some additional cabling, ringed layouts can address some of the security of supply issues of the radial design by incorporating a redundant path for the power flow with a string. The additional security comes at the expense of longer cable runs for a given number of wind turbines and higher cable rating requirements throughout the string circuit. A single-sided ring design, illustrated in Figure 4.3b, requires an additional cable run from the last wind turbine (i.e. G7) to the hub. This cable must be able to handle the full power flow of the string (e.g. 35 MW in a 5 MW seven-turbine string) in the event of a fault in the primary link to the hub end (denoted by the open breaker B1 in Figure 4.3b).

c. *Double-sided ring design*

Figure 4.3c illustrates a double-sided ring design (Sanino *et al.*, 2006; Quinonez-Varela *et al.*, 2007). In this configuration the last wind turbine in one string is interconnected to the last wind turbine in the next string (e.g. G7 to G8 as shown in Figure 4.3c). If the full output power of the wind turbines in one of the strings were to be delivered through the other string, then the cable at the hub end of the latter needs to be sized for the power output of double the number of wind turbines.

d. *Star design*

The star design shown in Figure 4.3d aims to reduce cable ratings and to provide a high level of security for the wind farm as a whole, since one cable outage only affects on wind turbine in general. A cost implication of this design is the more complex requirements at the wind turbine in the centre of the star (e.g. G5 in Figure 4.3d).

- *Power losses*

Key issues associated with power losses are the value of lost energy and the power factor at which the cables are operated. Further losses within the collector system may be introduced depending on the turbine technology.

- *Voltage levels*

Under normal operating conditions, the voltage levels throughout the collector system must be within permissible limits as defined by Grid Codes. In general, these are within $\pm 10\%$ of the rated voltage (e.g. 33 kV). Voltage regulation equipment shall act to adjust the busbar voltages by continuous modulation of the wind farm reactive power, within its reactive power range, or by tap-changing of the collector hub transformers.

- *Redundancy*

The purpose of redundancy in a wind farm collector system is as a contingency to keep as many wind turbines as possible connected during a fault event or maintenance downtime. Experience has demonstrated that repairing times required in offshore wind farms are significantly longer than those in onshore installations. For example the estimated repair time of a failed subsea power cable may vary from 720 h during summer to 2160 h in the winter.

- *Costs*

When compared with onshore wind farms, the investment costs and final cost of electricity produced in an offshore installation are higher. The additional expenditure of submarine foundations, turbine installation, subsea plant and grid connection constitute the major differences. Typically, the collector system has been seen as a minor contributor to the total investment cost of wind farms, especially in small-scale installations with radial collectors.

4.3.1   Wind Farm Clusters

The Wind Farm Cluster concept was created and developed as a natural evolution for wind energy management (Gesino *et al.*, 2011). In the past, wind turbines were grouped into wind farms, and nowadays wind farms are being grouped into Wind Farm Clusters. The aim of this structure is to allow the TSOs to manage wind energy more like a conventional power plant, avoiding some natural aspects of wind energy spread, such as the fluctuating nature of the wind, the sparse distribution of the wind farms and the existence of different generator technologies, amongst other issues.

A Wind Farm Cluster is a logical aggregation of existing physical wind farms, which are connected to the same grid node. The main goal of this structure is to allow the large-scale management of wind energy and the operation of wind farms as conventional power plants. The proper management and control of Wind Farm Clusters require advanced techniques and control strategies combined with high-tech wind energy forecast techniques. These advanced control mechanisms allow wind farm clusters to provide grid operators with active and reactive power control, wind power reserve and congestion management (facilitating the current and future Grid Code requirements).

4.4   Offshore Transmission

The offshore transmission network is used to transfer power from the wind farm to the power network located onshore. Most operational wind farms are connected through high-voltage alternating current (HVAC) networks with frequency of 50 Hz or 60 Hz; however, high-voltage direct current (HVDC) may be more cost effective and have lower electrical losses over longer distances. Key factors that influence the transmission technology are the distance between the offshore wind farm and mainland, and the amount of power transmitted (Volker *et al.*, 2008; CIGRE, 2009). Figure 4.4 shows transmission technology choices for HVAC and HVDC in relation to distance and power (Ackermann, 2005).

4.4.1   HVAC Transmission

HAVC transmission is used for most wind farm applications and provides a simple and economical connection from large offshore wind farms to the onshore grid, but there is a limitation for the cable length. Offshore wind turbine collector networks are medium voltage, typically 33 kV (Hopewell *et al.*, 2006), then at the offshore substation the voltage level is stepped up to the transmission level (e.g. 150 kV), (Zubiaga *et al.*, 2009). To improve the transmission efficiency and capacity, reactive power compensators such as static compensators (STATCOMs) and static VAr compensators (SVCs) may be incorporated into the system (Chaudhary *et al.*, 2008; Singh *et al.*, 2009).

The main components for an offshore HVAC transmission system are (Ackermann, 2005) (Figure 4.5):

- Offshore wind farm.
- Offshore substation (step-up transformer).
- Three-core polyethylene submarine cables (XPLE) to shore.

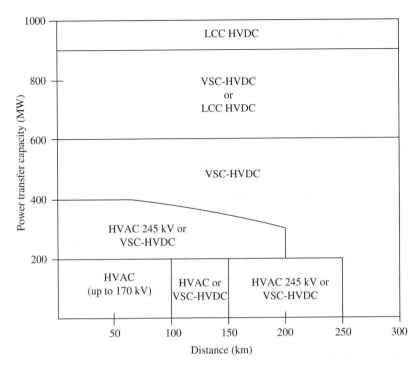

Figure 4.4   Choice of transmission technology for different wind farm capacity and distance from shore (Ackermann, 2005).

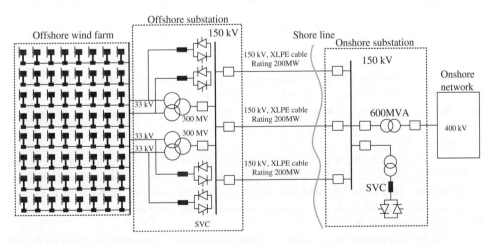

Figure 4.5   Basic configuration of HVAC interconnection of offshore wind farm to the grid (Ackermann, 2005).

- Static VAR compensator.
- Onshore substation.
- Ac submarine cables.

Some issues associated with HVAC transmission in offshore wind applications are (Bresesti *et al.*, 2007):

- Over long distances there are high power losses and large reactive power compensation devices are required at both ends of the line.
- The wind farm is synchronously coupled with the grid and any fault on the side of the wind farm will affect the grid and vice versa.
- High capacitance of the cables, may affect voltage shape.

The Thanet Wind Farm, located 12 km off the cost of Thanet, England, is the largest offshore wind farm connected via HVAC and started producing electricity in September 2010 (Thanet Offshore Wind Limited, 2005). The wind farm covers 35 km² and is located in average water depths of 20–25 m. There are 100 3-MW wind turbines giving 300 MW of generation capacity. The Thanet wind farm uses 33 kV as a collector voltage, in an offshore transformer station the voltage is raised to 132 kV.

4.4.2   HVDC Transmission

HVDC transmission systems in operation worldwide have been built using two different converter types:

a. Current Source Converter (CSC), which uses line-commutated switching devices (thyristors).
b. Voltage Source Converter (VSC), which uses self-commutated switching devices such as IGBT, MOSFET, and GTO.

Depending on the function and location of the converter stations, various HVDC configurations can be identified. These configurations are generic and applicable to both, current source and voltage source converter schemes (Agelidis *et al.*, 2006; Flourentzou, *et al.*, 2009).

- *Monopolar HVDC system*
  In the monopolar configuration, two converters are connected by a single pole line and either a positive or negative dc voltage is used. In Figure 4.6a there is only one insulated transmission conductor installed and the ground or sea provides the path for the return current. Alternatively, a metallic return conductor may be used as the return path.
- *Bipolar HVDC system*
  The bipolar configuration consists of two monopolar systems, one with positive polarity and the other with negative polarity, as shown in Figure 4.6b. Each terminal has two converters of equal rated voltage connected in series on the dc side. Normally the currents in the two poles are equal and there is negligible ground current. A bipolar scheme is the most commonly used configuration in applications where overhead lines are used to transmit power, as it

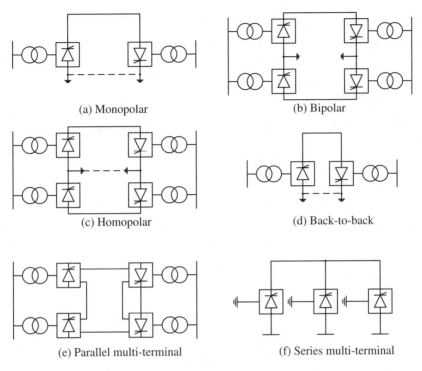

(a) Monopolar   (b) Bipolar

(c) Homopolar   (d) Back-to-back

(e) Parallel multi-terminal   (f) Series multi-terminal

Figure 4.6 Schematic diagrams of different HVDC circuit arrangements (Agelidis *et al.*, 2006; Flourentzou *et al.*, 2009).

is economically attractive to install an additional pole conductor. In case of failure of one pole, power transmission can continue on the other pole, increasing system reliability.

- *Homopolar HVDC system*
  In the homopolar configuration, shown in Figure 4.6c, two or more conductors have the negative polarity and can be operated with ground or a metallic return. With two poles operated in parallel, the homopolar configuration reduces the insulation costs, but the earth return current is the major disadvantage.

- *Back-to-back HVDC system*
  This is the common configuration for connecting two asynchronous ac systems. Two converter stations are located at the same site and a transmission line or cable is not needed. A block diagram of a back-to-back HVDC system is shown in Figure 4.6d. The two interconnected ac systems may have the same or different nominal frequencies.

- *Multi-terminal HVDC system*
  In this configuration there are more than two terminals; in practice no scheme has so far been operated with more than three terminals. Multi-terminal systems are generally divided into series and parallel systems as shown in Figure 4.6e and Figure 4.6f. The multi-terminal configuration imposes a significant complexity to achieve the control and communication between different converter substations, especially in the case of current source converter schemes.

Table 4.1 Examples of existing (and under construction) HVDC transmission systems using current source converter technology.

Project	Rating (MW)	DC voltage (KV)	Distance (km)	Commissioning year
Sumatra-Java (Indonesia)	3000	±500	700 (cable)	2013
Madeira (Brazil)	3150	±600	2500 (Overhead line)	2012
Xianjiaba-Shanghai (China)	6400	±800	2071 (Overhead line)	2011
SAPEI (Italy)	1000	±500	435 (Overhead line)	2010
Outaouais (Canada)	1250	±175	0	2009
NorNed (Norway-Netherlands)	600	±500	2X580	2008
Neptune (USA)	660	±550	105 (cable)	2007
Three Gorges-Shanghai (China)	3000	±500	1060 (Overhead line)	2006
Vizag II (India)	500	±88	0	2005
Rapid city (USA)	200	±13	0	2003
Moyle (USA)	500	2X250	700 (cable)	2001
SwePol (Sweden-Poland)	600	±450	245 (cable)	2000
Welch Monticello (USA)	600	450	0	1998
Chandrapur-Ramagundum	1000	2X205	0	1997
Kontek (Denmark-Germany)	600	400	119 (Overhead line) + 52 (cable)	1995
Sakuma (Japan)	300	±125	0	1993
Hydro Quebec(Canada-USA)	2250	±450	1100 (Overhead line)	1991
Vindhyachal (India)	500	2X69.7	0	1989
Zhoushan (China)	50	100	12 (Overhead line) + 42 (cable)	1987
Cross Channel 2(UK-France)	2000	±450	72 (cable)	1986
Highgate (USA)	200	±56	0	1985
Chateauguay (Canada–USA)	1000	2X140	0	1984

4.4.3 CSC-HVDC Transmission

The classic current source converter (CSC) is a mature technology and offers various benefits for bulk, long distance power transmission. Most existing dc transmission systems are based on this technology (Arrillaga and Watson, 2007). Table 4.1 summarises the most recent projects using CSC-HVDC technology. The first classic HVDC transmission link was installed in 1954. It used line-commutated current-source converters (LCC) with mercury-arc valves. The next development was in the 1960s by replacing the mercury-arc valves with semiconductor thyristor valves. Thyristor-based converters have now become the standard equipment for dc current source converter stations. The CSC always absorbs reactive power either in rectifier or conversion operation; therefore, reactive power compensation is always required.

The operation of a conventional HVDC transmission system is mainly affected by the network strength since the HVDC can be considered as a special load with a voltage and load angle. The strength of the system can be measured using the short circuit ratio (SCR) or the equivalent impedance at the bus. According to the IEEE Standard 1204–1997 (IEEE, 1997), the system SCR and Effective Short-Circuit Ratio (ESCR) are defined as follows:

$$SCR = \frac{S}{P_{dc}} \tag{4.1}$$

$$ESCR = \frac{S - Q_c}{P_{dc}} \tag{4.2}$$

where S is the ac system three-phase symmetrical short-circuit level in MVA at the converter terminals (ac bus), calculated at rated terminal voltage (1.0 pu), P_{dc} is the rated dc power of the terminals in MW, and Q_C is the three-phase fundamental MVAr at rated P_{dc} and rated terminal voltage. This includes ac filters and shunt capacitors. The system strength is classified as follows:

- System is strong if $SCR{>}3$ or $ESCR{>}2.5$
- System is weak if $3{>} SCR {>}2$ or $2.5{>} ESCR {>}1.5$
- System is very weak (very low SCR) if $SCR {<}2$ or $ESCR {<}1.5$.

HVDC transmission systems based on current source converter technology can be built as line-commutated converters (LCC) or capacitor-commutated converters (CCC).

4.4.3.1 Line-Commutated Converter (LCC)

The current source converter is capable of transferring the current in only one direction and the power flow reversal is achieved by dc voltage polarity reversal. A large dc inductor is connected in series at the dc terminals to get a smoother dc current. The current source converter can operate as a rectifier, where the firing angle of the converter is less than 90°, or as an inverter, in which case the firing angle is more than 90°. The basic building block of the converter station is the well-known six-pulse thyristor bridge or the 12-pulse thyristor configuration. With a six-pulse bridge the lowest order current harmonics are the 5th and 7th, while for the 12-pulse converter, the 11th and 13th are the lowest (Kundur, 1994; Arrillaga, 1998). Figure 4.7 shows the schematic of a line-commuted converter.

4.4.3.1.1 LCC-HVDC Control System
The main functions of the control system in an LCC-HVDC transmission system are to control the power flow between the sending and receiving end converters, and to protect the equipment against over-current and voltage stress. The basic desired features of an LCC-HVDC control system are (Kundur, 1994; IEEE Guide, 1997; Rashid, 2001; Agelidis *et al.*, 2006; Flourentzou *et al.*, 2009):

1. Limit maximum dc current: using constant-current control in the rectifier to protect the converter valves.
2. Maintain maximum dc voltage: using constant voltage control in the inverter to reduce the losses in transmission line and converter valves.

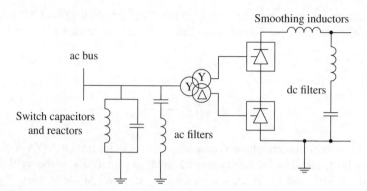

Figure 4.7   Line-commutated converter station.

3. Keep the ac reactive power demand low at either converter terminal. This means keeping the operating angles at the converters at the lowest level possible by using the minimum firing angle.
4. Prevent commutation failures at the inverter thus improving the stability of the power transmission.

The best way to understand the control principles of HVDC transmission is to study the voltage-current characteristics shown in Figure 4.8 (Kundur, 1994). These characteristics are used to illustrate the relationship between the dc voltage and current in steady state during power flow control on a dc line.

Under steady-state condition the control system of the inverter is designed to maintain the dc voltage by keeping the extinction angle (γ) constant. In this case the dc voltage droop increases as the dc current increases, as indicated by line DXFE and Eq. (4.3):

$$V_{dc} = \frac{3\sqrt{2}}{\pi} V_{l-l} \cos\gamma - \frac{3I_{dc}\omega L_c}{\pi} \tag{4.3}$$

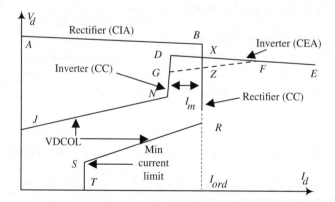

Figure 4.8 V_d-I_d characteristics for a two-terminal LCC-HVDC system (Kundur, 1994).

The inverter can operate with constant dc voltage as indicated by line GF, meaning that the extinction angle must increase beyond its minimum limit (always set as 18°). If the inverter is operated at either constant voltage or constant extinction angle then line BXZR is the normal operation of the rectifier, where the delay angle α is varied in order to keep the dc current constant. The delay angle α cannot be below its minimum set point (α_{min} always set as 5°). If α reaches its minimum point then the rectifier will operate at minimum constant ignition angle which is indicated by line AB, and defined as:

$$V_{dc} = \frac{3\sqrt{2}}{\pi} V_{l-l} \cos \alpha - \frac{3 I_{dc} \omega L_c}{\pi} \qquad (4.4)$$

The rectifier valves have limited thermal capability and cannot carry a large current; hence a maximum current limit is used in the control system to protect the valves. I_{dc} minimum is defined in the rectifier such as to ensure a minimum dc current to avoid the possibility of dc current extinction caused by the valve current dropping below the hold-on thyristor current. The operating point of the HVDC system is represented by points X or Z (the intersection of the rectifier and inverter characteristics). The operating point is achieved with the help of the converter transformer tap-changers. For the rectifier, the transformer tap-changer is used to keep the delay angle α at its normal operating range to meet the constant current setting I_{ref}. For inverter operation, the tap-changer is adjusted to meet the desired level of extinction angle control.

Another possible operating mode for the inverter is the constant current mode line DGN. The current demanded by the inverter is usually less than the current demanded by the rectifier by a current margin I_m that is typically about 10%. The current margin is chosen to be large enough to prevent the interaction between rectifier and inverter constant current modes due to current harmonics in the dc current. The current control at the inverter is activated when α reaches its minimum point. If for any reason such as low commutation voltage the line AB falls below either point X or Z, the operating point will shift to somewhere along the line DGN, then the inverter will control the dc current and the rectifier will control the dc voltage.

During low voltage conditions, it is not possible to maintain the dc current at its desired level. The low ac voltage will result in an increase in the reactive power consumption and in a low dc voltage, therefore increasing the risk of commutation failure. A voltage-dependant current-order limit (VDCOL) is used to reduce the dc current order when the voltage drops below a predetermined value. VDCOL is represented by line RST in the rectifier and line JN in the inverter. To prevent the inverter from operating in the rectifier region, and hence having power reversal, a minimum alpha limit is imposed, typically about 100–110°.

4.4.3.1.2 Control System Design

The control of power flow in an LCC-HVDC system is achieved by synchronizing the firing pulses generated by the trigger unit (Iwatta *et al.*, 1996). There are many control techniques but all of them use a voltage-controlled oscillator (VCO), and a ring counter to decouple the firing pulses from the commutation voltage. During normal operation the rectifier is operated with constant current control (CC) by changing the delay angle until α is less than the minimum limit, then the rectifier will operate at constant-ignition angle (CIA).

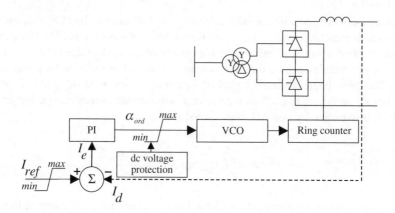

Figure 4.9   Block diagram of the rectifier control system.

Figure 4.9 shows the block diagram of the rectifier control system. The actual dc current flowing in the dc link is compared with the reference value I_{ord}. The resultant error signal I_e is then passed through a PI controller, the output of the PI controller is the alpha order α_{ord} which controls the frequency output of the VCO. The train of pulses generated by the VCO is fed to the ring counter, which has six or 12 stages (according to the converter type). Only one stage is ON at any time, therefore a complete set of six or 12 pulses is generated by the ring counter at equal intervals in one full cycle. A limit is used to prevent the dc current from exceeding the specific value to provide over-current protection to converter valves.

The control system of the inverter comprises of a dc voltage controller, extinction angle control, and current regulator. A control angle selector is placed to choose the desired control signal from the above controllers. The control system is similar to the rectifier control system including the comparing loop, PI controller, VCO, and counter ring. The block diagram of the inverter control system is shown in Figure 4.10.

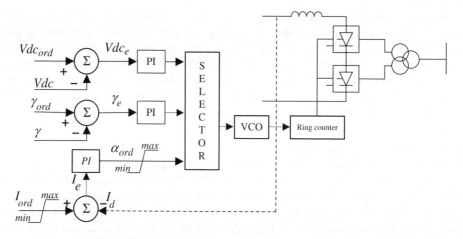

Figure 4.10   Block diagram of the inverter control system.

Some issues associated with LCC-HVDC transmission systems are (Kundur, 1994; Arrillaga, 1998):

- They require large reactive power for filtering purposes and to aid thyristor commutation. As a result, the converter station may be bulky and have a large footprint.
- They require start-up generators or synchronous compensator devices, which provide the necessary commutation voltage and reactive power compensation.
- They produce large amounts of harmonics and this makes the use of harmonic filters essential.
- They do not provide independent active and reactive power control.
- Power flow reversal necessitates that the dc link voltage polarity is reversed.
- There is a theoretical minimum power flow of 10% in the dc link to guarantee satisfactory operation.
- Only suitable for connection of ac networks with a high short-circuit ratio.
- Suffers from commutation failure.
- They require complex control strategies to achieve a multi-terminal configuration; as a result the number of terminals is limited.

4.4.3.2 Capacitor-Commutated Converter (CCC)

The capacitor-commutated converter is aimed to extend the application of CSC-HVDC transmission to systems with low short-circuit ratio. Figure 4.11 shows the schematic of a capacitor-commuted converter. The capacitor is connected in series with the converter transformer to reduce the commutation impedance of the converter, which in turn reduces the reactive power requirement of the converter. Increasing the size of the series capacitor can furthermore even allow operation at leading power factor (Meisingset and Gole, 2001). Other advantages over the LCC-HVDC include the ability to operate with weaker ac systems, and to deliver more power than an LCC-HVDC of similar size and cost. The negative impact of reduced commutation impedance is that additional stresses are imposed on the valves and converter transformers (higher di/dt).

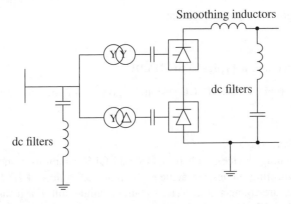

Figure 4.11 Capacitor-commutated converter station.

4.4.4   VSC-HVDC Transmission

High-voltage dc transmission systems based on voltage source converters (VSC-HVDC) were developed to address the shortcomings associated with CSC-HVDC systems. VSC converters use semiconductor switching devices which can be switched ON and OFF at any time in the waveform cycle such as IGBT, GTO and IGCT. This means that in VSCs, the current can be made to lead or lag the ac voltage, and so the converter can consume or supply reactive power to the connected ac network eliminating the need for reactive power compensation devices. Some advantages of the VSC technology are (Bahrman and Johnson, 2007; Cole and Belmans, 2009; Flourentzou *et al.*, 2009):

- In VSC-HVDC the converter is controlled using a pulse-width modulation (PWM) switching strategy that allows independent control of the magnitude and phase angle of the ac-side voltage. This makes it possible to control the real and reactive power independently.
- The use of PWM with a switching frequency in the range of 1 to 2 kHz is sufficient to separate the fundamental voltage from the sidebands, and suppress the harmonic components around and beyond the switching frequency components. This significantly reduces filtering requirements.
- Power flow can be reversed without the need to reverse the polarity of the dc link voltage.
- As the voltage source converter is capable of generating leading and lagging reactive power, the converter station can be used to provide voltage support to the ac network whilst transmitting the active power.
- Since the VSC actively controls the output current, the VSC-HVDC contribution to the fault current during ac faults is limited to rated current or less, and the converter can remain in operation to provide voltage support to the ac networks, provided the ac network remains stable.
- Black-start capability is provided, which is the ability to start/or restore power to a 'dead network'. This feature eliminates the need for a start-up generator in applications where footprint is critical or expensive such as with offshore wind farms.
- In cases where there is no need to transmit active power between the two ends, a VSC-HVDC can operate in 'sleep mode' (zero active power), and both converter stations operate as two independent STATCOMs to regulate the ac network voltages.
- VSC-HVDC can be controlled to provide fast frequency and/or damping support to the ac networks through active power modulation.

4.4.4.1   Components of a Typical VSC-HVDC

The schematic of a typical VSC-HVDC system is shown in Figure 4.12 (Arrillaga, 2007; Cole and Belmans, 2009).

4.4.4.1.1   VSC Converter

A VSC uses switching devices with turn ON and OFF capabilities such as IGBTs, with anti-parallel freewheeling diodes. To achieve the required ratings of HVDC transmission, a number of switches are connected in series to build a single unit. For instance, 3.3 kV valves are typically used in HVDC systems. The number of series-connected IGBTs required to

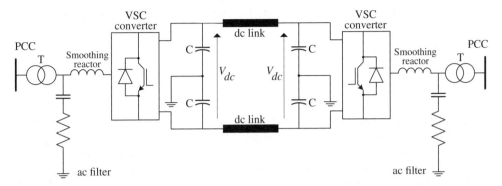

Figure 4.12   Schematic of a typical VSC-HVDC system (Arrillaga, 2007; Cole and Belmans, 2009).

withstand the converter rating is increased to provide a suitable redundancy if any sudden failure occurs in some of the IGBT switches.

The VSC switching devices can be connected in many ways to build different types of converter configurations such as: two-level converter, three-level converter or multi-level converter (more than three levels) as explained in detail in Appendix A.

4.4.4.1.2   Coupling Transformers
In a VSC-HVDC system, the ac filters are installed between the smoothing reactor and the transformer. The transformer is not exposed to harmonics or dc voltages like transformers in a conventional HVDC system. Hence, a normal transformer without special arrangement/specifications is used. In general, the functions of the transformers are:

- To transform the voltage of the ac system to a level suitable for the converter.
- To provide a reactance between the converter and the ac system to limit and control the ac current.

4.4.4.1.3   Smoothing Reactors
The smoothing reactors facilitate controlling the active and reactive power flow by regulating the currents through them. They also act as harmonic filters to reduce the harmonics in voltage and current waveforms, and to limit the short-circuit current.

4.4.4.1.4   Ac Harmonic Filters
The switching of the IGBTs using high-frequency PWM techniques produces only high-order harmonic components around and beyond the switching frequency components. Hence, only passive high-pass filters are needed to eliminate high-order harmonic components to prevent them from causing malfunction of the ac system equipment, or from generating radio and telecommunication disturbances.

4.4.4.1.5   Dc Capacitors
Two capacitors of the same size are located at the end of the dc sides. The main objective of the dc capacitors is to keep the power balance during transient conditions and to reduce the

harmonics on the dc side. Eq. (4.5) is used to calculate the dc capacitor size (Cuiqing, *et al.*, 2005):

$$\tau = \frac{0.5CV_{dc}^2}{S_n} \tag{4.5}$$

where τ is the time constant, C is the dc capacitor, V_{dc} is the dc voltage and S_n is the nominal power of the converter. The time constant is equal to the time needed to charge the capacitor from zero to the rated voltage V_{dc} if the converter is supplied with a constant active power equal to S_n. To obtain a smooth and sinusoidal voltage waveform, the dc voltage must be greater than the ac voltage (RMS value) for PWM by a factor of 2, and by a factor of 1.74 for PWM with third-harmonic injection (Yazdani and Iravani, 2010).

4.4.4.1.6   Dc Cables
The dc cables are typically XLPE (Cross-Linked Poly-Ethylene) polymer extruded cables. Polymeric cables are the best choice for HVDC transmission because of their mechanical strength, flexibility and low weight. The insulation system is triple-extruded, that is, the conductor screen, the insulation and the insulation screen are all extruded simultaneously (Weimers, 2000; Sood, 2004).

4.4.4.2   VSC-HVDC Steady-State Model

The equivalent ac/dc/ac circuit of a VSC-HVDC as that in Figure 4.12, is shown in Figure 4.13 (the ac harmonics filter is neglected). To simplify the description, the interface transformers are represented with their leakage reactance, X_T, in series with the winding resistance R_T, while the smoothing reactor is represented by its reactance X_F. v_c is the VSC converter output voltage and v_s is the terminal voltage of the ac network. V_{dc} is the converter dc voltage; R_{dc} and L_{dc} are the resistance and inductance of the dc link, respectively.

The ac system voltage and current relationships at the sending and receiving-ends are shown in the vector diagrams of Figure 4.14. In these diagrams δ is the phase angle difference between the ac system voltage and the VSC converter voltage, θ_s is the phase angle difference between the ac system voltage and the ac current flowing from or into the converter, while θ_c is the phase angle difference between the VSC converter voltage and the ac current. The subscripts

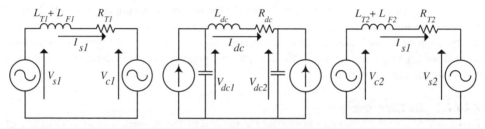

Figure 4.13   Equivalent circuit of a VSC-HVDC transmission system.

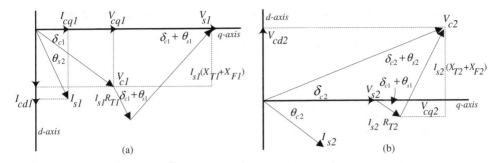

Figure 4.14 Vector diagram representation of a VSC-HVDC system: (a) sending end; (b) receiving end.

'1' and '2' represent the association with the sending-end and receiving-end converter stations, respectively.

The following expressions apply to the active and reactive powers at the sending and receiving ends of the link at fundamental frequency:

$$P_1 = \frac{v_{s1} v_{c1}}{X_{T1} + X_{F1}} \sin \delta_1 \tag{4.6}$$

$$P_2 = \frac{v_{s2} v_{c2}}{X_{T2} + X_{F2}} \sin \delta_2 \tag{4.7}$$

$$Q_{sc1} = \frac{v_{s1}(v_{s1} - v_{c1} \cos \delta_1)}{X_{T1} + X_{F1}} \tag{4.8}$$

$$Q_{sc2} = \frac{v_{s2}(v_{s2} - v_{c2} \cos \delta_2)}{X_{T2} + X_{F2}} \tag{4.9}$$

The following relationships are obtained from the vector diagram in Figure 4.14a (the sending-end VSC converter acting as rectifier):

The voltage at the VSC converter terminal is:

$$V_{c1} \cos \delta_1 = V_{s1} - I_{s1} R_{T1} \cos(\theta_1 + \delta_1) - I_{s1} X_{T1} \sin(\theta_1 + \delta_1) \tag{4.10}$$

From the vector diagram, the following relation is derived:

$$I_{s1} X_{T1} \cos(\theta_1 + \delta_1) - I_{s1} R_{T1} \sin(\theta_1 + \delta_1) = V_{c1} \sin \delta_1 \tag{4.11}$$

Using trigonometric equations and making some rearrangements, the load angle is

$$\tan \delta_1 = \frac{I_{s1} X_{T1} \cos \theta_1 - I_{s1} R_{T1} \sin \theta_1}{v_{c1} + I_{s1} X_{T1} \sin \theta_1 + I_{s1} R_{T1} \cos \theta_1} \tag{4.12}$$

The dq components of the current and voltage at the VSC converter terminal are:

$$I_{sd1} = -I_{s1} \sin(\delta_1 + \theta_1) \quad \& \quad I_{sq1} = I_{s1} \cos(\delta_1 + \theta_1) \tag{4.13}$$

$$v_{cd1} = -V_{c1} \sin \delta_1 \quad \& \quad v_{cq1} = V_{c1} \cos \delta_1 \tag{4.14}$$

Similarly, the relationship for the receiving-end (VSC acting as inverter) is derived from Figure 4.14b as follows.

The voltage at the VSC terminals is:

$$V_{c2} \cos \delta_2 = V_{s2} + I_{s2}R_{T2} \cos(\theta_2 + \delta_2) + I_{s2}X_{T2} \sin(\theta_2 + \delta_2) \tag{4.15}$$

The load angle is expressed as:

$$\tan \delta_2 = \frac{I_{s2}X_{T2} \cos \theta_2 - I_{s2}R_{T2} \sin \theta_2}{v_{c2} + I_{s2}X_{T2} \sin \theta_2 + I_{s2}R_{T2} \cos \theta_2} \tag{4.16}$$

4.4.4.3 VSC-HVDC Dynamic Model

Figure 4.15 shows the typical basic structure of a VSC-HVDC transmission system. The AC source voltages denoted by (v_{sa}, v_{sb}, v_{sc}) are three ideal sinusoidal waveforms displaced by 120° in the time domain. The three-phase currents denoted by (i_{sa}, i_{sb}, i_{sc}) are source currents flowing from the ac source to the VSC converter or in the reverse direction. The voltages and currents at the VSC terminal are denoted by (v_{ca}, v_{cb}, v_{cc}) and (i_{ca}, i_{cb}, i_{cc}) respectively; R_T and L_T are the resistance and inductance of the coupling transformer, R_F and L_F are the resistance and inductance of the coupling reactor. The dc side is modelled as series inductive (L_{dc}) and resistive (R_{dc}) circuits, while C, V_{dc} and I_{dc} are the dc link capacitor, dc link voltage and dc link current, respectively. In Figure 4.15 the left converter (VSC$_1$) is assumed to operate as a rectifier transforming the power from the ac source side to the VSC dc side. The right converter (VSC$_2$) is operated as an inverter and transforms the power from the VSC dc side to the ac side. The subscripts '1' and '2' denote the variable belonging to the sending-end station and receiving-end station, respectively.

Applying Kirchhoff voltage laws to the VSC-HVDC circuit in Figure 4.15 while adopting the time t in seconds and all quantities in per unit, the dynamic equations for the rectifier side are:

$$\frac{di_{sabc}}{dt} = -\frac{R_1}{L_1}i_{sabc} + \frac{1}{L_1}\left(v_{sabc} - v_{cabc}\right) \tag{4.17}$$

where $R_1 = R_{T1} + R_{F1}$ and $L_1 = L_{T1} + L_{F1}$.

The dynamic equations for the inverter station are:

$$\frac{di_{sabc2}}{dt} = -\frac{R_2}{L_2}i_{sabc2} + \frac{1}{L_2}\left(v_{cabc2} - v_{sabc2}\right) \tag{4.18}$$

where $R_2 = R_{T21} + R_{F2}$ and $L_2 = L_{T2} + L_{F2}$.

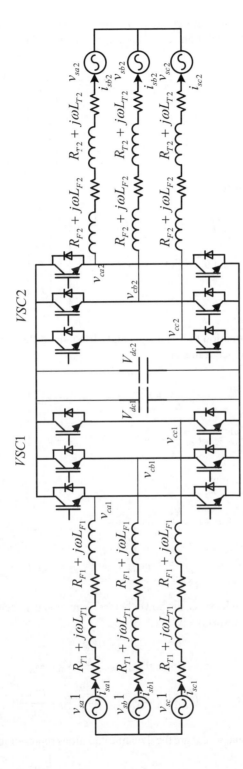

Figure 4.15 VSC-HVDC circuit.

To simplify the design of the control system, the three-phase quantities are expressed in the dq reference frame using the Park transformation matrix given in Eq. (4.19) as:

$$
\begin{bmatrix} f_d \\ f_q \\ f_0 \end{bmatrix} = \begin{bmatrix} \cos\theta & \cos\left(\theta - \dfrac{2\pi}{3}\right) & \cos\left(\theta - \dfrac{4\pi}{3}\right) \\ \sin\theta & \sin\left(\theta - \dfrac{2\pi}{3}\right) & \sin\left(\theta - \dfrac{4\pi}{3}\right) \\ \dfrac{1}{2} & \dfrac{1}{2} & \dfrac{1}{2} \end{bmatrix} \begin{bmatrix} f_a \\ f_b \\ f_c \end{bmatrix} = [T] \begin{bmatrix} f_a \\ f_b \\ f_c \end{bmatrix} \tag{4.19}
$$

And the inverse Park transformation is:

$$
\begin{bmatrix} f_a \\ f_b \\ f_c \end{bmatrix} = \begin{bmatrix} \cos\theta & \cos\left(\theta - \dfrac{2\pi}{3}\right) & \cos\left(\theta - \dfrac{4\pi}{3}\right) \\ \sin\theta & \sin\left(\theta - \dfrac{2\pi}{3}\right) & \sin\left(\theta - \dfrac{4\pi}{3}\right) \\ \dfrac{1}{2} & \dfrac{1}{2} & \dfrac{1}{2} \end{bmatrix}^{-1} \begin{bmatrix} f_d \\ f_q \\ f_0 \end{bmatrix} = [T]^{-1} \begin{bmatrix} f_d \\ f_q \\ f_0 \end{bmatrix} \tag{4.20}
$$

Applying Park's transformation to Eqs. (4.17) and (4.18), the differential equations for the rectifier side in the dq reference frame are:

$$
\begin{aligned}
\frac{di_{sd1}}{dt} &= -i_{sd1}\frac{R_1}{L_1} + \omega i_{sq1} + \frac{1}{L_1}\left(v_{sd1} - v_{cd1}\right) \\
\frac{di_{sq1}}{dt} &= -i_{sq1}\frac{R_1}{L_1} - \omega i_{sd1} + \frac{1}{L_1}\left(v_{sd1} - v_{cd1}\right)
\end{aligned} \tag{4.21}
$$

Similarly, for the inverter side:

$$
\begin{aligned}
\frac{di_{sd2}}{dt} &= -i_{sd2}\frac{R_2}{L_2} + \omega i_{sq2} + \frac{1}{L_2}\left(v_{cd2} - v_{sd2}\right) \\
\frac{di_{sq2}}{dt} &= -i_{sq2}\frac{R_2}{L_2} - \omega i_{sd2} + \frac{1}{L_2}\left(v_{cd2} - v_{sd2}\right)
\end{aligned} \tag{4.22}
$$

If the converter is assumed lossless, the power balance equation between the dc and ac sides in the rectifier and inverter, respectively, are:

$$
\begin{aligned}
V_{dc1}I_{dc} &= \frac{3}{2}\left(v_{cd1}i_{sd1} + v_{cq1}i_{sq1}\right) - CV_{dc1}\frac{dV_{dc1}}{dt} \\
V_{dc2}I_{dc} &= \frac{3}{2}\left(v_{cd1}i_{sd1} + v_{cq1}i_{sq1}\right) + CV_{dc2}\frac{dV_{dc2}}{dt}
\end{aligned} \tag{4.23}
$$

Equations (4.21), (4.22) and (4.23) are the dynamic equations representing the VSC-HVDC.

In a converter with sinusoidal PWM the relationship between the modulation index M, dc link voltage and the *dq* components of the ac voltage is given by:

$$v_{cd} = \frac{1}{2}MV_{dc}\cos\delta \tag{4.24}$$

$$v_{cq} = \frac{1}{2}MV_{dc}\sin\delta \tag{4.25}$$

The instantaneous total power in the time domain is expressed as:

$$S(t) = v_a(t).i_a(t) + v_b(t).i_b(t) + v_c(t).i_c(t) \tag{4.26}$$

Applying the park transformation to Eq. (4.26) yields:

$$p(t) = \frac{3}{2}[v_d(t).i_d(t) + v_q(t).i_q(t)] \tag{4.27}$$

$$q(t) = \frac{3}{2}[-v_d(t).i_q(t) + v_q(t).i_d(t)] \tag{4.28}$$

4.4.4.4   VSC-HVDC Control System

Using PWM modulation, the VSC controller can adjust the modulation index, M, and converter phase angle, θ, to give any combination of voltage and phase shift and hence, control the active power, reactive power, ac voltage and dc link voltage.

Vector control is the most popular technique to modify M and θ in order to achieve the desired control function. In vector control the three-phase rotating voltage and current are transformed to the *dq* reference before calculating the modulation index and the phase angle. It is then synchronized with the ac three-phase voltages using a Phase-Locked Loop (PLL). The controlled variable is forced to follow the reference values to generate the desired values of M and θ, which are then used in the PWM technique to control the VSC switching.

The control system of the VSC converters is composed of an inner and outer controller. The main function of the inner controller is to control the currents in the complete range of operation, and to ensure that the converter is not overloaded to protect the converter from transient currents during system disturbances. The outer controller is responsible for supplying the reference values to the inner controller. Figure 4.16 shows an example of a VSC control system including: the inner and outer controller, and the PLL.

There are four possible control modes to choose from for the outer controller of the VSC converter (Cuiqing *et al.*, 2005):

- dc voltage control mode
- active power control mode
- ac voltage control mode
- reactive power control mode.

The choice of controller to be added to the VSC converter (not all controllers can be used at the same time), depends on the application of the VSC-HVDC, but in all cases an option for

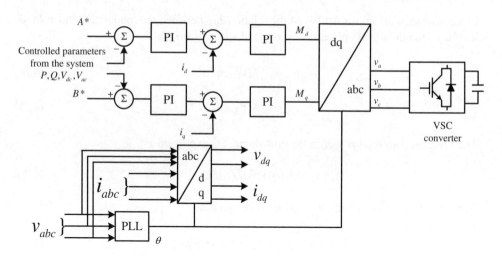

Figure 4.16   Block diagram of the VSC-HVDC control system.

the dc voltage controller is necessary to ensure the power balance between the sending and receiving converters. In this discussion, the operation mode of the converters is chosen such that VSC_2 controls the dc voltage and the ac voltage at B_2, whilst VSC_1 controls the active power and the ac voltage at B_1.

4.4.4.4.1   Inner Controller Design

In order to design an inner current controller the cross-coupling terms in Eq. (4.21) are decoupled as follows (Giddani *et al.*, 2009):

$$v_{sd1} = u_{d1} - \omega L_1 i_{q1} + v_{d1} \tag{4.29}$$

$$v_{sq1} = u_{q1} + \omega L_1 i_{d1} + v_{q1} \tag{4.30}$$

Substituting (4.29) and (4.30) into Eq. (4.21) then

$$\frac{d}{dt}\begin{bmatrix} i_{sd1} \\ i_{sq1} \end{bmatrix} = \begin{bmatrix} -R_1/L_1 & 0 \\ 0 & -R_1/L_1 \end{bmatrix}\begin{bmatrix} i_{sd1} \\ i_{sq1} \end{bmatrix} + \begin{bmatrix} 1/L_1 & 0 \\ 0 & 1/L_1 \end{bmatrix}\begin{bmatrix} u_{d1} \\ u_{q1} \end{bmatrix} \tag{4.31}$$

where u_d and u_q are new variables obtained from the inner controllers which regulate the dq-axes currents.

The values of u_d and u_q are defined as:

$$u_{d1} = k_{pi}\left(i^*_{d1} - i_{d1}\right) + k_{ii}\int\left(i^*_{d1} - i_{d1}\right)dt \tag{4.32}$$

$$u_{q1} = k_{pi}\left(i^*_{q1} - i_{q1}\right) + k_{ii}\int\left(i^*_{q1} - i_{q1}\right)dt \tag{4.33}$$

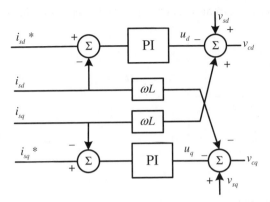

Figure 4.17 Block diagram of the inner controller.

where k_{pi} and k_{ii} are the proportional and integral gains of the current controller, and the superscript $*$ refers to the reference value. The inner current loop for converter VSC$_1$ is designed according to Eqs. (4.29) and (4.30) and the block diagram of the inner controller is shown in Figure 4.17.

Replacing the integral part of Eqs. (4.32) and (4.33) by new auxiliary control variables z_d and z_q, the following equations are obtained:

$$u_d = k_{pi} \left(i_{sd}^* - i_{sd} \right) + k_{ii} z_d \tag{4.34}$$

$$u_q = k_{pi} \left(i_{sq}^* - i_{sq} \right) + k_{ii} z_q \tag{4.35}$$

where,

$$\frac{dz_d}{dt} = -i_{sd} + i_{sd}^* \tag{4.36}$$

$$\frac{dz_q}{dt} = -i_{sq} + i_{sq}^* \tag{4.37}$$

Substituting Eqs. (4.34)–(4.37) into Eq. (4.31) then

$$\frac{d}{dt}\begin{bmatrix} i_{sd} \\ z_d \\ i_{sq} \\ z_q \end{bmatrix} = \begin{bmatrix} -(R_1 + k_{pi})/L_1 & K_{ii}/L_1 & 0 & 0 \\ -1 & 0 & 0 & 0 \\ 0 & 0 & -(R_1 + k_{pi})/L_1 & K_{ii}/L_1 \\ 0 & 0 & -1 & 0 \end{bmatrix} \begin{bmatrix} i_{sd} \\ z_d \\ i_{sq} \\ z_q \end{bmatrix} + \begin{bmatrix} K_{pi}/L_1 & 0 \\ 1 & 0 \\ 0 & K_{pi}/L_1 \\ 0 & 1 \end{bmatrix} \begin{bmatrix} i_{sd}^* \\ i_{sq}^* \end{bmatrix}$$

$$\tag{4.38}$$

From Eq. (4.38) the *dq* currents are defined in the Laplace domain as follows:

$$
\begin{bmatrix} i_{sd}(s) \\ z_d(s) \\ i_{sq}(s) \\ z_q(s) \end{bmatrix} = \cfrac{1}{s^2 + \cfrac{(R_1 + K_{pi})}{L_1}s + \cfrac{K_{ii}}{L_1}} \begin{bmatrix} s & K_{ii}/L_1 & 0 & 0 \\ -1 & s + (R_1 + K_{pi})/L_1 & 0 & 0 \\ 0 & 0 & s & K_{ii}/L_1 \\ 0 & 0 & -1 & s + (R_1 + K_{pi})/L_1 \end{bmatrix} \begin{bmatrix} i_{sd}^* \\ i_{sq}^* \end{bmatrix}
$$

$$(4.39)$$

Therefore, the Laplace transfer function of the inner current controller is:

$$
G(s) = \cfrac{\cfrac{K_{pi}}{L_1}\left(s + \cfrac{K_{ii}}{K_{pi}}\right)}{s^2 + \cfrac{(R_1 + K_{pi})}{L_1}s + \cfrac{L_{ii}}{L_1}}
$$

$$(4.40)$$

From the characteristic equation of the current controller in the denominator of Eq. (4.40), the proportional and integral gains of the current controller are defined as follows (Douangsyla *et al.*, 2004):

$$k_{pi} = 2\zeta\omega_n L_1 - R_1 \tag{4.41}$$

$$K_{ii} = \omega_n^2 L_1 \tag{4.42}$$

where, ω_n and ζ are the natural frequency and damping factor, respectively. These values must be selected carefully to ensure stable operation over the full operating range.

4.4.4.4.2 Outer Controller Design
The outer controller consists of the dc voltage controller, ac voltage controller, active and reactive power controller and the frequency controller. In this section different outer controllers are described.

4.4.4.4.2.1 Active and Reactive Power Controller
Based on the *dq* power Eqs. (4.27) and (4.28) the active and reactive powers delivered to the point of common coupling (PCC) are:

$$p(t) = \frac{3}{2}\left[v_{sd}(t)i_{sd}(t) + v_{sq}(t)i_{sq}(t)\right] \tag{4.43}$$

$$q(t) = \frac{3}{2}\left[-v_{sd}(t)i_{sq}(t) + v_{sq}(t)i_{sd}(t)\right] \tag{4.44}$$

where v_{sd} and v_{sq} are the dq voltages at the PCC. For balanced steady-state operation $v_{sq} = 0$, therefore Eqs. (4.43) and (4.44) can be expressed as:

$$p(t) = \frac{3}{2}v_{sd}(t).i_{sd}(t) \tag{4.45}$$

$$q(t) = \frac{-3}{2}v_d(t).i_q(t) \tag{4.46}$$

Hence the *dq* current reference values are:

$$i_{sd}^* = \frac{2P^*}{3v_{sd}} \tag{4.47}$$

$$i_{sq}^* = \frac{-2Q^*}{3v_{sd}} \tag{4.48}$$

Figure 4.18 shows the schematic block diagram of the active and reactive power controller.

4.4.4.4.2.2   dc and ac Voltage Controller

In the dc voltage controller, the measured dc voltage is compared with the reference voltage, the error signal between the voltage reference and the measured value is fed to a PI controller to generate the *d*-axis reference current (i_{sd}^*). Eq. (4.49) is the mathematical model of the dc current controller:

$$i_{sd}^* = k_{pdc}(V_{dc}^* - V_{dc}) + k_{idc}\int (V_{dc}^* - V_{dc})dt \tag{4.49}$$

Similarly, the mathematical model of the ac current controller is:

$$i_{sq}^* = k_{pac}(V_{ac}^* - V_{ac}) + k_{iac}\int (V_{ac}^* - V_{ac})dt \tag{4.50}$$

where k_{pdc}, k_{idc}, k_{pac}, k_{iac} are the proportional and integral gains of the dc voltage controller and the ac voltage controller, respectively. Figure 4.19 shows the schematic block diagram of the dc and ac voltage controller.

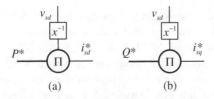

Figure 4.18   Block diagram of the active and reactive controller.

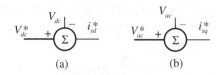

Figure 4.19   Block diagram of the dc and ac voltage controller.

4.4.5   Multi-Terminal VSC-HVDC Networks

Although most existing dc schemes have been realised as point-to-point connections, CSC-HVDC or VSC-HVDC transmission can also be configured in multi-terminal arrangements. VSC-HVDC technology seems to present suitable options for multi-terminal configurations. The active power reversal in a multi-terminal system based on voltage source converters is achieved by changing the dc current direction without the need to change the dc link voltage polarity and operating mode of the converter. This results in instantaneous power reversal without increasing the stress on the dc cable and interfacing transformer, as it occurs in CSC-HVDC (Long *et al.*, 1990; Weixing and Boon Teck, 2002; Wulue *et al.*, 2006).

The active power flow between converters in a multi-terminal VSC-HVDC system is controlled using combinations of direct active power control in the converters that control active power, and dc voltage droop in the converters used to control the dc link voltage (Lie *et al.*, 2008; Adam *et al.*, 2009). This increases control flexibility of multi-terminal VSC-HVDC systems to respond to any network alteration resulting from active power imbalance (due to add/loss of major load or generation), and ac faults. A reliable coordinated control system is required between the converter stations in addition to robust dc circuit breakers to improve the chance of system recovery from dc faults or loss of converters. Some other features of multi-terminal VSC-HVDC transmission systems are (Arrillaga and Watson, 2007; Adam *et al.*, 2009):

- Increased power flow control flexibility along desired routes.
- Improved system reliability and stability. During the outage of a VSC converter, or loss of a dc link cable, re-routing of the power is available through other paths of the system; therefore the impact of losing a major generating unit is minimised.
- Suitable for creation of transmission systems connecting different ac networks and integrating renewable energy resources. The converter in the dc network could operate as rectifier or inverter to transmit the power to/from a specific point in the dc network.

4.4.5.1   Configurations of Multi-Terminal dc Transmission Systems

There are different configurations for multi-terminal systems for wind farm integration, which differ in the number VSC converters and their arrangement (Padiyar, 2005; Hongbo and Ekstrom, 1998; Haileselassie *et al.*, 2008; Jiebei and Booth, 2010).

Figure 4.20 shows two options for using multi-terminal VSC-HVDC for offshore wind farm integration: double-input single-output and double-input double-output configurations. The power flow control takes into consideration that the power direction is from the wind farm into the ac grid (no power reversal).

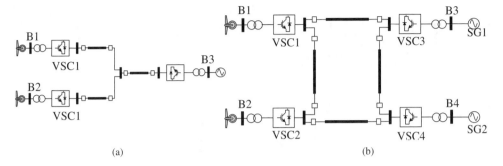

Figure 4.20 Examples of possible multi-terminal configurations for wind farm integration: double-input single-output configuration. (b) double-input double-output configuration.

Multi-terminal VSC-HVDC can be used instead of conventional ac transmission to accommodate the power between ac networks. In this case the HVDC system is used to create a large power pool that may facilitate the following objectives (Padiyar, 2005):

- Increased connection of renewable power to the ac network without the need for bulky storage systems.
- Improved resilience of the ac networks attached to this power pool against power and frequency instability problems during local loss of generation or major load.
- Power trading between these regional networks in real time with full control over power flows.
- Inherent reactive power capability in each converter station to provide voltage control at the point of common coupling.
- Flexibility of asynchronous operation.

Figure 4.21 shows the example of a four-terminal VSC-HVDC network, which may be suitable to interconnect offshore wind generation to different countries (i.e. transnational dc network).

Offshore oil and gas platforms use electricity generated from gas turbines located at offshore sites. Multi-terminal VSC-HVDC systems can be used to supply the oil and gas platforms and to transmit the offshore wind power to the onshore ac network instead of gas turbines (Figure 4.22). The potential benefits of a multi-terminal dc system configuration in this case are reduced operational costs, increased system reliability and reduced CO_2 emissions (Haileselassie *et al.*, 2008).

4.5   Offshore Substations

Offshore substations are used to increase the voltage before power is transmitted to the onshore grid. The major purpose of the offshore substation is to increase the voltage so that electrical losses can be avoided during power transfer via cables. Figure 4.23 shows the schematic of an offshore substation (Bazargan, 2007).

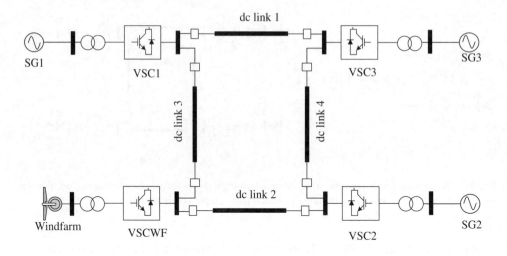

Figure 4.21 Example of a four-terminal network.

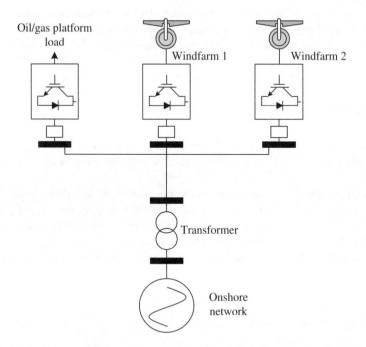

Figure 4.22 Multi-terminal HVDC system for offshore wind farms, oil/gas platform and onshore network (Haileselassie *et al.*, 2008).

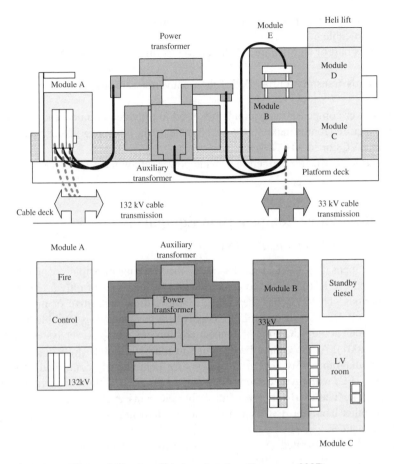

Figure 4.23   An offshore substation (Bazargan, 2007).

An offshore substation is typically delivered as one component after contract by supplier and weighs generally from 1800 tonnes to 2200 tonnes. The platform level is generally 25–30 metres above sea level and has an area of 800–1200 m^2 (Bazargan, 2007). One substation can typically support up to 500 MW of output from the wind farm. With an increase in wind farm size, the number of substations may increase too. These offshore substations are generally not service-based but still have a small workshop available inside it.

The major components of an offshore substation are divided into three categories:

- Components related to electrical systems.
- Components related to facilities.
- Components related to structure.

The main components related to the electrical systems are as follows:

- The major component of the offshore substation is the transformer. The transformer steps up the voltage according to requirements for optimum power transfer and minimisation of losses.

- Back-up diesel generator is also made available on the substation in case of loss of power via the export cable.
- Switchgear is part of an offshore substation to discriminate between the export cables and array cables.
- If the onwards transmission is HVDC then converters are also installed in the offshore substation.
- Reactive power compensation equipment is used to provide optimum reactive power compensation required for the maximum power transfer to the onshore grid (e.g. bank of reactors, capacitors, SVCs or STATCOMs).
- The substation is properly earthed to provide power safety in case of safety hazard or short circuit.

4.6   Reactive Power Compensation Equipment

4.6.1   Static Var Compensator (SVC)

Static Var Compensators (SVCs) are based on the working principles of a variable shunt susceptance and use fast thyristor controllers with settling times of only a few fundamental frequency periods. The SVC has the ability to either draw capacitive or inductive current from the network. By suitable control of its equivalent reactance, it is possible to regulate the voltage magnitude at the SVC point of connection, thus enhancing power system performance (Acha *et al.*, 2001). An SVC may be designed in many different ways. Figure 4.24 shows the schematic diagrams of typical arrangements for continuously controlled SVCs, namely fixed capacitor (FC) with thyristor-controlled reactor (TCR), and thyristor-switched capacitor (TSC) with TCR. The thyristors are controllable elements enabling smooth control of the TCR. On the other hand, the TSC is a fast-switched element that achieves voltage regulation in a stepwise manner.

When the SVC is operated in a voltage control mode, it presents fast-settling times of almost one period in the case of the FC/TCR arrangement. In the case of switched elements (TSC/TCR), the response time is usually 30–60 μsec depending upon the SVC configuration and system strength. To achieve such a high-speed response, it is necessary to assess the type and size of the power components as well as the control scheme according to the specific network configuration and operation requirements.

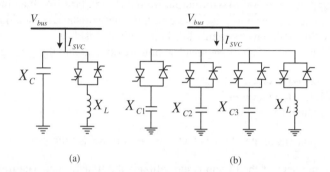

(a)   (b)

Figure 4.24   Typical structures of and SVC: (a) a TCR with a fixed capacitor. (b) TCR with a TSC.

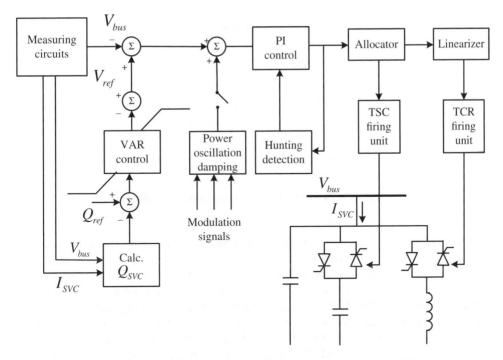

Figure 4.25 Simplified voltage control block diagram of a typical SVC.

Figure 4.25 shows a simplified block diagram of a voltage control scheme for a typical SVC application (Song and Johns, 1999; Hingorani and Gyugyi, 2000). Essential elements in this control scheme are: the measuring circuits, the voltage regulator, the allocator, the lineariser, and the TCR and TSC firing circuits.

The *measuring circuits* measure the voltages and currents at different points of the network providing the necessary information to the SVC for control and protection purposes. The measured signals are conditioned to provide suitable control to the other blocks of the control system. The *voltage regulator* performs the closed-loop voltage control and uses a PI that provides the total SVC susceptance reference required to minimise the error. When a *power oscillation damping* function is included it acts on the voltage regulation to provide damping of slow electro-mechanical swings in the power system. The *allocator* converts the susceptance reference from the voltage regulator into specific information, which is then processed to determine the number of reactive power banks that must be switched on and the required firing angle. The lineariser converts the susceptance from the allocator to a firing angle α. To maintain the same control response over the entire SVC operating range, the angle α is determined as a nonlinear function of the susceptance reference order. This function is normally given as a table that is derived from Eq. (4.51).

$$1 - X_L B(\alpha) = \alpha + \frac{\sin(\pi\alpha)}{\pi} \tag{4.51}$$

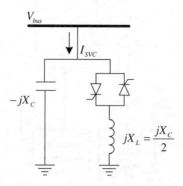

Figure 4.26   SVC capacitance and inductance values.

where $B(\alpha)$ is the susceptance of the TCR fired at the angle α. The stability controller supervises the operation of the voltage controller incorporating *hunting detection and gain adjustment* functions. Unstable operation (hunting), which may take place during weak system operating conditions, will be detected and the gain of the PI would be reduced by half, aiming to achieve stable operation. The *TSC-TCR firing units* compute the angles α and generate firing pulses for the TSC and TCR thyristor valves. Other control elements that can be added to the control circuit are the dc and reactive power controller. With the addition of these elements the control system becomes more robust and efficient; however, its complexity increases considerably.

Figure 4.26 shows the SVC capacitance and inductance values for a FC/TCR topology. These parameters can be determined according to the reactive power compensation requirements, Q_{SVC}, for a given application, as follows:

$$X_C = \frac{V_{bus}^2}{Q_{SVC}} \tag{4.52}$$

$$X_L = \frac{X_C}{2} \tag{4.53}$$

Each SVC has its own reactive power-firing angle characteristic, $Q_{SVC}(\alpha)$, which is a function of the inductive and capacitive reactances. The effective reactance X_{SVC}, of the SVC as a function of the firing angle, α, can be calculated using the fundamental frequency equivalent reactance of the TCR, X_{TCR}, using:

$$X_{TCR} = \frac{\pi X_L}{(\sigma - \sin \sigma)} \tag{4.54}$$

and

$$\sigma = 2(\pi - \alpha) \tag{4.55}$$

where X_L is the reactance of the inductor and σ and α are the conduction and firing angles of the thyristor, respectively. At $\alpha = 90°$ the TCR conducts fully and the equivalent reactance X_{TCR}

becomes X_L. At $\alpha = 180°$ the TCR is blocked and its equivalent reactance becomes infinite. The total effective reactance of the SVC, including the TCR and capacitive reactances, is determined as:

$$X_{SVC} = \frac{X_C X_{TCR}}{X_C + X_{TCR}} \tag{4.56}$$

that, as a function of the conduction angle, σ, can be expressed by:

$$X_{SVC} = \frac{\pi X_C X_L}{X_C (\sigma - \sin \sigma) - \pi X_L} \tag{4.57}$$

and then as a function of the firing angle, α, it is given as:

$$X_{SVC} = \frac{\pi X_C X_L}{X_C [2 (\pi - \alpha) + \sin 2\alpha] - \pi X_L} \tag{4.58}$$

Using Eq. (4.58), Q_{SVC} as a function of the firing angle can be calculated using the following fundamental relationship:

$$Q_{SVC} = V_{bus}^2 \frac{X_C [2 (\pi - \alpha) + \sin 2\alpha] - \pi X_L}{\pi X_C X_L} \tag{4.59}$$

4.6.2 Static Compensator (STATCOM)

A STATCOM is a VSC-based device, with a linear capacitor acting as voltage source (Song and Johns, 1999; Hingorani and Gyugyi, 2000; Acha *et al.*, 2001). A STATCOM is used for provision of dynamic reactive power compensation to power transmission systems, providing voltage support and increased transient stability margins. The most basic configuration of a STATCOM consists of a two-level VSC with a capacitor acting as a dc energy storage device, a coupling transformer connected in shunt with the ac system, and associated control circuits as shown in Figure 4.27. As the voltage source is created from a capacitor the STATCOM has very little real power capability.

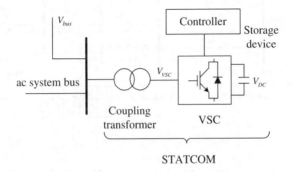

Figure 4.27 Schematic diagram of a STATCOM connected to an ac system.

Table 4.2 Power exchange between a STATCOM and the ac system.

Voltage relation	Power exchange		
	STATCOM	⇔	ac system
$|V_{VSC}| > |V_{bus}|$	Q	⇒	
$|V_{VSC}| < |V_{bus}|$		⇐	Q
$\delta < 0$	P	⇒	
$\delta > 0$		⇐	P

A summary of the power exchange between a STATCOM and the ac network is shown in Table 4.2.

The size of the capacitor in a STATCOM may be selected by analytical methods (Acha *et al.*, 2001) taking into account dc voltage ripple constraints and power rating. In addition, the dc capacitor size has a direct impact on the performance of the closed-loop control system and there will always exist a compromise between the VSCs harmonic generation and the speed of response of the control system (Xu *et al.*, 2001). The response time of a STATCOM can be shorter than that of an SVC, partly because the natural response (i.e. the response without control action) will be to counteract the change in the ac voltage amplitude, and partly because the VSC uses switching at a frequency higher than the fundamental frequency. Since the reactive power from a STATCOM falls linearly with the ac voltage (as the current can be maintained at the rated value even for a low ac voltage), a STATCOM also provides better reactive power support at low ac voltages than a SVC.

The use of a STATCOM to provide voltage control and fault-ride through solution for a large wind farm based on FSIG wind turbines is shown in Figure 4.28 (Wu *et al.*, 2003; Cartwright *et al.*, 2004b.

The use of a STATCOM in combination with energy storage (super capacitors) has also been proposed to facilitate the integration of a FSIG-based wind farm (Banos *et al.*, 2006)

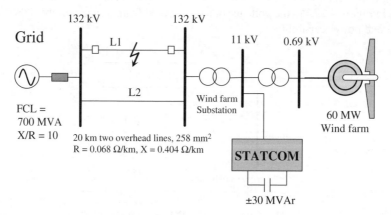

Figure 4.28 A large FSIG-based wind farm connected to the system with a STATCOM (Wu *et al.*, 2003; Cartwright *et al.*, 2004b).

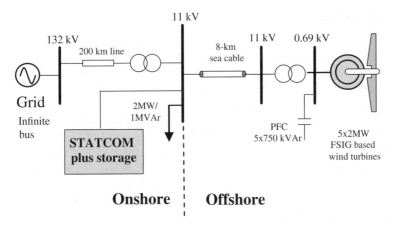

Figure 4.29 Network set up used to illustrate the contribution of a STATCOM plus energy storage to facilitate the integration of an FSIG-based wind farm (Banos *et al.*, 2006).

(Figure 4.29). Used with the storage device, the STATCOM provides additional benefits to wind farm operation such as increased capability to damp electro-mechanical oscillations, and increased power quality and reliability of the electricity supply.

The use of a hybrid LCC-HVDC inter-connector comprising a thyristor-based converter and a VSC-based (STATCOM) (rated at ±100 MVar) for transmitting power from a 500 MW offshore wind farm based on FSIGs has been presented in (Andersen and Xu, 2004; Cartwright *et al.*, 2004) (Figure 4.30). For the offshore converter, the reactive power required is provided by filters and the STATCOM. For the onshore station, conventional filters are used for reactive power compensation. The STATCOM provides both, the necessary commutation voltage to the HVDC converter and reactive power compensation to the network during steady state, dynamic and transient conditions.

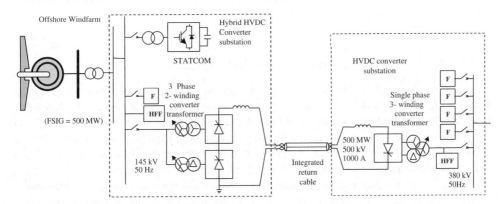

Figure 4.30 Diagram of hybrid HVDC system connecting a large FSIG-based wind farm (Cartwright *et al.*, 2004b; Andersen and Xu, 2004).

4.7   Subsea Cables

To transfer power from offshore wind farms to the onshore grid either ac or dc subsea transmission systems can be used. The selection of subsea cables depends on factors such as: distance to shore, voltage and transmitted power capacity. Before installation, a comprehensive survey is necessary in order to establish cable routes (Contact, 2008).

Offshore submarine cables have various types of insulation systems such as extruded polymer insulation, mass impregnated cables and self-contained fluid filled cables. There are two types of extruded insulated cables: cross-linked polyethylene (XLPE) and ethylene propylene rubber (EPR). These are lighter than other cable types and because of that they are easier to handle in installation on site. In addition the cables are environmentally friendly due to lack of oil and associated risk of leakage. Mass-impregnated cables are suitable for voltages up to 600 kV dc and power of 1000 MW. Self-contained fluid-filled cables are suitable for high voltages up to 1000 kV ac and 600 kV dc systems.

4.7.1   Ac Subsea Cables

Normally for HVAC systems, XLPE subsea cables are used for offshore applications. Currently, available options for offshore wind farms are three-core cables carrying from 60 to 225 kV or single-core design carrying up to 420 kV. Three-core designs have three cores where each is separately insulated and combined together in one cable, which can be used to transfer smaller amounts of power. To transmit a large amount of power three separate single-core subsea cables can be used, each cable will carry one phase of the three-phase system. A fourth cable is often added in parallel in case of fault/damage on one of the cables (Contact, 2008).

4.7.2   Dc Subsea Cables

Offshore wind power is tending towards higher voltages, with HVDC cables being more efficient for long distances for offshore wind farms than HVAC cables. For LCC-HVDC mass-impregnated cables are usually installed, and for VSC-HVDC systems extruded insulated cables will be installed (Contact, 2008).

4.7.3   Modelling of Underground and Subsea Cables

To model onshore and offshore electrical power systems successfully, it is necessary to have accurate impedance models of the underground and subsea cables otherwise, it is impossible to reliably predict system resonances and the effects of any generated harmonics from power electronic converters. As the oil and gas industry and offshore wind generation expand toward deeper sea, the cable circuits tend to increase in length. At present, cable circuits are being employed with lengths in the order of 100 km. To develop accurate impedance models, a good understanding of the physical phenomena that goes into the makeup of the cable impedance is necessary. Also, for the power system studies, the steady-state and harmonic generation of cables must be known. Over the years, research has been tailored to determine a suitable model for calculating the impedance of cables (Ametani, 1980; Kane *et al.*, 1995; Nagaoka and Ametani, 2000; Mugala and Eriksson, 2007). Subsea cables which have different layers of heavy armour on the outside to give added strength both for laying and protecting against mechanical damage, especially with the growing trend of offshore wind power generation feeding oil drilling production fields. Because of the heavy armour, the electromagnetic effects between the layers within the subsea cable need to be considered carefully when developing impedance models. Subsea cable arrangements and structures are diverse with inductive and capacitive elements, which suggests that each cable type will generate a harmonic resonance characteristic (Chien and Bucknall, 2007; Amornvipas and Hofmann, 2010). A unified approach which matches overhead transmission lines is difficult to achieve for underground and submarine cables due to their extremely complex construction and layout.

Acknowledgements

The material presented in this chapter is adapted from the PhD thesis 'Control Design and Stability Assessment of VSC-HVDC networks for Large-Scale Offshore Wind Integration', University of Strathclyde, 2009, by Giddani Osman Addalan Kalcon, and used with his permission.

Some material in Section 4.6 is adapted from that prepared by the Dr Anaya-Lara and originally published in Acha *et al.*, 2001.

References

ABB Technical Description (2008) *Article: CCC-Capacitor Commutated Converters,* Doc. No. 9AKK104295D3439, Available at: http://www.abb.com/abblibrary/downloadcenter/

Acha, E., Agelidis, V.G., Anaya-Lara, O. and Miller, T.J.E. (2001) *Electronic Control in Electrical Power Systems,* Butterworth-Heinemann, ISBN 0-7506-5126-1.

Ackermann, T. (2005) *Wind Power in Power Systems,* John Wiley & Sons, Ltd.

Adam, G.P. *et al.* (2009) Multi-terminal DC transmission system based on modular multilevel converter. Universities Power Engineering Conference (UPEC), 2009 Proceedings of the 44th International, pp. 1–5.

Agelidis, V.G., Demetriades, G.D. and Flourentzou, N., *et al.* (2006) Recent Advances in High-Voltage Direct-Current Power Transmission Systems. Industrial Technology, 2006. ICIT 2006. IEEE International Conference on, 2006, pp. 206–213.

Ametani, A. (1980) A general formulation of impedance and admittance of cables. *IEEE Transactions on Power Apparatus and Systems,* **PAS-99**(3), 902–910.

Amornvipas, C. and Hofmann, L. (2010) Resonance analysis in transmission systems: experience in Germany. Power and Energy Society General Meeting, 2010 IEEE, 25–29 July 2010. 1–8.

Anaya-Lara, O., Tande, J.O., Uhlen, K. and Adaramola, M. (2013) Control challenges and possibilities for offshore wind farms. SINTEF Report, TR A7258, ISBN: 978-82-594-3604-7, 2013.

Andersen, B.R. and Xu, L. (2004) Hybrid HVDC for power transmission to island networks. *IEEE Transactions on Power Delivery,* **4**, 1884–1890.

Arrillaga, J. (1998) *High Voltage Direct Current Transmission,* 2nd edn, The Institution of Electrical Engineers.

Arrillaga, J. Liu, Y. H. and Watson, N. R. *et al.* (2007) *Flexible Power Transmission Systems-The HVDC Options*: John Wiley & Sons Ltd, Publication.

Arrillaga, Y.H.L.L.J. and Watson, N.R. (2007) *Flexible Power Transmission Systems-The HVDC Options,* John Wiley & Sons Ltd, Publication.

Ault, G.W., Gair, S. and McDonald, J.R. (2005) Electrical system designs for the proposed 1 GW Beatrice offshore wind farm. Fifth International Workshop on Large-Scale Integration of Wind Power and Transmission Networks for Offshore Wind Farms, Glasgow, Scotland.

Bahrman, M.P. and Johnson, B.K. (2007) The ABCs of HVDC transmission technologies. *Power and Energy Magazine, IEEE*, **5**, 32–44.

Banos, C., Aten, M., Cartwright, P. and Green, T.C. (2006) Benefits and control of STATCOM with energy storage in wind power generation. Proceedings of the 8th IEE International Conference on AC and DC Power Transmission, 2006.

Bazargan, M. (2007) Renewables offshore wind: offshore substation. *Power Engineering*, **21**(3), 26–27.

Bresesti, P. *et al.* (2007) HVDC connection of offshore wind farms to the transmission system. *Energy Conversion, IEEE Transactions on*, **22**, 37–43.

Cartwright, P., Anaya-Lara, O., Wu, X. *et al.* (2004a) Grid compliant offshore wind power connections provided by FACTS and HVDC solutions. Proceedings of the European Wind Energy Conference EWEC 2004, London, U.K.

Cartwright, P., Xu, L. and Sasse, C. (2004b) Grid integration of large offshore wind farms using hybrid HVDC transmission. Proceeding of the Nordic Wind Power Conference NWPC '04.

Chaudhary, S.K.T., Teodorescu, R. and Rodriguez, P. (2008) Wind Farm Grid Integration Using VSC Based HVDC Transmission – An Overview. Energy 2030 Conference, 17–18 November 2008. ENERGY 2008, IEEE, pp. 1–7.

Chien, C.H. and Bucknall, R.W.G. (2007) Analysis of harmonics in subsea power transmission cables used in VSC-HVDC transmission systems operating under steady-state conditions. *IEEE Transaction on Power Delivery*, **22**(4), 2489–2497.

CIGRE (2009) Integration of large-scale wind generation using HVDC and power electronics. CIGRE Working Group, B4.39 Technical brochure 2009.

Cole, S. and Belmans, R. (2009) Transmission of bulk power. *Industrial Electronics Magazine, IEEE*, **3**, 19–24.

Cuiqing, D., Bollen, M. H. J. *et al.* (2005) Analysis of the control algorithms of voltage-source converter HVDC. Power Tech, 2005 IEEE Russia, pp. 1–7.

Cuiqing, D., Bollen, M. H. J., Agneholm, E. and Sannino, A *et al.* (2007) A new control strategy of a VSC-HVDC system for high-quality supply of industrial plants. *Power Delivery, IEEE Transactions on*, **22**, 2386–2394.

Douangsyla, S., Kanthee, A., Kando, M., Kittiratsatcha, S. and Kinnares, V *et al.* (2004) Modeling for PWM voltage source converter controlled power transfer. Communications and Information Technology, 2004. ISCIT 2004. IEEE International Symposium on, vol. 2, pp. 875–878.

E.ON Climate and Renewables (2012) E.ON Offshore Wind Energy Factbook, www.eon.com (accessed on 4 September 2013).

Flourentzou, N. Agelidis, V.G. and Demetriades, G.D. *et al.* (2009) VSC-based HVDC power transmission systems: an overview. *IEEE Trans on Power Electronics*, **24**, 592–602.

Giddani, O.A., Adam, G.P., Anaya-Lara, O. and Lo, K.L, *et al.* (2009) Grid integration of a large offshore wind farm using VSC-HVDC in parallel with an AC submarine cable. Universities Power Engineering Conference (UPEC), 2009 Proceedings of the 44th International, pp. 1–5.

Gesino, A.J., Lange, B. and Rohrig, K. (2011) Large-scale integration of offshore wind power through wind farm clusters. Fraunhofer IWES, Renewable Energy World Conference & Expo, Europe, 2011.

Haileselassie, T.M. and Molinas, M. (2008) Multi-Terminal VSC-HVDC System for Integration of Offshore Wind Farms and Green Electrification of Platforms in the North Sea. Nordic Workshop on Power and Industrial Electronics, June 2008.

Hingorani, N. G., Gyugyi, L. (2000) *Understanding FACTS: Concepts and Technology of Flexible AC Transmission Systems*, John Wiley & Sons Ltd, ISBN: 0780334558, 9780780334557.

Hongbo, J. and Ekstrom, A. (1998) Multiterminal HVDC systems in urban areas of large cities. *Power Delivery, IEEE Transactions on*, **13**, pp. 1278–1284.

Hopewell, P.D. *et al.* (2006) Optimising the Design of Offshore Wind Farm Collection Networks. Universities Power Engineering Conference, 2006. UPEC '06. Proceedings of the 41st International, 2006, pp. 84–88.

IEEE Guide for Planning DC Links Terminating at AC Locations Having Low Short-Circuit Capacities, *IEEE Std* 1204-1997, p. i, 1997.

Iwatta, Y., Tanaka, S., Sakamoto, K., Konishi, H., Kawazoe, H. (1996) Simulation study of a hybrid HVDC system composed of self-commutated converter and a line-commutated converter, in Sixth International Conference on AC and DC Power Transmission, pp. 381–386.

Jiebei, Z. and Booth, C. (2010) Future multi-terminal HVDC transmission systems using Voltage source converters. Universities Power Engineering Conference (UPEC), 2010 45th International, pp. 1–6.

Kane, M., Ahmad, A. and Auriol, P. (1995) Multiwire shielded cable parameter computation. *IEEE Transactions on Magnetics*, **31**, 1646–1649.

Kundur, P. (1994) Power System Stability and Control, Electric Power Research Institute, Power System Engineering Series.

Lie, X. *et al.* (2005) VSC transmission operating under unbalanced AC conditions – analysis and control design. *Power Delivery, IEEE Transactions on*, **20**, 427–434.

Lie, X. *et al.* (2008) Multi-terminal DC transmission systems for connecting large offshore wind farms. Power and Energy Society General Meeting – Conversion and Delivery of Electrical Energy in the 21st Century, 2008 IEEE, pp. 1–7.

Long, W.F. *et al.* (1990) Application aspects of multiterminal DC power transmission. *Power Delivery, IEEE Transactions on*, **5**, 2084–2098.

Meisingset, M. and Gole, A.M. (2001) A comparison of conventional and capacitor commutated converters based on steady-state and dynamic considerations. AC-DC Power Transmission, 2001. Seventh International Conference on (Conf. Publ. No. 485), pp. 49–54.

Mugala, G., Eriksson, R. and Pettersson, P. (2007) Dependence of insulated power cable wave propagation characteristics on design parameters. Dielectrics and Electrical Insulation, IEEE Transactions on, **14**, 393–399.

Musial, W., Ram, B. (2010) Large-scale offshore wind power in the United States, NREL, NREL/TP-500-4075.

Nagaoka, N. and Ametani, A. (2000) *Modelling of Frequency-Dependent Lines and Cables by Means of Algebra Processing Program*, IEEE.

Padiyar, K.R. (2005) *HVDC Power Transmission Systems: Technology and System Interactions*, New Age International (P) Limited, Publishers, New Delhi.

Quinonez-Varela, G., Ault, G.W., Anaya-Lara, O. and McDonald, J.R. (2007) Electrical collector system options for large offshore wind farms, *IET Renewable Power Generation*, **1**(2), 107–114.

Rashid, M.H. (2001) *Power Electronics Handbook*, Academic Press.

Sanino, A., Liljestrand, L., Breder, H. and Koldby, E. (2006) On some aspects of design and operation of large offshore wind parks. Sixth International Workshop on Large-Scale Integration of Wind Power and Transmission Networks for Offshore Wind Farms, Delft, The Netherlands, pp. 85–94.

Singh, B. *et al.* (2009) Static synchronous compensators (STATCOM): a review. *Power Electronics, IET*, **2**, 297–324.

Song, Y.-H. and Johns, A.T. (1999) *Flexible AC Transmission Systems (FACTS)*, IET, ISBN: 0852967713, 9780852967713.

Sood, V.K. (2004) *HVDC and FACTS Controllers Application of Static Converter in Power System*: Kluwar Academic Publisher.

Thanet Offshore Wind Limited (2005) Thanet Offshore Wind Farm Environmental Statement Non Technical Summary. Available: http://www.warwickenergy.com/pdf/ThanetNTSlr.pdf, last accessed 18 December 2013.

Volker, T., Mehler, C., Raffel, H. and Orlik, B. *et al.* (2008) New HVDC-Concept for power transmission from offshore wind farms. Wind Power to the Grid – EPE Wind Energy Chapter 1st Seminar, 2008. EPE-WECS 2008, 27–28 March, pp. 1–6.

Weimers, L. (2000) New markets need new technology. Power System Technology, 2000. Proceedings. PowerCon 2000. International Conference on, vol. 2, pp. 873–877.

Weixing, L. and Boon Teck, O. (2002) Multi-terminal HVDC as enabling technology of premium quality power park. Power Engineering Society Winter Meeting, 2002. *IEEE*, Vol. 2, pp. 719–724.

Wulue, P., Yong, C. and Hairong, C., *et al.* (2006) Hybrid Multi-terminal HVDC System for Large Scale Wind Power. Power Systems Conference and Exposition, 2006. PSCE '06. 2006 IEEE PES, pp. 755–759.

Wu, X., Arulampalam, A., Zhan, C. and Jenkins, N. (2003) Application of a static reactive power compensator (STATCOM) and a dynamic braking resistor (DBR) for the stability enhancement of a large wind farm. *Wind Engineering*, **27**(2), 93–106.

Xu, L., Agelidis, V.G. and Acha, E. (2001) Development considerations of a DSP-controlled PWM VSC-based STATCOM. IEE Proceedings, Electric Power Applications.

Yazdani, A. and Iravani, R. (2010) *Voltage-Sourced Converters in Power Systems: Modeling, Control, and Applications*, John Wiley & Sons, Inc., Hoboken, New Jersey.

Zubiaga, M. Abad, G., Barrena, J. A., Aurtenetxea, S. and Carcar, A. *et al.* (2009) Spectral analysis of a transmission system based on AC submarine cables for an offshore wind farm. Industrial Electronics, 2009. IECON '09. 35th Annual Conference of IEEE, 2009, pp. 871–876.

5

Grid Integration of Offshore Wind Farms – Case Studies

This chapter presents case studies showing the connection of large-scale offshore wind farms to ac onshore grids via VSC-HVDC transmission. Other offshore transmission schemes such as HVAC subsea transmission, and HVAC in parallel with HVDC transmission, are also analysed and discussed. In addition, two examples that illustrate the use of VSC-MTDC networks are presented and explained; one case focuses on wind farm connection to the mainland grid, and the other case on large power exchange between transnational ac power grids. Some of the examples consider ac faults to illustrate the feature of HVDC transmission that prevents fault propagation.

5.1 Background

It is expected that several large offshore wind farms will be located far away from the mainland ac grids, and therefore transmission of such large power over subsea cables introduces various technical challenges. In HVAC subsea transmission the electrical losses caused by high charging currents in the submarine cable increase as transmission distance increases. This characteristic limits the interconnection to less than 100 km from shore even if a number of large dynamic voltage compensating devices, such as SVC or STATCOM, are installed for safe and stable operation. Another limitation of the ac transmission option is its synchronous connection between the wind farms and the ac mainland grids (Adam *et al.*, 2010b; Haileselassie *et al.*, 2010; Haileselassie *et al.*, 2011; Ahmed *et al.*, 2012a).

Dc subsea transmission systems provide an attractive alternative to HVAC transmission. They have reduced transmission losses and also result in an asynchronous connection between the mainland ac grid and the wind farm; the latter can potentially prevent the adverse impact on the wind farm of a fault in the ac grid. Also, if VSC-HVDC transmission is used then no extra reactive power compensation devices are needed (Bahrman *et al.*, 1980; Banks and Williams, 1983; Baker *et al.*, 1989; Bakken and Faanes, 1997; Andersen and Lie, 2004; Bahrman, 2006;

Offshore Wind Energy Generation: Control, Protection, and Integration to Electrical Systems, First Edition.
Olimpo Anaya-Lara, David Campos-Gaona, Edgar Moreno-Goytia and Grain Adam.
© 2014 John Wiley & Sons, Ltd. Published 2014 by John Wiley & Sons, Ltd.
Companion Website: www.wiley.com/go/offshore_wind_energy_generation

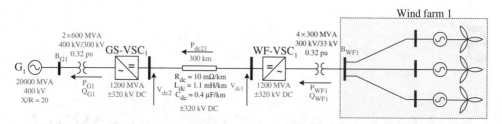

Figure 5.1   Integration of a large-scale wind farm using point-to-point HVDC transmission.

Barthold, 2006; Bahrman and Johnson, 2007; Bahrman, 2008; Basu, 2009; Buijs *et al.*, 2009; Azimoh *et al.*, 2010; Ahmed *et al.*, 2011a; Barker and Whitehouse, 2012; Adam *et al.*, 2012a; Abdel-Khalik *et al.*, 2013). For short distances (less than 80 km) subsea dc transmission tends to be more expensive than its ac counterpart, but for longer distances and transmission of larger power blocks it tends to become more competitive (Banks and Williams, 1983; Alexandridis and Galanos, 1988; Baker *et al.*, 1989; Bakken and Faanes, 1997; Al-Dhalaan *et al.*, 1998; Andersen and Barker, 2000; Andersen and Lie, 2004; Bahrman, 2006; Bahrman and Johnson, 2007; Bahrman, 2008; Adam *et al.*, 2009; Basu, 2009; Buijs *et al.*, 2009; Bai *et al.*, 2010; Ahmed *et al.*, 2011b; Ahmed et al., 2012a; Ahmed *et al.*, 2012b; Aik and Andersson, 2013).

5.2   Offshore Wind Farm Connection Using Point-to-Point VSC-HVDC Transmission

Figure 5.1 illustrates the use of VSC-HVDC transmission in the connection of a large-scale wind farm with a rated output power of 1200 MW. All wind turbines within the wind farm are of the variable-speed type with fully-rated converters. Since the fully-rated converter decouples the wind generator from the collection network, only the grid-side converter is modelled in detail. For further simplicity, one equivalent unit represents the entire offshore wind farm. In this illustrative example, the grid-side converter (GS-VSC$_1$) of the VSC-HVDC is configured to regulate the dc link voltage level at 640 kV and the ac voltage at the PCC (B$_{G1}$). The wind farm side converter of the VSC-HVDC keeps a stiff ac bus at the wind farm main platform (B$_{WF1}$). Figure 5.2 shows the control system of the wind farm side converter and Figure 5.3

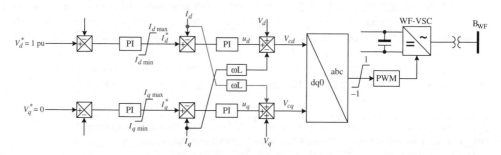

Figure 5.2   Control system of the wind farm-side converter (WF-VSC).

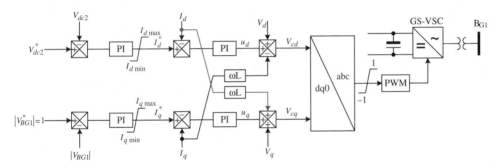

Figure 5.3 Control system of the grid-side converter (GS-VSC).

shows the control system of the grid-side converter, both for the VSC-HVDC link. Controlling the wind farm side converter in this manner allows it to transfer the entire power generated by the wind farm to the grid without creating a power imbalance between the ac and dc sides. Additionally, it allows the control of the wind farm side converter independently of the wind turbine generator technology. Figure 5.1 shows the main parameters of the system (the directions of power are positive).

Figure 5.4 shows the representation of the fully-rated converter wind generator for this example; where P_g represents the active power of the wind farm at the main platform, and Q_g represents its reactive power output. As part of the exercise, a noise signal is injected into the active power reference (P_g^*) of the wind turbine generator to reflect the effect of the wind variability in the output power, hence illustrating mitigation effect of the VSC-HVDC transmission system, while Q_g^* is set to zero for unity power operation.

Figure 5.5 shows the waveforms obtained from this case study, which starts with a wind farm output power of only 600 MW (0.5 pu), but at $t = 2$s its power output has increased to 1080 MW (0.9 pu). Observe that as the power from the wind farm to the main platform increases, the wind farm and grid side converters of the VSC-HVDC link (WF-VSC and GS-VSC) keep the voltage constant at B_{WF} and B_{G1} (see Figure 5.5 (a)–(d)). Also notice that

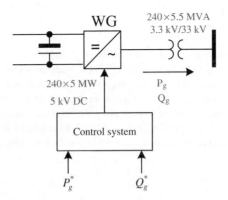

Figure 5.4 Simplified representation of the fully-rated converter when only its grid-side converter is modelled.

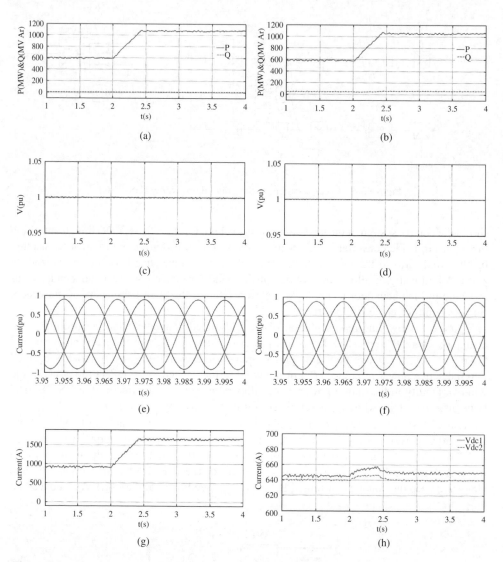

Figure 5.5 Waveforms illustrating steady-state and dynamic behaviour when a large scale-wind farm is integrated into the ac grid using point-to-point VSC-HVDC transmission (a) Active and reactive powers WF-VSC presents at wind farm main platform (B_{WF1}) (b) Active and reactive powers GS-VSC injects into point of common coupling (B_{G1}) (c) Voltage magnitude at wind farm main platform (B_{WF1}) (d) Voltage magnitude at point of common coupling (B_{G1}) (e) Zoomed version of the current waveforms at wind farm main platform (B_{WF1}) (f) Zoomed version of the current waveforms at point of common coupling (B_{G1}) (g) dc link current (h) dc link voltages of wind farm and grid-side converters WF-VSC and GS-VSC.

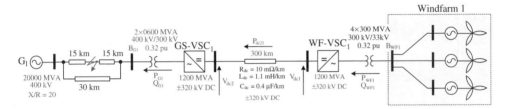

Figure 5.6 Test system used to illustrate the improved ac fault ride-through capability using a VSC-HVDC to connect a large scale offshore wind farm to the mainland grid.

the control over the storage elements of the VSC-HVDC contribute to the mitigation of the effect of wind speed variability injected into reference power for the wind turbine generator as it can be seen that the voltage at the PCCs is not significantly affected (Figure 5.5 (e) and (f)). Figure 5.5 (g) and (h) show how the dc link current and dc voltage of the wind farm side converter vary with input power as expected. This case study illustrates that controlling the wind farm side converter of the VSC-HVDC link as indicated, ensures power balance between the wind farm collection network and the dc side of the wind farm side converter (WF-VSC).

To illustrate the improved ac fault ride-through behaviour of the wind farm when integrated into the mainland grid using a VSC-HVDC link, a symmetrical ac three-phase fault to ground, through a 2.5 Ω fault resistance, is applied to one of the tie lines that connects bus B_{G1} to the grid G_1 (Figure 5.6). Assume that the system operating conditions remain the same as the previous example, and the fault is applied at time $t = 3s$ and cleared after 140 ms. Observe that the decoupling feature of the VSC-HVDC link prevents propagation of the ac fault from the grid towards the wind farm therefore no electrical or mechanical stresses are expected on the wind turbine generators. This is clearly illustrated as the power, voltage and current waveforms at the wind farm main platform remain unaffected, despite similar quantities in the grid side exhibiting significant changes (increased currents and depressed voltages, see Figure 5.7 (a)–(e)). However, the plots of the dc current and dc voltages across both converters have shown that as the fault in the grid side depresses the voltage at the point of common coupling (B_{G1}), the ability of the grid side converter to transfer power to the grid reduces (Figure 5.7 (g) and (h)). This causes a power imbalance between the ac and dc sides of the HVDC link, and excessive active power that cannot be transferred to the ac grid will be trapped in the dc side and absorbed by the dc link capacitors. As a result, the dc link voltages of the grid and wind farm side converter will rise. This shows that VSC-HVDC transmission remains vulnerable to severe voltage depression at the grid side close to the point-of-common coupling (PCC), or sudden loss of ac grid (Adam *et al.*, 2010a). Some measures presently used by the industry to address this issue include the use of a dc chopper at the dc link of the VSC-HVDC link, or in the dc link of individual wind turbine generators (Adam *et al.*, 2010a; Adam *et al.*, 2012b).

5.3 Offshore Wind Farm Connection Using HVAC Transmission

The option of connecting offshore wind farms to the mainland ac grids using ac transmission looks attractive because of its apparent simplicity, well-understood dynamics during ac network

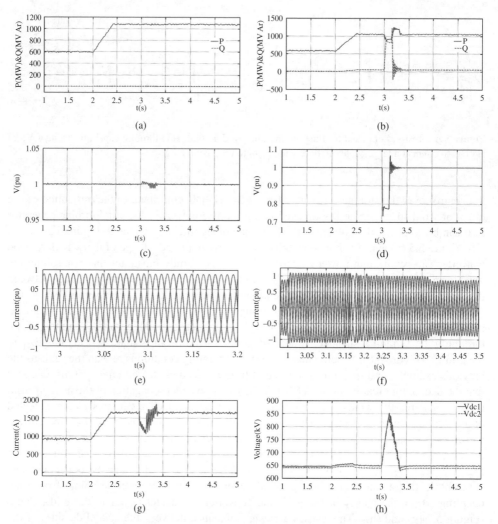

Figure 5.7 Waveforms illustrating improved ac fault ride-through of an offshore wind farm using VSC-HVDC to connect to the mainland grid (a) Active and reactive powers at the main wind farm platform bus B_{WF1} (b) Active and reactive powers at the point of common coupling B_{G1} (c) Voltage magnitude at B_{WF1} (d) Voltage magnitude at $_{BG1}$ (e) ac current waveforms at BWF1 zoomed around fault period (f) ac current waveforms at BG1 zoomed around fault period (g) dc current (h) dc voltages of the grid and wind farm side converters.

faults and proven protection strategies. Currently these factors continue to outweigh others in the possible options for power transmission from offshore wind farms even when the connection distance exceeds 80 km.

Figure 5.8 shows the schematic diagram of a large offshore wind farm connected to the mainland grid using HVAC transmission. Under this scenario, the wind farm reactive power capability must be put into practice for voltage support at the wind farm main platform.

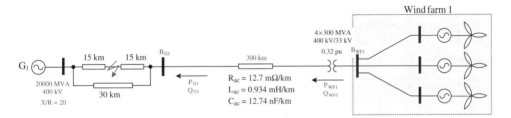

Figure 5.8 Integration of a large-scale offshore wind farm using HVAC transmission system.

Therefore, the grid side converter of the fully-rated converter wind turbine generator should control the ac voltage instead of controlling reactive power for power factor as in the previous example. In this case study, the reactive power capability of the fully-rated converter is assumed to be ±0.872 pu. Otherwise, in order to improve ac fault ride-through dynamic reactive power compensating devices must be installed at the wind farm terminal.

Figure 5.9 shows the results obtained from this case study. The operating conditions are the same as in the example in the previous section. Observe that the use of HVAC transmission results in synchronous connection that permits the impact of an ac fault in the grid to propagate to the wind farm (see Figure 5.9 (a)–(f)). Additionally, despite the reactive power capability of the grid-side converter of the wind turbine generator being increased to a level comparable with its active power capability, the point-of-common coupling (B_{G1}) suffers from voltage depression when the wind farm operates near its rated output power as can be seen in Figure 5.9 (d). Such levels of voltage depression are not allowable in all known Grid Codes, hence a large STATCOM or SVC will be required. In addition, the power loss is expected to increase significantly due to the flow of reactive current component besides that responsible for active power transfer. The grid-side converter of each individual wind turbine generator is also exposed to excessive voltage and current stresses during ac faults in the onshore grid side. This example has shown one of the reasons VSC-HVDC transmission may be more attractive than HVAC for connection of large offshore wind farms.

5.4   Offshore Wind Farm Connected Using Parallel HVAC/VSC-HVDC Transmission

Figure 5.10 shows the grid connection of a large-scale offshore wind farm using HVAC link in parallel with VSC-HVDC link. In this hybrid connection, the wind farm has a synchronous connection to the grid, thus ac faults in the grid will propagate toward the wind farm. However, the parallel power path provided by the VSC-HVDC link also gives voltage support at both ends of the HVAC link, and therefore the wind farm ability to ride-through an ac fault in the onshore grid is improved. To ensure full power control in the proportion of power transferred through the VSC-HVDC, the wind farm-side converter (WF-VSC$_1$) must be configured to directly regulate active power and ac voltage at the wind farm main platform bus (B_{WF1}).

In order to avoid the problem of loop-flow, the reference active power to the WF-VSC must not exceed the wind farm output power. Additionally, the converter stations of the VSC-HVDC link must be used to regulate ac voltage as illustrated in the phasor diagram in Figure 5.11, where $\vec{V}_1$, and $\vec{I}_1$ represent voltage and current vectors at bus B_{WF1}, while, $\vec{V}_2$ and $\vec{I}_2$ represent

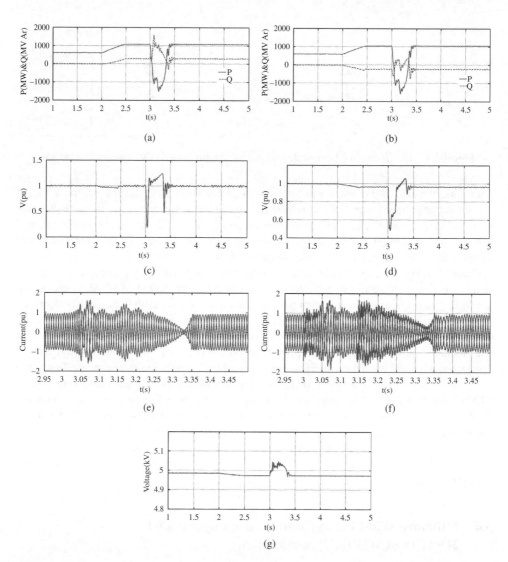

Figure 5.9 Waveforms illustrating fault ride-through of a large-scale wind farm connected to grid using HVAC transmission system (a) Active and reactive powers measured at the wind farm main platform bus (B_{WF1}) (b) Active and reactive powers measured at point-of-common coupling with the mainland ac grid (B_{G1}) (c) Voltage magnitude wind farm at main platform bus (B_{WF1}) (d) Voltage magnitude at point-of-common coupling with the mainland ac grid (B_{G1}) (e) Snapshots of the current waveforms at B_{WF1} around the fault period (f) Snapshots of the current waveforms at B_{G1} around the fault period (g) Dc link voltage of the grid-side converter of the fully-rated converter wind turbine generator.

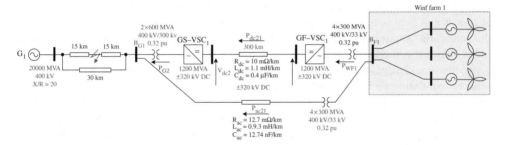

Figure 5.10 Integration of a large-scale wind farm using a VSC-HVDC link in parallel with HVAC transmission system.

those at B_{G1}. δ_{c1} and δ_{c2} are angles of the converter terminal voltages relative to the voltages at B_{WF1} and B_{G1}, while φ_{c1} and φ_{c2} are power factor angles at B_{WF1} and B_{G1}. The phasor diagram of the hybrid connection in Figure 5.11 illustrates that the voltage support at both ends of the ac cable are achieved at no extra cost.

To illustrate the steady-state and ac fault ride-through performance of the hybrid connection, the system in Figure 5.10 is simulated with the same operating conditions as that in previous

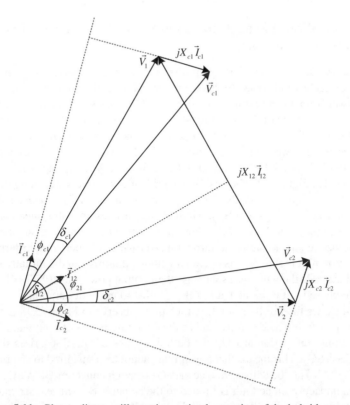

Figure 5.11 Phasor diagram illustrating optimal operation of the hybrid connection.

examples. In this example, initially, active power referenced to WF-VSC is set to zero, and at time $t = 4$ begins increasing gradually to 720 MW (0.6 pu). Figure 5.12 (a)–(d) show the wind farm output power, and the power transmitted through the VSC-HVDC and HVAC links. Notice that the hybrid connection with the wind farm side converter directly regulates active power and offers full control over the portions of power transmitted in both links. Also, observe that during an ac fault both converters of the VSC-HVDC link use their STATCOM functionalities to inject reactive power in an attempt to support ac voltage at BWF1 and BG1.

Figure 5.12 (e) and (f) show that the voltage at the wind farm main platform bus experiences less dip than that at the grid side point-of-common coupling (BG1), thanks to the voltage support from the VSC-HVDC converter stations. The plots for the current waveforms in Figure 5.12 (g)–(j) show that the wind generator output current waveforms do not exhibit any significant over-current, and this is because the wind farm side converter of the VSC-HVDC is able to fully compensate for the voltage sag developing in the grid side. This response shows that the hybrid connection has the potential to improve ac fault ride-through of the wind farm even under extreme cases of ac faults in the grid side. Figure 5.12 (h) shows that with hybrid connection the amount of active power which cannot be transferred to the grid due to extreme voltage collapse at the PCC (B_{G1}) can be minimised. Thus, the rise of the dc link voltages at the VSC-HVDC and wind turbine generators can also be minimised. A HVAC link also offers an alternative power path during a dc side fault in the VSC-HVDC link.

5.5 Offshore Wind Farms Connected Using a Multi-Terminal VSC-HVDC Network

Figure 5.13 shows an example of using an offshore multi-terminal HVDC network for connection of large-scale wind farms. In this case, a four-terminal HVDC network can minimise the loss of power from a wind farm in case of a scheduled outage for maintenance, or a forced outage due to a fault in one of the converter stations or in an ac or dc line. The presence of two grid side converters offers the possibility of power exchange between the two ac grids G_1 and G_2, besides receiving power from wind farms 1 and 2. In this example, the two wind farm-side converters WF-VSC$_1$ and WF-VSC$_2$ control the stiffness of the ac voltages at buses B_{WF1} and B_{WF2}, as depicted in Figure 5.2, whereas the two grid-side converters are configured to regulate their dc voltage. The latter allows the flow of power into ac grids G_1 and G_2, and in addition, regulates the ac voltage at B_{G1} and B_{G2}. All the wind turbines use fully-rated converters as in the previous examples, and Figure 5.13 shows the system parameters. In an attempt to illustrate the steady-state and ac fault ride-through performance of such a multi-terminal HVDC network, the system in Figure 5.13 is operated with wind farms 1 and 2 initially injecting 600 MW each but at $t = 2$s the wind farms increase their output powers to 1080 MW each. At time $t = 3$s, a three-phase fault is applied at bus B_{G1}, with 140 ms fault duration.

Figure 5.14 shows the results from the simulation. Observe in Figure 5.14 that prior to the fault the system operates successfully, and as the operating points of both wind farms remain unchanged during fault period at BG_1, the dc-link voltage of GS-VSC$_1$ rises slightly higher than that of GS-VSC$_2$. This means that power that cannot be transferred to ac grid G_1 through converter GS-VSC$_1$ is to be delivered to ac grid G_2 through converter GS-VSC$_2$. It should be noted that the impact of an ac fault is limited to the ac network. This causes transient power flow in the dc side that leads to the rise of all dc link voltages and redirection of a small amount

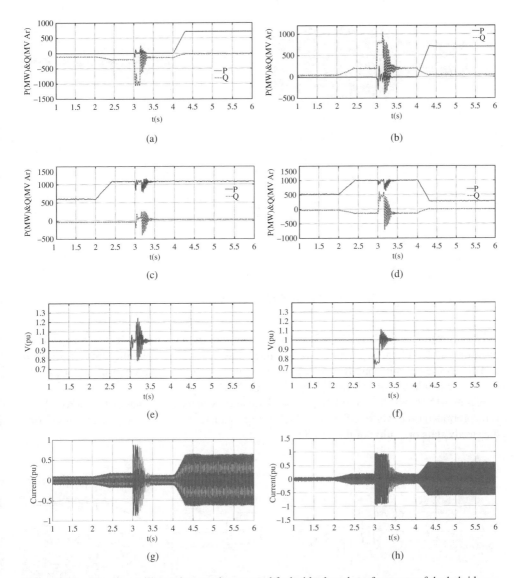

Figure 5.12 Waveforms illustrating steady-state and fault ride-though performance of the hybrid connection (VSC-HVDC in parallel with HVAC) (a) Active power the sending-end converter of the VSC-HVDC link (WF-VSC1) exports to the grid, and its reactive power exchange with B_{WFI} (b) Active and reactive power the receiving-end converter of the VSC-HVDC (GS-VSC) injects into the point-of-common coupling (BG1) (c) Wind farm output active and reactive powers delivered into the main platform bus (B_{WFI}) (d) Active and reactive powers exchanged at the receiving-end of the HVAC link (e) Voltage magnitude at the wind farm main platform bus (BWF1) (f) Voltage magnitude at the point of common coupling (BG1) (g) Current the wind farm-side converter (WF-VSC1) exchanges with the main platform bus BWF (h) Current the grid-side converter (GS-VSC1) injects to BG1 (i) Currents at the receiving end of the HVAC link (j) Currents the wind farm generator injects into the main platform bus BG1 (k) Dc voltages of the sending- and receiving-end converters of the VSC-HVDC link. (*Continued*)

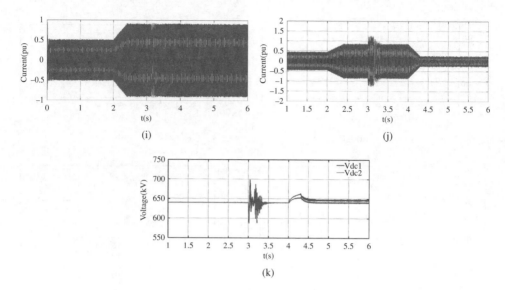

Figure 5.12   (*Continued*)

of power from GS-VSC$_1$ to GS-VSC$_2$. Notice also that the use of two grid side converters, operating in dc voltage regulation mode, reduce power despatch flexibility into ac grids G$_1$ and G$_2$. However, this condition is improved by incorporating droop control into one of the dc voltage regulated converters GS-VSC$_1$ and GS-VSC$_2$.

Figure 5.15 shows the control system block diagram when a dc voltage-active power droop is incorporated in converter GS-VSC$_1$. Such droop characteristic can be derived by considering the steady state operation of the dc grid.

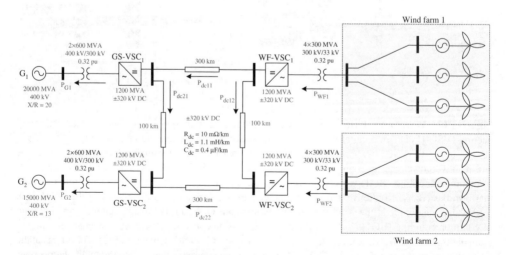

Figure 5.13   Integration of large-scale wind farms into power systems using voltage source converter based multi-terminal HVDC transmission system.

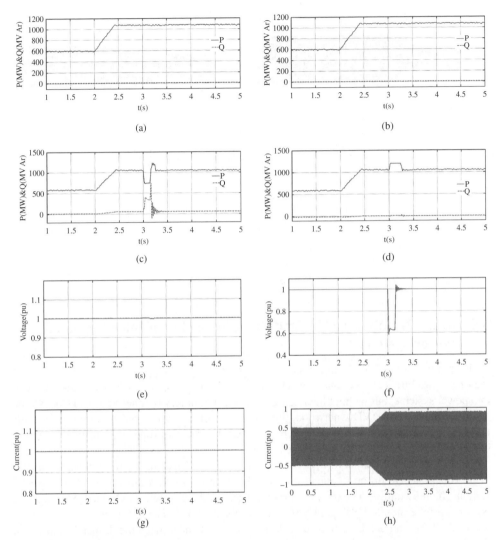

Figure 5.14 Waveforms illustrating the steady-state and ac fault ride-through of a multi-terminal VSC-HVDC transmission network connecting two offshore wind farms to two isolated ac grids (a) Active and reactive power the wind farm-side converter (WF-VSC$_1$) injects into the HVDC network (b) Active and reactive powers the wind farm-side converter converter (WF-VSC$_2$) injects into the HVDC network (c) Active and reactive powers the grid-side converter (GS-VSC$_1$) injects into ac grid G$_1$ (d) Active and reactive powers the grid-side converter (GS-VSC$_2$) injects into ac grid G$_2$ (e) Voltage magnitude at B$_{WFl}$ (f) Voltage magnitude at B$_{G1}$ (g) Voltage magnitude at B$_{G2}$ (h) Current waveforms WF-VSC$_1$ injects into B$_{WFl}$ (i) Current waveforms GS-VSC$_1$ injects into G$_1$ (j) Current waveforms GS-VSC$_2$ injects into G$_2$ (k) dc voltage across all converter terminals. (*Continued*)

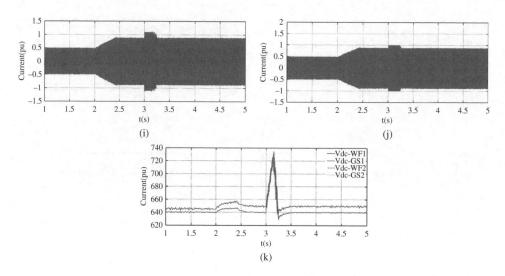

Figure 5.14   (*Continued*)

5.6   Multi-Terminal VSC-HVDC for Connection of Inter-Regional Power Systems

Multi-terminal VSC-HVDC transmission is also potentially applicable for connection of regional power grids. This connection structure can facilitate active power exchange between two power systems, but without interfering with their regulatory regimes and Grid Codes. The Multi-terminal VSC-HVDC networks provide a platform for the realisation of super grids currently advocated across Europe that may cover all parts of Europe, and this vision aims to harness all forms of renewables in these regions.

Figure 5.16 shows a radial form of four-terminal VSC-HVDC that helps to illustrate the use of dc voltage-active droop to enhance power despatch of the dc voltage regulated converter stations. In this example, G_1 to G_4 represent ac grids, and converters VSC_1 and VSC_2 regulate power, while terminals VSC_3 and VSC_4 regulate dc voltage. The droop control operates in terminal VSC_3, and can be activated or deactivated on demand from the despatch centre. All the system parameters are displayed in Figure 5.16, and an example of a droop characteristic

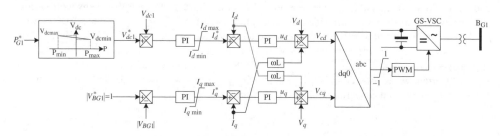

Figure 5.15   Control system block diagram when droop control is implemented in converter GS-VSC1 to improve power dispatch flexibility into AC network.

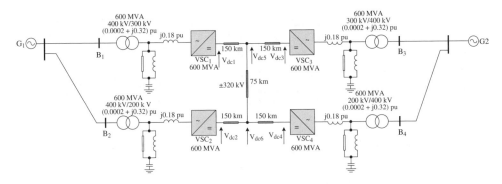

Figure 5.16 Example of four-terminal VSC-HVDC transmission.

is shown Figure 5.17, where the dc voltage can be related to active power as: $V^*_{dc3} = K(P^*_3 - P_{3\,min}) + V_{dc3\,max}$, where $V_{dc3\,min}$, $V_{dc3\,max}$ are the dc voltages corresponding to maximum and minimum power capability of converter 3 ($P_{3\,max}$ and $P_{3\,min}$). Such droop control can also be implemented using dc currents instead of voltage as can be found in the literature. However, this chapter favours the use of dc voltage over current, because dc link currents tend to be sensitive to small dc voltage variations. Initially, VSC_1 and VSC_2 are commanded to export 420 MW and 540 MW into G2, while both VSC_3 and VSC_4 regulate their dc link voltage. At $t = 3s$, the droop control activates to keep steady power injection from VSC_3 into G_2 at 460 MW, independent of the level of power injection in other terminals, and at t = 5s VSC_1 is commanded to increase its power export from 420 MW to 480 MW. At $t = 6s$, a load, comprised of 180 MW and 200 MVAr, is connected to B3.

Figure 5.18 presents the results obtained. Figure 5.18 (a) and (b) show the active and reactive power converter stations exchange with G_1 and G_2. Observe that with activation of droop control at VSC_3, VSC_3 power exchange with G2 remains unaffected as the level

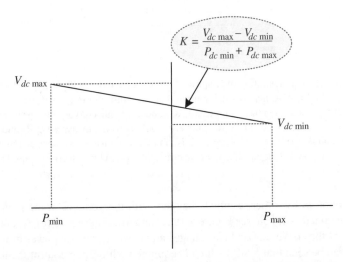

Figure 5.17 Example of dc voltage-active power droop characteristic.

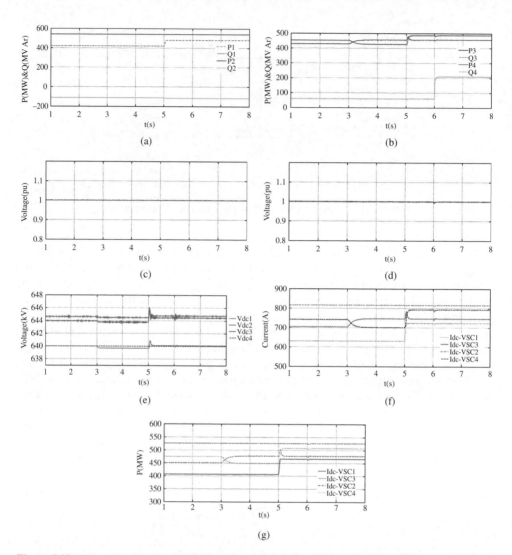

Figure 5.18 Waveforms illustrating the improved power control flexibility can be achieved with droop in multi-terminal VSC-HVDC network using more than one converter to regulate dc voltage for reliable operation (a) Active power and reactive powers VSC_1 and VSC_2 inject to dc side, measured at B_1 and B_2 (b) Active power and reactive powers VSC_3 and VSC_4 deliver to G_2, measured at B_3 and B_4 (c) Voltage magnitude at B_1 and B_2 (d) Voltage magnitude at B_3 and B_4 (e) Converter stations dc link voltages (f) Dc currents measured at the dc link terminals of converter stations (g) Dc powers measured at the converter stations dc links.

of power injection in the other terminal changes, with the voltages at the point-of-common couplings maintained at 1.0 pu, see Figure 5.18 (c) and (d). Figure 5.18 (e), (f) and (g) showing the variations of the converters dc link voltages and currents, and dc power flows with power injections. It can be seen that VSC_3, which is equipped with droop control, adjusts its dc link voltage, hence the current flow into it in an attempt to maintain its power exchange with B_3.

These results have shown that the use of dc voltage-active power droop is favoured over the dc current-active power as no uncontrolled current surges are observed in the dc side.

Acknowledgements

The material presented in this chapter is partially based on research conducted by the Authors under EPSRC Supergen Wind.

References

Abdel-Khalik, A.S., Massoud, A.M., Elserougi, A.A. and Ahmed, S. (2013) Optimum power transmission-based droop control design for multi-terminal HVDC of offshore wind farms. *IEEE Transactions on Power Systems*, **28**, 3401–3409.

Adam, G.P., Ahmed, K.H., Finney, S.J. *et al.* (2012a) New breed of network fault-tolerant voltage-source-converter HVDC transmission system. *IEEE Transactions on Power Systems*, **28**, 335–346.

Adam, G.P., Ahmed, K.H., Finney, S.J. and Williams, B.W. (2010a) AC fault ride-through capability of a VSC-HVDC transmission systems. Energy Conversion Congress and Exposition (ECCE), 2010 IEEE, 12–16 September 2010, pp. 3739–3745.

Adam, G.P., Anaya-Lara, O., Burt, G. *et al.* (2009) Comparison between Two VSC-HVDC Transmission Systems Technologies: Modular and Neutral Point Clamped Multilevel Converter. IEEE 13th Annual conference of the Industrial Electronic Society IECON2009. Porto-Portugal, 3–5 November 2009.

Adam, G.P., Finney, S.J., Bell, K. and Williams, B.W. (2012b) Transient capability assessments of HVDC voltage source converters. Power and Energy Conference at Illinois (PECI), 2012 IEEE, 24–25 February 2012, pp. 1–8.

Adam, G.P., Finney, S.J., Williams, B.W. *et al.* (2010b) Control of Multi-Terminal DC Transmission System Based on Voltage Source Converters. IET, the 9th International Conference on AC and DC Power Transmission, London, UK.

Ahmed, N., Haider, A., Angquist, L. and Nee, H.P. (2011a) M2C-based MTDC system for handling of power fluctuations from offshore wind farms. Renewable Power Generation (RPG 2011), IET Conference on, 6–8 September 2011, pp. 1–6.

Ahmed, N., Haider, A., Van Hertem, D. *et al.* (2011b) Prospects and challenges of future HVDC SuperGrids with modular multilevel converters. Power Electronics and Applications (EPE 2011), Proceedings of the 2011–14th European Conference on, 30 August 2011–1 September 2011, pp. 1–10.

Ahmed, N., Norrga, S., Nee, H.P. *et al.* (2012a) HVDC SuperGrids with modular multilevel converters – The power transmission backbone of the future. Systems, Signals and Devices (SSD), 2012 9th International Multi-Conference on, 20–23 March 2012, pp. 1–7.

Ahmed, W. and Manohar, P. (2012b) DC line protection for VSC-HVDC system. Power Electronics, Drives and Energy Systems (PEDES), 2012 IEEE International Conference on, 16–19 December 2012, pp. 1–6.

Aik, D.L.H. and Andersson, G. (2013) Analysis of voltage and power interactions in multi-infeed HVDC systems. *IEEE Transactions on Power Delivery*, **28**, 816–824.

Al-Dhalaan, S., Al-Majali, H.D. and O'kelly, D. (1998) HVDC converter using self-commutated devices. *IEEE Transactions on Power Electronics*, **13**, 1164–1173.

Alexandridis, A.T. and Galanos, G.D. (1988) Design of a reduced order observer for optimal decentralized control of HVDC systems. *IEEE Transactions on Power Systems*, **3**, 963–969.

Andersen, B. and Barker, C. (2000) A new era in HVDC? *IEE Review*, **46**, 33–39.

Andersen, B.R. and Lie, X. (2004) Hybrid HVDC system for power transmission to island networks. *IEEE Transactions on Power Delivery*, **19**, 1884–1890.

Azimoh, L.C., Folly, K., Chowdhury, S.P. *et al.* (2010) Investigation of voltage and transient stability of HVAC network in hybrid with VSC-HVDC and HVDC Link. Universities Power Engineering Conference (UPEC), 2010 45th International, 31 August 2010–3 September 2010, pp. 1–6.

Bahrman, M., Larsen, E., Piwko, R.J. and Patel, H.S. (1980) Experience with HVDC - turbine-generator torsional interaction at Square Butte. *IEEE Transactions on Power Apparatus and Systems*, **PAS-99**, 966–975.

Bahrman, M.P. (2006) Overview of Hvdc transmission. Power Systems Conference and Exposition, 2006. PSCE '06. 2006 IEEE PES, 29 October 2006–1 November 2006, pp. 18–23.

Bahrman, M.P. (2008) HVDC transmission overview. Transmission and Distribution Conference and Exposition, 2008. T&D. IEEE/PES, 21–24 April 2008, pp. 1–7.

Bahrman, M.P. and Johnson, B.K. (2007) The ABCs of HVDC transmission technologies. *Power and Energy Magazine, IEEE*, **5**, 32–44.

Bai, J.-F., Xu, J.-Z. and Luo, L.-F. (2010) Characteristics of power transmission and dynamic recovery of FCC-HVDC with different SCR. Power System Technology (POWERCON), 2010 International Conference on, 24–28 October 2010, pp. 1–6.

Baker, A.C., Zaffanella, L.E., Anzivino, L.D. *et al.* (1989) A comparison of HVAC and HVDC contamination performance of station post insulators. *IEEE Transactions on Power Delivery*, **4**, 1486–1491.

Bakken, B.H. and Faanes, H.H. (1997) Technical and economic aspects of using a long submarine HVDC connection for frequency control. *IEEE Transactions on Power Systems*, **12**, 1252–1258.

Banks, R.S. and Williams, A.N. (1983) The public health implications of HVDC transmission lines: an assessment of the available evidence. *IEEE Transactions on Power Apparatus and Systems*, **PAS-102**, 2640–2648.

Barker, C.D. and Whitehouse, R.S. (2012) A current flow controller for use in HVDC grids. AC and DC Power Transmission (ACDC 2012), 10th IET International Conference on, 4–5 Dec. 2012, pp. 1–5.

Barthold, L.O. (2006) Technical and Economic Aspects of Tripole HVDC. Power System Technology, 2006. PowerCon 2006. International Conference on, 22–26 October 2006, pp. 1–6.

Basu, K.P. (2009) Stability enhancement of power system by controlling HVDC power flow through the same AC transmission line. Industrial Electronics & Applications, 2009. ISIEA 2009. IEEE Symposium on, 4–6 October 2009, pp. 663–668.

Buijs, P., Cole, S. and Belmans, R. (2009) TEN-E revisited: Opportunities for HVDC technology. Energy Market, 2009. EEM 2009. 6th International Conference on the European, 27–29 May 2009, pp. 1–6.

Haileselassie, T.M. and Uhlen, K. (2010) Primary frequency control of remote grids connected by multi-terminal HVDC. Power and Energy Society General Meeting, 2010 IEEE, 25–29 July 2010, pp. 1–6.

Haileselassie, T.M. and Uhlen, K. (2011) Power flow analysis of multi-terminal HVDC networks. PowerTech, 2011 IEEE Trondheim, 19–23 June 2011, pp. 1–6.

6

Offshore Wind Farm Protection

During fault conditions the transient current and associated under/over voltage pose a significant threat to a wind turbine. The limited current handling capability of power electronics is a significant issue when considering the DFIG and FRC technologies. Studies investigating the common causes of failure for wind turbines in Sweden, Germany and Finland show that failures in the electrical system lead to significant downtime (Ribrant and Bertling, 2007). To address these issues appropriate protection technology must be incorporated within the individual wind turbines and coordinated over the whole wind farm.

Compliance with grid code requirements in terms of fault ride-through capability and grid support provided by a wind generator poses a particular challenge to the wind industry. Over the past decade the grid codes of European countries with high penetration of wind power have had to be revised to cope with the impact of sudden disconnection of the wind power. To enhance wind generators fault ride-through capability the turbine protection must attempt to enable it to ride through a fault before triggering the main or backup protection relays. Thus, the protection coordination process becomes more challenging than average protection coordination procedures.

6.1 Protection within the Wind Farm ac Network

Figure 6.1 shows a typical electrical layout of a wind farm divided into different protection zones. Any short circuits within the wind generator protection zone would normally be detected and isolated as fast as possible by the protections within the zone. The wind generator protection zone may have additional backup protection within the wind generator zone or the feeder protection zone.

The protection within the distribution feeder zone must be able to detect and disconnect the feeder during faults within itself or in any of the wind generator units connected to it. The type of protection within the feeder zone must be selective; not operating for out-of-zone faults unless working as a backup protection for the wind generator protection zone.

Offshore Wind Energy Generation: Control, Protection, and Integration to Electrical Systems, First Edition.
Olimpo Anaya-Lara, David Campos-Gaona, Edgar Moreno-Goytia and Grain Adam.
© 2014 John Wiley & Sons, Ltd. Published 2014 by John Wiley & Sons, Ltd.
Companion Website: www.wiley.com/go/offshore_wind_energy_generation

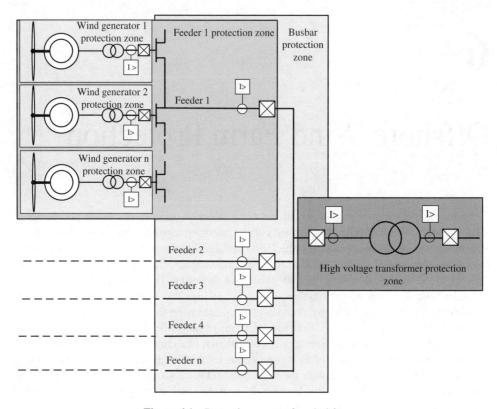

Figure 6.1   Protection zones of a wind farm.

A fault in the busbar must be cleared as fast as possible, given the large magnitude of the fault currents associated with this type of fault. However, the busbar protection must be selective to avoid tripping for a fault in a feeder unless programmed to work as a backup protection for a feeder fault.

6.1.1   Wind Generator Protection Zone

6.1.1.1   Generator Protection

A wind generator zone is usually protected against the following events (Cardenas *et al.*, 2010; Vikesjö and Messing, 2011):

1. Instant over current for the low-voltage side, including:
   * Generator stator short circuit. The pickup current must be selected to differentiate between a short circuit within the generator zone and the transient current response of the generator in case of voltage level drop caused by a fault outside the protection zone.
   * Generator stator earth fault.
   * Short circuit within the converter.

- Earth fault within the converter.
- Transformer short circuit.
- Transformer earth fault.
2. Generator negative sequence over current (asymmetric current).
3. Generator thermal overload.
4. Generator over current protection in case of overloading.
5. Over/under voltage: action depending of grid code requirements.
6. Over/under frequency: action depending on grid code requirements.

The level of redundancy of the protection can be based on the N-1 criterion; or based on the evaluation between risk and investment of the protection zone.

6.1.1.2 Power Electronic Protections

Upon the detection of a fault in the generator or in the ac network close to the wind turbine, the first protection mechanism is usually the immediate blockade of the pulses driving the commutation of the power electronic converters connected to the turbine. Consequently the power electronic switches are protected from the large transient currents due to the fault by forcing all the switches to the OFF state. This protection measurement is very fast (less than 1 ms (Davies *et al.*, 2009)) with activation before the initial maximum peak of the fault current. However, additional protection schemes are required to safeguard the integrity of the power electronic devices in wind turbines, according to the wind turbine type and the characteristics of the power electronic devices.

6.1.1.2.1 DFIG Power Electronic Protection
As described in Chapter 2, the power rating of the B2B converter of a DFIG wind turbine is just a fraction of the power rating of the generator. Consequently the overcurrent capacity of the power electronic devices for the B2B of the DFIG is much smaller than the transient fault currents induced in the rotor windings of the DFIG.

During a fault, transient currents that may be as large as the ac grid fault currents will be induced in the rotor windings (due to the law of lux conservation). The immediate blockade of the firing pulses for the B2B is not enough to protect the integrity of the power electronic devices, particularly in the case of the rotor-side converter (RSC). To protect the B2B converter protection schemes have been developed, such as the following:

1. Crowbar protection.
   The crowbar protection short circuits the rotor windings to a set of resistors whilst disconnecting the RSC from the circuit. The resistance is chosen to be large enough to reduce the transient currents back to a safe level in a short period of time (as detailed in Section 2.2.4). The electric circuit of the crowbar protection can be constructed using a three-phase full-bridge rectifier and a switching device (e.g. a GTO or an IGBT), connected in parallel between the rotor windings and the RSC terminals as shown in Figure 6.2.
2. dc chopper.
   This protection scheme consists of a braking resistor connected in parallel with the capacitor in the dc link of the B2B converter (Figure 6.3).

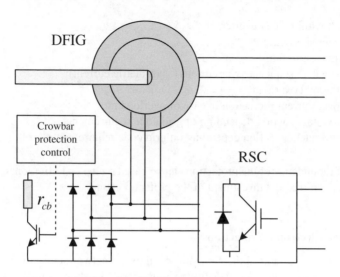

Figure 6.2   DFIG crowbar protection.

In the event of a fault, the dc link voltage increases due to a sudden increase of the RSC currents and the inability of the GSC to deliver power to the ac grid due to the low voltage. Thus the aim of the dc chopper is to limit the transient overvoltage of the dc capacitor and protect both the IGBTs of the B2B converter and the dc capacitor. The braking resistor also provides a way to dissipate the extra energy in the dc circuit (Pannell *et al.*, 2013). This protection scheme is also used for wind turbines with fully-rated converter topologies.

The protection triggers when the dc voltage rises above a pre-defined threshold. Under this condition the IGBT of the dc chopper is turned on and the braking resistor gets connected in

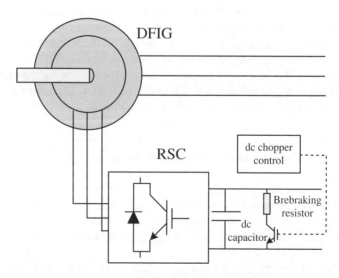

Figure 6.3   DFIG dc chopper protection.

parallel with the dc circuit. This causes the dc circuit energy to dissipate in the braking resistor and reduces the dc voltage level.

6.1.1.2.2 FRC-WT Power Electronic Protection

In case of a fault near a wind turbine with a fully-rated converter, no parts of the generator are directly connected to the grid but are instead interfaced via a B2B converter. Since the output current is controlled at any time by the converter, the turbine can be regarded as a current source with no transient over current response in case of a grid fault. However, the sudden decrease in grid voltage level during the fault affects the energy transfer capability from the turbine to the grid. This undelivered energy produces an increase in the wind turbine speed and in the dc voltage level of the B2B converter. Both situations can affect the integrity of the turbine. Thus protection inside the B2B converter is designed to deal with these problems:

1. dc chopper.
   Similar to the dc chopper of a DFIG wind turbine, a braking resistor is inserted in the dc circuit to dissipate the energy coming from the generator in such a way that the energy balance in the dc link is maintained. This prevents the excessive rise of the dc link voltage. A power electronic switch controls the braking resistor. The power dissipation of the braking resistor is controlled by the duty cycle of the power electronic switch (Mohan *et al.*, 1989). One way to control this duty cycle is by recording the dc current input of the converter working as a rectifier just before the fault and measure the value of the dc current of the converter working as an inverter during the fault. The comparison of these values is used to calculate the dc current to flow through the braking resistor (Conroy and Watson, 2007). Several manufacturers offer a wide variety of controlled braking resistors for motor drives and wind turbine applications.
   A diagram of the dc chopper protection for fully-rated converter wind turbines is shown in Figure 6.4.
2. dc series dynamic resistor.
   The dc series dynamic resistor topology for wind turbines is described in (Jin *et al.*, 2010b). The main characteristic of this device is that the series connection on the dc side can control the current magnitude directly. To avoid the loss of control in the rectifier the resistance will share any transient overvoltage in the dc circuit. A diagram of the dc series dynamic resistor is shown in Figure 6.5.

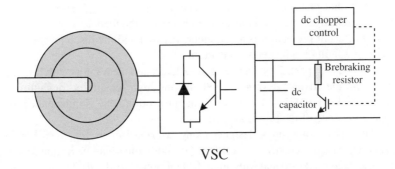

Figure 6.4 dc chopper protection for fully rated converter wind turbines.

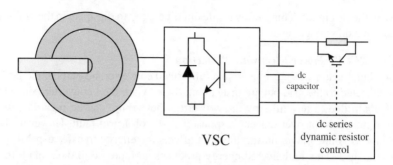

Figure 6.5 dc series dynamic resistor for fully-rated converter wind turbines.

6.1.2   Feeder Protection Zone

Any fault within the wind turbine unit would normally be detected and isolated by the protection within the wind generator zone. If this is not the case then the feeder protection has the potential to act as backup protection. However, it is highly desirable to have backup protection in the wind generator zone to prevent the disconnection of the entire feeder, along with the other wind generators connected to it.

The coordination of the wind generator zone backup protection may be achieved using an over current protection. Ideally, the pickup current has to be low enough to detect faults at the end of the line or in the low-voltage side of the wind turbine transformer, yet high enough to allow the uninterrupted operation under normal load conditions.

The setting for the pickup current I_{pickup} can be set to (Vikesjö and Messing, 2011):

$$1.2\frac{I_{max}}{k} \leq I_{pickup} \leq 0.7 I_{sc\,min} \tag{6.1}$$

where I_{max} is the maximum load current through the feeder, k is the reset ratio of the protection and $I_{sc\,min}$ is the smallest short circuit current measured at the feeder (that is, the short-circuit current at the low-voltage side of the wind turbine transformer).

There may be cases where Eq. (6.1) cannot be fulfilled because of the magnitude of I_{max}. If this is the case a solution could be the use of directional phase over current protection or under impedance protection.

When a directional phase over current protection is used, an I_{max} current setting smaller than the total current from the wind turbines connected to the feeder would be set. The use of directional over current protection is an efficient way to avoid the disconnection of a healthy feeder in case of a fault on a feeder adjacent to it.

The under impedance protection can be used to detect and clear faults with low short-circuit current level with the advantage of selectively having high sensitivity for fault currents and low sensitivity for load currents. The reach of the protection zone is independent of the fault type and operating state of the power system.

The impedance protection can be designed using two zones, encircling the feeder impedance and the load area. The impedance characteristic for both zones can be of quadrilateral shape, with some time delay (Vikesjö and Messing, 2011). Figure 6.6 shows the characteristic of the under impedance relay for feeder protection.

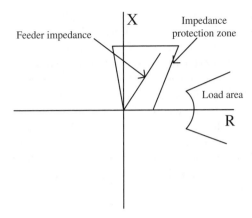

Figure 6.6 Characteristic of the under impedance relay for feeder protection (Vikesjö and Messing, 2011).

6.1.3 Busbar Protection Zone

As common practice, the busbar faults are required to be cleared as fast as possible because of the large fault currents that may be induced. However, the busbar protection system must be selective and avoid tripping in case the fault happens in a feeder protection zone. Figure 6.7 shows a commonly used protection principle for the busbar protection zone.

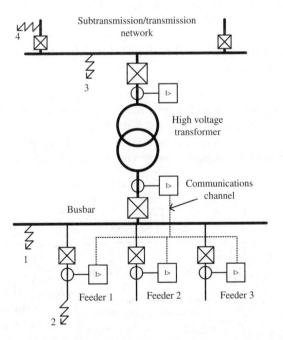

Figure 6.7 Busbar protection configuration (Vikesjö and Messing, 2011).

When a fault occurs in the low-voltage side of the high-voltage transformer (fault 1) the over current protection of the busbar will pick up and, after a short time delay, trigger the circuit breaker (CB) on the low-voltage side of the transformer clearing the fault. However, if a fault occurs in a feeder (fault 2) the trip of busbar protection must be prevented by a blocking signal coming from the protection devices within the faulted feeder. There is the risk, however, that a short circuit in a busbar may trigger the protections of the feeders and the trip signal of the busbar CB is blocked by the protection of the feeders. In order to avoid this problem, the feeders can use directional short circuit protection to differentiate between faults happening inside or outside its protection zone.

Finally, the trip of the busbar protection must be followed by the tripping of all CBs of the feeders connected to the busbar in order to stop any fault current contribution coming from the wind turbine generators.

6.1.4   High-Voltage Transformer Protection Zone

The substation protection in a wind farm follows the conventional protection schemes typically included in substations. These include bus differential protections for the medium-voltage and high-voltage buses, transformer differential protection, gas pressure protection and breaker failure protection, amongst others (Ackermann, 2012).

When a fault occurs on the subtransmission/transmission line (fault 4 in Figure 6.7) the usual procedure is to open the CBs at both ends of the line after a time delay. The protection on the high-voltage side of the transformer may be unable to distinguish between load current and the current at fault locations 3 and 4 in Figure 6.7. For this case the use of an under impedance protection is recommended (Vikesjö and Messing, 2011).

6.2   Study of Faults in the ac Transmission Line of an Offshore DFIG Wind Farm

In this section, two case studies are reviewed to show the behaviour of a DFIG wind farm under an ac transmission line fault. The system under test is presented in Figure 6.8.

The fault starts at t = 1s and the CBs of the faulted line open 200 ms after. The fault causes a voltage dip of 0.6 pu (0.4 pu retained) for 200 ms in the DFIG wind farm network. Prior to the fault the DFIG wind farm has an incoming wind speed of 11 m/s and is producing 0.71 pu active power.

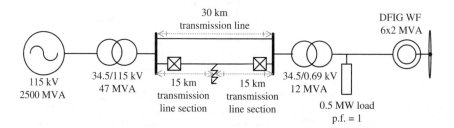

Figure 6.8   Single-line diagram for the study of faults in the ac transmission network.

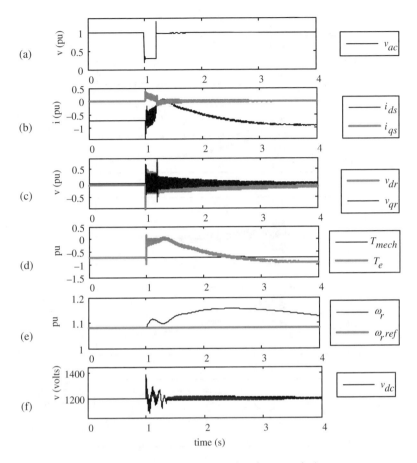

Figure 6.9 Simulation results of case study 1.

6.2.1 Case Study 1

At the instant when the fault transient i_s is induced, as shown in Figure 6.9b the magnitude of these transient currents is not enough to induce transient i_r or to trigger the crowbar protection. Consequently v_r in the rotor circuit never suffers an interruption, as shown in Figure 6.9c. This has a positive effect in the regulation of the energy flowing in the dc circuit, which even under the fault period, is controlled by the RSC and GSC to maintain the v_{dc} at its reference value. The voltage dip causes a sudden reduction of the generated T_e as illustrated in Figure 6.9d. Because of this, ω_r increases during the fault period and also when T_e recovers after the fault is released (Figure 6.9e). When the fault is cleared, the transient over speeding of ω_r is eventually controlled by the RSC and the DFIG wind farm rides through the fault successfully.

6.2.2 Case Study 2

The key results of the second case study are presented in Figure 6.10 and in Figure 6.11. For this case study the transmission line is reduced in order to have a voltage dip of 0.8 pu

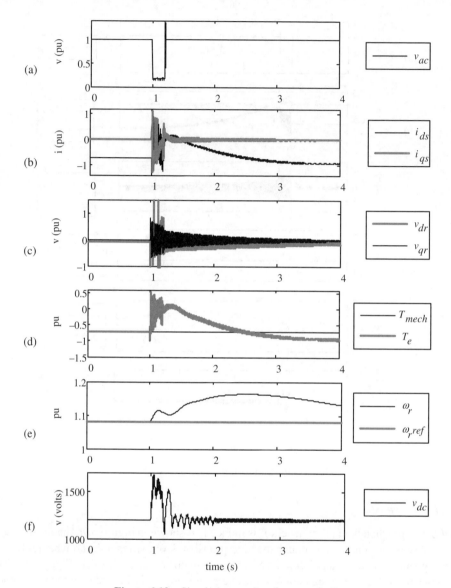

Figure 6.10 Simulation results of case study 2.

(0.2 pu retained). Figure 6.11 is a more detailed view of Figure 6.10 centered on the period of 0.9 to 1.3 seconds. It can be seen that the magnitudes of the transient currents are larger than in the previous case. Therefore, a transient i_r large enough to trip the crowbar protection is generated. The triggering of the crowbar protection results in v_r being equal to zero for a period of 10 ms (the time activation period of the crowbar protection, immediately after the fault, as clear in Figure 6.10c and shown in more detail in Figure 6.11c). After the crowbar protection deactivates, v_{dc} begins increasing due to the limited capacity of the GSC to deliver

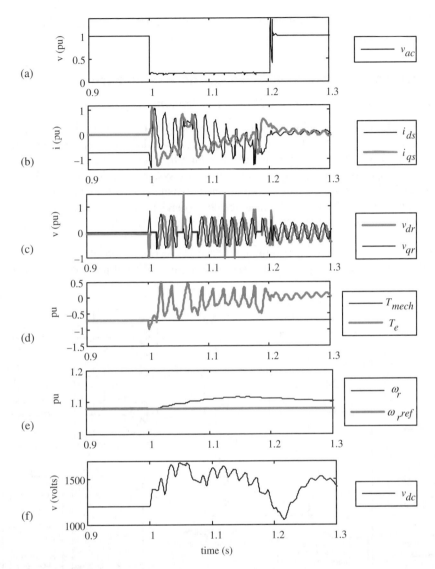

Figure 6.11 Simulation results of case of study 2 zoomed around the period between 0.9 to 1.3 seconds.

power to the ac network. When v_{dc} reaches a level of 1600 V, the crowbar protection triggers again and stops the power flow from the rotor circuit into the dc link of the B2B converter. Following the activation of the crowbar v_{dc} begins to decrease and the dc capacitor is protected from an overvoltage. As the fault continues the RSC attempts to regain control over the DFIG variables unsuccessfully and the crowbar protection activates one last time before the fault is cleared.

The second case study shows the over speeding of ω_r is slightly larger than in the previous case, this is because of the fault current magnitude and its effect on T_e. After the fault is cleared, the transient over speeding of ω_r is eventually controlled by the RSC and the DFIG wind farm accomplishes a successful fault ride-through.

6.3 Protections for dc Connected Offshore Wind Farms

6.3.1 VSC-HVDC Converter Protection Scheme

Figure 6.12 shows the protections included in a VSC-HVDC converter station. The philosophy of the protection within the station is to prompt the removal of any element of the electrical system from service in the event of a fault. The fault may be due to a short circuit or due to an abnormal operation of one of the elements of the system. The protection system makes use of ac CBs at the ends of the transmitting and receiving HVDC stations to de-energise the converter transformer. This will eliminate the dc current and voltage at the stations.

The actions taken by the HVDC station protection scheme can be:

1. Transient current limitation. This protective action is carried out by the temporary blocking of the converter control pulses on a per phase basis. This will prevent the circulation of

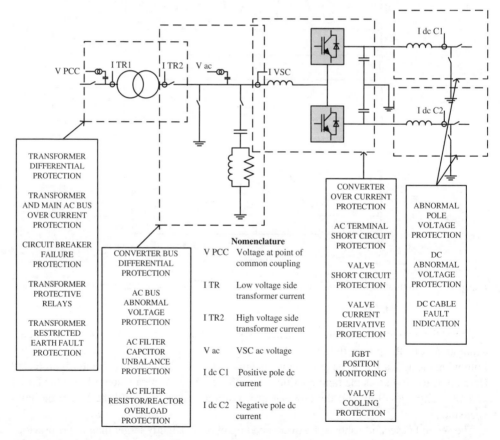

Figure 6.12 Protections included in a VSC-HVDC station (Source: ABB report – *HVDC Light, It's time to connect*. Available at www.abb.com/hvdc).

transient over currents that may damage the IGBTs of the converter. The control pulses can be re-established once the current returns to safe levels of operation. In the case of an overvoltage occurrence, the pulses are blocked for all three phases simultaneously.

2. Permanent blocking of the converter. This protective action sends a permanent turn-off control pulse to all the IGBTs of the station forcing the current circulating through them and the voltage levels at the terminals of the station to be zero.

3. ac circuit-breaker trip. This protection action disconnects the ac network from the converter equipment. This is carried out in order to prevent the ac system feeding a fault on the valve side of the converter transformer or in the dc side of the converter station.

4. Pole isolation. The pole isolation disconnects the dc side positive and negative poles from the dc cable during a short circuit in the dc side of the converter, or during a system malfunction. It should be noted that the pole isolation is carried out after the ac CBs have de-energised the converter stations.

6.3.2 Analysis of dc Transmission Line Fault

When a fault condition occurs on the dc line, the dc capacitors of the VSC-HVDC quickly discharge, failing to back bias the anti-parallel diodes of the IGBTs. As a consequence of this the fault current is driven towards the fault incidence point in the dc side of the converter (Figure 6.13). Under this situation, the converter station is not capable of interrupting the ac power from feeding the fault current through the diodes because the IGBT firing pulses, even if they are blocked, do not exert any control over the anti-parallel diodes of the valves. In such circumstances the interruption of the rectified fault current is not possible by means of using conventional circuit breakers (CB) in the dc line. This is because the dc current, contrary to its ac counterpart, lacks the zero crossing instant on every cycle used to interrupt the current stream. If a conventional ac breaker is used to try to interrupt the dc current, a damaging arc will be created burning the switchgear.

Until now, the approach conventionally used to interrupt the dc fault and restore the full power transmission in a VSC-HVDC link is to use ac CBs in both ac ends of the VSC-HVDC link. The process of opening the ac CBs, clearing the fault and then closing the breakers can take up to 2 seconds (Fairley, 2013). This time period could be prohibitive in applications where the amount of power transmitted is large enough to affect the ac network stability. In addition, another protective measure already in use consists of a press-pack thyristor, which is

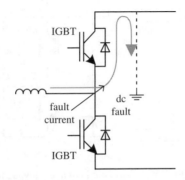

Figure 6.13   dc fault current path in a VSC-HVDC converter.

connected in parallel to the endangered anti-parallel diode and is fired in the event of a dc line-to-line fault. As a result, most of the fault current flows through the thyristor and not through the diode it protects. Press-pack thyristors have an inherent capability to withstand high surge currents (Davies *et al.*, 2009). However, the lack of dc fault current blocking capabilities of the VSC-HVDC converter and the delay in restoring the power flow is not the major problem for VSC-HVDC grids subject to a dc fault. A larger concern during a dc fault is the induced transient currents flowing through the valves of the VSC-HVDC station as discussed below.

In conventional HVDC transmission systems based on line-commutated converters, a large smoothing reactance is connected in series with the cables. In such cases the dc link has no over current problem caused by cable faults. However, in the case of VSC-HVDC converters, the dc circuit lacks this reactance and the fault currents are much higher in amplitude and speed of propagation. The characteristic impedance of an XLPE cable is typically well below 100 Ω, which is significantly lower than an overhead transmission line (Van Hertem and Ghandhari, 2010). A dc fault current interruption in the order of milliseconds is required to protect the anti-parallel freewheeling diodes of the IGBTs from the transient over currents, which in a matter of milliseconds have enough intensity to destroy them.

6.3.3 Pole-to-Pole Faults

The dc pole-to-pole fault is the most severe fault occurring in the dc side of the VSC-HVDC network. Despite this, it is rather unlikely for this type of fault to occur (since the positive and negative pole of the dc circuit travel in two independent insulated cables with some distance in-between). If it does occur, then the initial dc short circuit current can be determined by an equivalent circuit comprising the equivalent resistance and inductance of the dc cable connected in series with the pre-charged dc capacitor of the dc link (Figure 6.14). The expressions for the voltage v_{dc_sc} and current i_{dc_sc} of the equivalent circuit are (Jin *et al.*, 2010a, 2012):

$$v_{dc\_sc} = \frac{v_{dc0}\omega_0}{\omega}e^{-\delta t}\sin(\omega t + \beta) - \frac{i_{dc0}}{\omega C}e^{\delta t}\sin\omega t \tag{6.2}$$

$$i_{dc\_sc} = C\frac{dv_c}{dt} = \frac{i_{dc0}\omega_0}{\omega}e^{-\delta t}\sin(\omega t + \beta) + \frac{v_{dc0}}{\omega L}e^{-\delta t}\sin\omega t \tag{6.3}$$

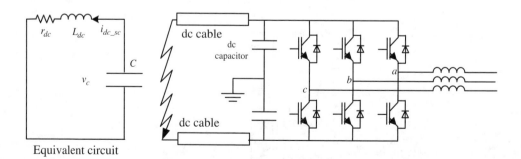

Equivalent circuit

Figure 6.14   dc pole-to-pole fault and equivalent circuit.

where v_{dc0} and i_{dc0} are the initial conditions of the dc voltage and current of the circuit, C, L_{dc} and r_{dc} are the capacitance, inductance and resistance of the equivalent dc circuit, $\delta = \frac{r_{dc}}{2L_{dc}}$ $\omega^2 = \frac{1}{L_{dc}C} - \left(\frac{r_{dc}}{2L_{dc}}\right)^2$, $\omega_0 = \sqrt{\delta^2 + \omega^2}$ and $\beta = \arctan \frac{\omega}{\delta}$.

Simulation studies have shown the magnitude of the initial fault current for a pole-to-pole fault to be up to 73 times the nominal dc current with a rise time <5 ms (Jin *et al.*, 2010a). After the dc capacitor is discharged, the inductor cable discharges through the anti-parallel freewheeling diodes of the IGBTs in the VSC. Also, since the dc voltage drops, these diodes become directly polarised and begin driving current from the ac network to the dc fault point. This current can be regarded as the steady-state fault current.

The IGBTs in the VSC-HVDC stations have very small over current and overvoltage tolerance. Consequently, the initial fault current in a pole-to-pole fault is more than enough to permanently damage the diodes of the IGBTs. To avoid such a destructive scenario a dc CB strong enough to interrupt the dc current in a very short period of time is required.

A complementary solution to use along the dc CB is a dc pole reactor, calculated to function as a current limiter (Fujin and Zhe, 2013). A high inductance in the dc circuit considerably reduces the rate of rise of the fault current during the initial capacitor discharge period. This enables the use of dc CBs with longer breaking time. For example, a 100 mH dc pole reactor allows for a maximum rise of the fault current of 3.5 kA/ms in a 320 kV dc network (Callavik *et al.*, 2012). Furthermore, a larger dc reactor also reduces the steady-state fault current level. Nevertheless the use of dc reactors results in additional complexity in the design and implementation of good performance dc voltage controllers.

6.3.4 Pole-to-Earth Fault

The magnitude of the dc short circuit current in a pole-to-earth fault depends on the grounding system of the converters. It usually consists in the neutral of the high-voltage transformer of the wind farm and the dc link midpoint. Initially, an arc generates between the pole and the sheath of the cable and a ground loop through the sheath and the next grounding point is established. Later on it is most likely that the arc currents will increase up to the point of destruction of the cable at the ground fault location. From that moment on, a low ohmic path is established between the pole and the ground. Here the soil resistivity and ionisation characteristics will define the fault resistance. This fault resistance cannot be ignored. Thus the equivalent circuit for the analysis of the initial magnitude of fault current can be drawn as shown in Figure 6.15.

In that way, the expressions for the voltage v'_{dc_sc} and current i'_{dc_sc} of the pole-to-earth fault equivalent circuit under the initial conditions of $v'_{dc\_sc} = V_0$, $i'_{dc\_sc} = I_0$ are (Jin *et al.*, 2010a, 2012)

$$i'_{dc\_sc} = C' \frac{dv'_{dc\_sc}}{dt} = A_1 p_1 e^{p_1 t} + A_2 p_2 e^{p_2 t} \tag{6.4}$$

$$v'_{dc\_sc} = A_1 e^{p_1 t} + A_2 e^{p_2 t} \tag{6.5}$$

Where $p_{1,2} = \frac{r_f + r'_{dc}}{2L_{dc}} \pm \sqrt{\left(\frac{r_f + r'_{dc}}{2L_{dc}}\right)^2 - \frac{1}{L'_{dc}C'}}$, $A_1 = \frac{1}{p_2 - p_1}\left(p_2 V_0 + \frac{I_0}{C'}\right)$ and $A_2 = \frac{1}{p_2 - p_1}\left(-p_1 V_0 - \frac{I_0}{C'}\right)$.

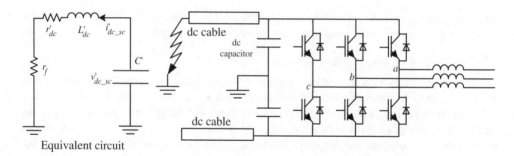

Equivalent circuit

Figure 6.15   dc pole-to-earth fault and equivalent circuit.

In this type of fault, the dc capacitor discharges down to a level set by the fault resistance. Simulations have shown the initial transient current to be up to 11 times the normal dc current with an initial rise time of 5 ms (Jin *et al.*, 2010a).

6.3.5   HVDC dc Protections: Challenges and Trends

The reason for choosing fast HVDC CBs instead of their ac counterpart comes from the need for a fast interruption of the faulted currents in order to avoid the damage of the freewheel diode of the IGBT (Jovcic *et al.*, 2011). The ac CBs are not fast enough to protect the IGBT (up to now, the minimum tripping time for the fastest ac CBs is around 20 ms). When electromechanical dc CBs (combined with a parallel commutation circuit and a parallel arrester), are used, it is necessary to produce an oscillation of the dc current which facilitates the arc extinction between the main contacts once a zero-cross point of the fault current occurs. However, such a process requires around 10–20 ms which is not fast enough for protecting the IGBT (Jovcic *et al.*, 2011). Current investigations are focusing on the research and development of solid-state dc breakers where the switching time can be as low as a few microseconds (Jovcic *et al.*, 2011). However, this technology still requires improvement and a reduction in cost. There are proposals in the open literature which focus on using special semiconductor materials for designing the CBs for the HVDC. Some propose the use of other power electronic based solutions like the hybrid configurations consisting of solid-state CB and a fast dc switch (Franck, 2011). A hybrid HVDC CB which consists of a semiconductor-based load commutation switch connected in series with a fast mechanical switch and a dc reactor, able to interrupt a major dc fault current within 5 ms has also been reported (Callavik *et al.*, 2012).

6.3.6   Simulation Studies of Faults in the dc Transmission Line of an Offshore DFIG Wind Farm

When the VSC experiences a dc fault, the voltage in the dc link quickly decreases, according to the type of fault; in any case large currents will circulate through the anti-parallel freewheeling diodes of each IGBT valve. When these over currents are detected, the IGBT valves are immediately blocked, and ideally, a pair of fast dc CBs will isolate the faulted cable before the transient currents damage the diodes of the IGBTs. The blocking of the IGBTs leads to a total loss of the voltage reference to the wind farm. A sudden loss of the voltage reference, just as in the occurrence of a three-phase fault, causes the largest transient currents in the stator windings of each DFIG of the wind farm. The transient i_s, in turn, induces large transient currents to the rotor windings, which are large enough to trigger the crowbar protection. This implies leaving the speed and the reactive power consumption of the wind turbine uncontrolled for a period of time.

As explained in Chapter 2 the induction machine requires a voltage reference at its terminals to produce T_e. If this is not available all the incoming T_{mech} from the wind will turn into kinetic energy, speeding up the rotor of the machine at the highest speed rate-of-change. If the slip speed of the IM increments beyond a critical stability point, the machine runs away and T_e collapses permanently. In order to avoid reaching the rotor speed critical stable point, the crowbar protection allows a large over speeding of the IM. However the higher ω_r is, when the ac voltage is restored, the larger the steady-state currents induced in the rotor windings will be and thus larger voltages are needed to control them.

When ω_r is above the synchronous speed, active power is generated in the rotor windings which goes into the RSC and is then delivered to the grid by the GSC. The fundamental aspect for achieving a successful fault ride-through is the capability of the RSC of the DFIG to control ω_r after the crowbar protection is deactivated. This capacity is, however, limited by the power rating of the converter.

Figure 6.16 shows a single-line diagram of a DFIG wind turbine connected to a multi-terminal HVDC network. This network is used to analyse the behaviour of the DFIG during a pole-to-pole fault happening near the VSC-HVDC station 2 at t = 17s.

One millisecond after the inception of the fault, a blocking signal is sent to the IGBTs of the three VSC-HVDC stations forcing them to be in off state. After 5 ms fast HVDC CBs isolate the faulted poles and prevent the transient dc currents from damaging the freewheeling diodes of the VSC-HVDC converter. Finally, after 200 ms, the dc voltage level of the HVDC network recovers and the blocking signal is removed from the IGBTs of stations 1 and 3. The power exchange between the DFIG wind farm and the ac grid is restored. The results of the simulation are presented in Figures 6.17 and 6.18. Figure 6.18 is a more detailed view of Figure 6.17 centred on the period 16.9 to 18 seconds.

As shown in Figures 6.17a and 6.18a, the grid voltage collapses almost immediately after the fault occurrence. Here the transient currents of the stator are similar to those of a three-phase fault near the stator terminals, as shown in Figures 6.17b and 6.18b. The magnitude of the transient rotor currents induced during the fault period are large enough to increase

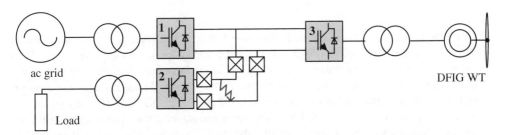

Figure 6.16 Multi-terminal VSC-HVDC network for a pole-to-pole fault case of study.

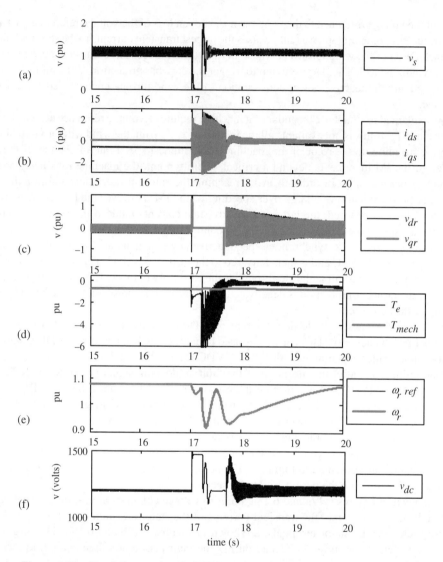

Figure 6.17   Simulation results of a DFIG wind turbine under a pole-to-pole dc fault.

abruptly the dc voltage level of the B2B converter before the crowbar protection disconnects the RSC from the rotor circuit (Figures 6.17f and 6.18f). The dc voltage level of the B2B converter remains above its reference value during all the fault period; this is due to the total inability of the GSC to deliver any energy to the DFIG ac grid. Because of this, the crowbar protection remains active during the fault period, and no voltage is injected to the rotor circuit (Figures 6.17c and 6.18c). The initial transient stator and rotor currents create a transient peak of T_e that slows down the rotor speed just after the fault occurrence (Figures 6.17d and 6.18d). When the dc fault is cleared, the transient currents create a T_e peak that affects ω_r and keeps

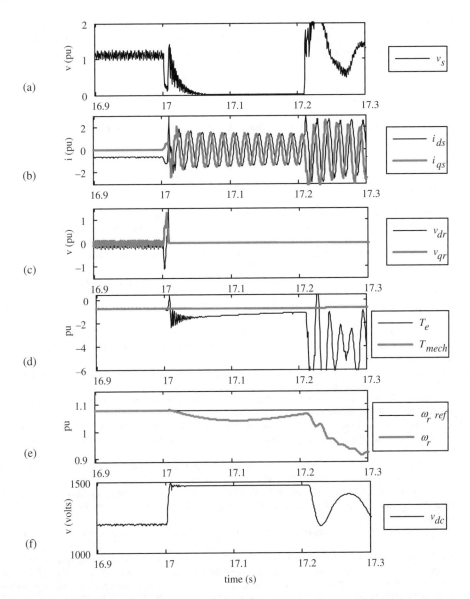

Figure 6.18 Simulation results of a DFIG wind turbine under a pole-to-pole dc fault zoomed around the period between 16.9 to 17.3 seconds.

the crowbar protection active for an extended period of time. However, once the dc fault is cleared and the VSC-HVDC converter connected to the DFIG starts generating ac voltage again, the GSC is able to control the dc voltage level in the B2B converter.

Even with the extended period of time that the crowbar protection is active during the dc fault, the DFIG eventually manages to regain control of its variables after the VSC-HVDC converter restores the voltage reference in the wind farm grid because the rotor speed did not reach a level beyond the control capacity of the B2B converter. Thus, as a result of using fast HVDC CBs the fault ride-through of the DFIG wind turbine was successful.

Acknowledgements

The material presented in this chapter is partially based on research conducted by the Authors under EPSRC Supergen Wind.

References

Ackermann, T. (2012) *Wind Power in Power Systems*, John Wiley & Sons Ltd.

Callavik, M., Blomberg, A., Häfner, J. and Jacobson, B. (2012) The Hybrid HVDC Breaker An innovation breakthrough enabling reliable HVDC grids. ABB Grid Systems, Technical Paper.

Cardenas, J., Muthukrishnan, V., Mcginn, D. and Hunt, R. (2010) Wind farm protection using an IEC 61850 process bus architecture. IET Conference Publications, 2010, p. 10.

Conroy, J.F. and Watson, R. (2007) Low-voltage ride-through of a full converter wind turbine with permanent magnet generator. *Renewable Power Generation, IET*, **1**, 182–189.

Davies, M., Dommaschk, M., Dorn, J. *et al.* (2009) *HVDC PLUS – Basics and Principle of Operation*, SIEMENS, Germany.

Fairley, P. (2013) Germany jump-starts the supergrid. *IEEE Spectrum*, **50**, 32–37.

Franck, C.M. (2011) HVDC circuit breakers: a review identifying future research needs. *IEEE Transactions on Power Delivery*, **26**, 998–1007.

Fujin, D. and Zhe, C. (2013) Design of protective inductors for HVDC transmission line within DC grid offshore wind farms. *IEEE Transactions on Power Delivery*, **28**, 75–83.

Jin, Y., Fletcher, J.E. and O'Reilly, J. (2010a) Multiterminal DC wind farm collection grid internal fault analysis and protection design. *IEEE Transactions on Power Delivery*, **25**, 2308–2318.

Jin, Y., Fletcher, J.E. and O'Reilly, J. (2010b) A series-dynamic-resistor-based converter protection scheme for doubly-fed induction generator during various fault conditions. *IEEE Transactions on Energy Conversion*, **25**, 422–432.

Jin, Y., Fletcher, J.E. and O'Reilly, J. (2012) Short-circuit and ground fault analyses and location in VSC-based DC network cables. *Industrial Electronics, IEEE Transactions on*, **59**, 3827–3837.

Jovcic, D., Van Hertem, D., Linden, K. *et al.* (2011) Feasibility of DC transmission networks. Innovative Smart Grid Technologies (ISGT Europe), 2011 2nd IEEE PES International Conference and Exhibition on, 5–7 Dec. 2011, pp. 1–8.

Mohan, N., Undeland, T.M. and Robbins, W.P. (1989) *Power Electronics: Converters, Applications, and Design*, Wiley.

Pannell, G., Zahawi, B., Atkinson, D.J. and Missailidis, P. (2013) Evaluation of the performance of a DC-link brake chopper as a DFIG low-voltage fault-ride-through device. *IEEE Transactions on Energy Conversion*, **28**, 535–542.

Ribrant, J. and Bertling, L. (2007) Survey of failures in wind power systems with focus on Swedish wind power plants during 1997–2005. Power Engineering Society General Meeting, 2007. IEEE, 24–28 June 2007, pp. 1–8.

Van Hertem, D. and Ghandhari, M. (2010) Multi-terminal VSC HVDC for the European supergrid: obstacles. *Renewable and Sustainable Energy Reviews*, **14**, 3156–3163.

Vikesjö, J. and Messing, L. (2011) *Wind Power and Fault Clearance*, Elforsk, Stockholm Sweden.

7

Emerging Technologies for Offshore Wind Integration

7.1 Wind Turbine Advanced Control for Load Mitigation

Both generator and pitch control have enabled greater power regulation and lighter blade construction due to a lower load spectrum and a lighter gearbox (as a consequence of the reduction in torque peaks). Difficulties arise in turbulent winds when excessive loading occurs, which causes premature wear on turbine components. Limitations in current technology to mitigate more efficiently these excessive loads have led the research to new wind turbine control approaches (Bossanyi, 2000; Johnson *et al.*, 2008; Wright and Fingersh, 2008; Leithead *et al.*, 2009).

Controllers for load mitigation in large commercial wind turbines are typically designed using classical control techniques such as PI control. A loop is often added to the generator torque control below-rated wind speed to actively damp the drive train torsion mode of the turbine. To actively damp the tower fore-aft motion a single-input single-output (SISO) control loop is implemented into the pitch control above-rated wind speed.

7.1.1 Blade Pitch Control

Another way to mitigate turbine loads is through individual pitch control (IPC) (Bossanyi, 2005; Johnson *et al.*, 2008). With IPC each blade is pitched independently and both classical control and multivariable control may be used. Classical control methods use multiple control loops to add active damping to several flexible turbine modes or to minimise the effects of asymmetric wind variations. These controls must be designed with great care to avoid loop interactions that may cause turbine instability (especially as turbines become larger and more flexible, and the degree of coupling between individual control loops increases) (Leithead *et al.*, 2003; Dominguez and Leithead, 2006). Advanced multi-input multi-output (MIMO) multivariable control design methods can be used to meet these objectives using

Offshore Wind Energy Generation: Control, Protection, and Integration to Electrical Systems, First Edition.
Olimpo Anaya-Lara, David Campos-Gaona, Edgar Moreno-Goytia and Grain Adam.
© 2014 John Wiley & Sons, Ltd. Published 2014 by John Wiley & Sons, Ltd.
Companion Website: www.wiley.com/go/offshore_wind_energy_generation

all the available actuators and sensors in a reduced number of control loops and maximising load alleviation. Cyclic pitch control is another advanced technique that varies the blade pitch angles with a phase shift of 120° to alleviate the load variations caused by rotor tilt and yaw errors.

There are three concerns when considering individual pitch control: first, the entire blade must still be pitched; second, the pitching mechanism may be unable to act fast enough to relieve the oscillating loads due to wind gusts; third, there is a concern that individual blade pitch will result in over-use of the pitching mechanism. It is important to design turbines to use individual pitch from the start; retrofitting current turbines with individual pitch control will lead to premature failure of the pitch mechanism due to the resulting high duty cycle. Challenges with implementation include response time requirements to counter load perturbations, the need for larger pitch motors, and the power required to operate the system under a new control strategy.

7.1.2   Blade Twist Control

One concept for controlling fatigue loads on a wind turbine blade is to use passive blade bend-twist coupling (Johnson *et al.*, 2008). The aero-elastic tailored blade is designed so that the twist distribution changes as the blade bends due to aerodynamic loads. This is now possible through the advent of composite materials, which can be implemented in a deliberate fashion to control flap-twist. The transient loads due to wind gusts theoretically could be reduced because the blade would twist towards lower angles of attack, thereby mitigating the loads and potentially reducing pitch activity as well. Some of the challenges with this concept include reduced energy capture, higher costs, and blade integrity issues. First, reduced energy capture may occur due to altering a blade that is designed for optimum energy capture at rated speed by causing it to twist. Second, higher costs associated with material and manufacturing techniques may make the concept uneconomical. Third, the fabrication technique may lead to decreased stiffness and additional material may be required to counteract additional blade deflection.

7.1.3   Variable Diameter Rotor

Variable diameter rotors operate by extending/retracting a *tip* blade out of a *root* blade to increase/decrease the diameter (Johnson *et al.*, 2008). This concept is capable of improving energy capture in low wind speeds and reducing loads on the rotor in high wind conditions. During low wind speed, a large rotor diameter provides more capture area, which results in larger aerodynamic loads and an increase in energy capture. However, this operation generates larger blade root and tower base bending loads. In higher wind speeds, the rotor diameter can be decreased to avoid excessive loads. The tip blade would extend and retract independently of the pitching mechanism and it would respond to gross changes in the wind speed; the pitch control would still be used to regulate power. The variable diameter rotor has the potential for increasing energy production for a given load spectrum. Some engineering challenges that need addressing to make this turbine concept successful include complex control strategies, the need to maintain a high aerodynamic efficiency, increased blade weight, and general issues with durability and reliability of the system as a whole.

7.1.4   Active Flow Control

Active Flow Control (AFC) is the control of the local airflow surrounding the blade (Johnson *et al.*, 2008). The purpose of flow control is often to improve the aerodynamic performance of an aerofoil or lifting surface. However, for utility-scale wind turbines the main focus is to reduce extreme loads, which occur during high wind activity, and to mitigate fatigue loads, which vary along a blade and can occur randomly. To do this, active load control devices or 'smart' devices must include actuators and sensors located along the span of the blade. The system must be able to sense changes in the local flow conditions and respond quickly to counter any negative impact on blade loading. This arrangement provides active 'smart' control over the rotor. By definition, a smart structure involves distributed actuators and sensors and one or more microprocessors that analyse the responses from the sensors and use integrated control theory to command the actuators to apply localised strains/displacements to alter system response. Numerous investigations on the use of AFC devices show that significant load reduction is possible.

7.2   Converter Interface Arrangements and Collector Design

To gain understanding on how wind turbine generator technology may influence control approaches, it is necessary to consider the electrical system as a whole, that is, the turbine topology (converter interface arrangement), wind farm collector, and the offshore transmission type (e.g. ac or dc). Two cases are discussed next in the context of the converter interface arrangement and location (Parker and Anaya-Lara, 2011, 2012, 2013).

7.2.1   Converters on Turbine

7.2.1.1   Ac String

The conventional ac string arrangement is shown in Figure 7.1. Turbines feature a squirrel-cage induction generator (SCIG), or a permanent magnet generator (PMG) connected to a fully-rated converter, or alternatively a doubly-fed induction generator (DFIG) and partially-rated converter. The output of the converter is stepped up to the collection network voltage and the turbines are connected together in strings. The number of turbines on a string is determined by the ratings of both the turbine and the cable being used. Voltage is limited by water-treeing effect with wet insulation cables, while dry insulation cables with a lead sheath around the insulation would be too expensive. A higher voltage also requires a higher voltage rating for the transformer, which increases the cost and size. Available current ratings are also limited; this is because the skin depth of the ac current means that conductors with larger areas will be less effective. For this reason, the cost of ac cables tends to increase exponentially with current capacity.

7.2.1.2   Dc String

An arrangement using dc in the collection strings is shown in Figure 7.2. In this system, the turbines output a dc voltage, which is then stepped up to the transmission voltage at the collection platform. In most studies, the turbine produces a voltage of around 40–50 kV dc, which requires an ac-dc converter capable of producing such a voltage, featuring many

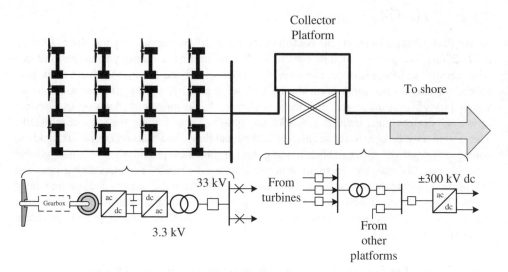

Figure 7.1 Conventional ac strings (Parker and Anaya-Lara, 2013).

switching devices in series, or a lower voltage ac-dc converter and step-up dc-dc converter. A solution involving a lower voltage converter and a dc voltage of 5 kV is also possible, which has the advantage of eliminating the turbine transformer and using a conventional 3.3 kV three-level converter. However, the currents in the strings will be extremely high, requiring thick cable and leading to high losses. Dc systems are attractive because they could reduce the number of conversion steps between ac and dc; however, converters with a voltage boost ratio will require a transformer, requiring conversion to ac and back.

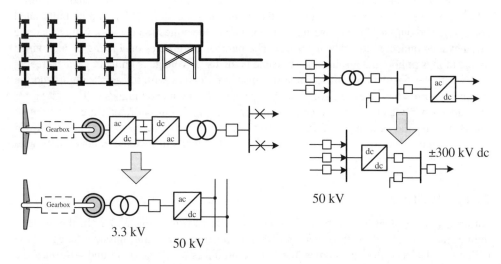

Figure 7.2 Dc strings connection.

As dc cables do not suffer from water-treeing degradation, higher voltages could be used without needing dry-insulation cables; furthermore, the current in a dc cable can use the entire surface area of the conductor. The implication of this is that the cable cost will increase linearly with current capacity rather than exponentially as with ac. Because of these factors, it could be possible to implement longer turbine strings much more cheaply with dc than with ac collection. However, this is difficult to quantify as currently there are no commercially available cables with the required configuration and voltage rating (previous studies of the cable cost have extrapolated the cost for multi-core dc collection cables from the costs for single-core HVDC transmission cables with a significantly higher voltage rating).

Another issue with dc collection networks is with fault protection: due to the fact that the current does not continually reverse as with ac, the switching arc will not be automatically extinguished when the current reverses following a circuit breaker being opened. Various dc circuit breaker designs have been proposed, but these become increasingly expensive at higher voltage ratings. Dc collection and transmission networks have been designed considering the use of power converters which are capable of stepping down the voltage as well as stepping up, and these can be used to limit the fault current, but at the cost of extra complexity.

7.2.1.3 Dc Series

An alternative dc collection architecture is to use series dc connection of the turbines as shown in Figure 7.3. Here the dc outputs of the turbines are connected in series, and the turbines connected in a loop. This allows the high collection voltage to be achieved without using high voltage converters, although the wind generator converter would need to be isolated with respect to ground. An isolation transformer would need to be used, or a generator capable of handling a high voltage offset. Another option is to use a transformer-isolated converter in the turbine, where the high-voltage side of the converter only consists of a passive diode rectifier, which is much easier to isolate.

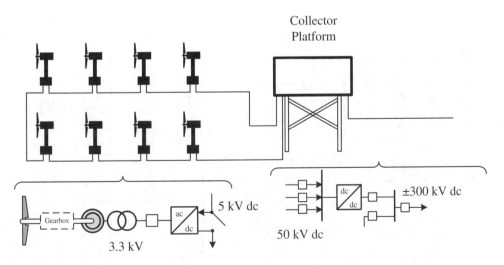

Figure 7.3   Series dc connection.

This arrangement could reduce the cable costs because it only uses a single-core cable loop, although there is no scope to taper the current rating of the cable. In the event of a turbine fault, the faulty turbine could be bypassed using a mechanical switch; however, any cable faults will mean that none of the turbines on the loop would be able to export power.

A related idea is to increase the turbine output voltage and the length of the strings so that the full transmission voltage is produced, eliminating the need for the collection platform. This system has been shown to have the lowest losses due to the high collection network voltage, and the lowest cost due to the elimination of the collection platform. Several strings could be used in parallel to increase fault tolerance. The disadvantage of this system is that the transformer and converter in the turbine must be capable of isolating the full transmission voltage, and high voltage transformers with a low enough power rating are not commercially available.

7.2.2   Converters on Platform

7.2.2.1   Ac Cluster

An increasingly popular idea is connecting turbines with fixed-speed induction generators to a variable-frequency ac collection grid, with strings of turbines being connected through a single converter. This places the converters on the collection platform, allowing them to be more easily repaired in the event of a fault, and a single large converter could potentially be cheaper than several small ones. An ac or dc collection system could be used within the collection platform as shown in Figure 7.4.

The speed of all the turbines in the string can be varied together to track the maximum power point for the current wind speed, but speed control over the individual turbines is lost. The speed of each turbine will be able to vary by a small amount relative to the others due to the slip of the induction generator, with an increase in turbine speed leading to an increase in slip and an increase in torque. Depending on the number of turbines connected to each converter, this will result in a reduction in the amount of power extracted.

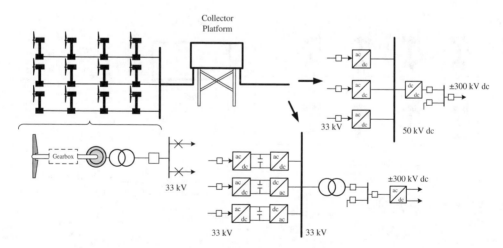

Figure 7.4   Cluster ac connection.

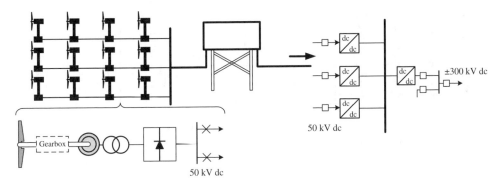

Figure 7.5   Parallel dc cluster collection.

This system could also have an impact on the drive train loads experienced by the turbines: a turbine experiencing a gust would not be able to speed up to absorb the excess power, leading to a high transient torque, putting strain on the drive train and blade roots. Research on the reliability of turbines in service has shown that the transition to variable speed turbines has reduced the level of blade failure compared with fixed-speed turbines.

7.2.2.2   Parallel dc Cluster

This method, shown in Figure 7.5, uses a permanent magnet generator and passive rectifier in the turbine, with a dc-dc converter for each string of turbines. The speed of the turbine will be determined by the dc voltage of the string; thus, the system will behave in a similar way to the ac cluster connection system described previously, with similar issues of drive-train torque transients during gusts. It is considered that the passive filter will have considerably greater reliability than an active converter.

For a given dc voltage, the amount of possible speed variation of the turbine will depend on the generator inductance, with a higher inductance giving a greater variation in speed. The passive rectifier is unable to supply the generator with reactive power; if the generator inductance is too high as well, the maximum torque will be reduced. Inductance is typically much higher in low speed machines, found in direct-drive turbines; in these cases, capacitors can be used between the generator and rectifier to supply the reactive power requirements. The main advantage of dc over ac clustering is the greater efficiency of the permanent-magnet generator, compared with the induction generator used in the ac system. The greater current and voltage capability of the dc cables could also lead to larger cluster sizes and a reduction in cable cost; however, this could also reduce the power capture. A dc system could also reduce the number of conversion steps, increasing efficiency.

7.2.2.3   Series dc Cluster

A variation of the parallel cluster arrangement is to connect the turbines in series in a loop, with each loop controlled by a single converter, as shown in Figure 7.6. In this case, the converter will control the current within the loop, which will determine the generator torque within the

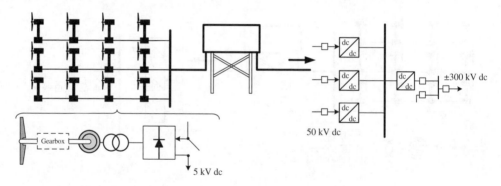

Figure 7.6   Series dc cluster connection.

turbine and will be much more analogous to the conventional turbine control method. As the turbine speeds will be capable of varying individually, transient torque spikes should not be a problem. This emerging connection method is in its initial stage of investigation.

7.2.3   Ac Collection Options: Fixed or Variable Frequency

Ac fixed-frequency operation of the offshore network is normal practice and possible with both synchronous (HVAC) and asynchronous (HVDC) connection of the offshore wind farm. Ac variable frequency operation at the collection network would be cost-effective only when the wind farm is connected to the grid through a HVDC transmission system (Hyttinen and Bentzen, 2002; Jovcic, 2006; Jovcic and Milanovic, 2006; Feltes and Erlich, 2007). This is because the offshore HVDC rectifier can control the offshore frequency independently from the onshore grid, whereas for a synchronous ac transmission link an additional ac-ac or ac-dc-ac conversion system would need to be installed. Examples of wind farm configurations using dc collection as proposed in the literature are described next (Lundberg, 2003; Meyer *et al.*, 2007).

7.2.3.1   Examples of Variable-Frequency Collection Configurations

The use of variable frequency operation in the collection network with DFIGs and a HVDC transmission link pursue one of the following objectives (Feltes and Erlich, 2007):

- Reduce the rating of DFIG converters.
- Extend the speed range to maximise the power capture without increasing the rating of the DFIG converters.

A multi-terminal configuration based on VSC-HVDC transmission that allows variable frequency operation in the offshore collection network is under investigation (Jovcic, 2006). The proposal is similar to the collecting network configuration in (Jovcic and Milanovic, 2006) but with a multi-terminal HVDC link based on current source converters that use forced-commutated devices such as IGBTs. In the latter arrangement, the generator transformers are not needed to step up the voltage to transmission level because the current source converters are connected in series. In both schemes, no additional converters are required at each WTG terminal, and 2MW permanent magnet generators are synchronised together in a group connected to a centralised converter and all are operating at the same speed.

A wind turbine topology that allows variable frequency operation with squirrel-cage induction generator over a wide range of operating conditions is also under investigation (Meier et al., 2004a, 2004b; Meir, 2005; Meyer et al., 2007). The topology replaces the fully-rated back-to-back converter with a single stage cyclo-converter to decouple the frequency at the offshore substation from the wind turbine side. The scheme has the following features and potential advantages:

- The 50 Hz three-phase transformers within each wind turbine generator and at offshore substations are replaced by medium frequency (400–500 Hz) single-phase transformers. The transformer insulation must be designed to cope with increased voltage stress, and high dv/dt resulting from step changes in the voltage due to the snubber capacitors of the converter.
- Since only a single-phase converter is required at the offshore substation instead of a three-phase converter, the number of series-connected devices required decreases significantly resulting in significant reduction in cost, conduction and switching losses.
- Soft switching of the cyclo-converter and offshore converter of the VSC-HVDC reduces the switching losses significantly.
- The use of thyristors rather than IGBTs in the cyclo-converter reduces the power loss and cost.

7.2.3.2   Ac Variable-Frequency Collection Evaluation

Variable-frequency operation at the offshore collection network in conjunction with a HVDC transmission link and WTGs of the DFIG type can maximise the power extraction and reduce the overall wind farm cost according to reference (Feltes and Erlich, 2007). However, the consequences of having variable frequency at the offshore network regarding switchgear and protection, transformer operation, voltage and current rating of the equipment located at the offshore network need to be thoroughly investigated. Standard power transformers are designed for a specific frequency of operation (50 Hz or 60 Hz), and the normal tolerance of frequency variation is around +5%, that is, for a 50 Hz unit, limits of 47.5 to 52.5 Hz. For a lower frequency design of transformer, a larger core would be needed in order to maintain the required voltage ratio for a specific current rating and a reasonable flux density avoiding saturation. However, this can be overcome by reducing the voltage in proportion to a reduction in frequency. Another aspect to be investigated is how the reactive power flow through the transformers and cables changes with variable frequency.

Operating VSC-HVDC at variable frequency has been demonstrated practically in the gas platform Troll A (Hyttinen and Bentzen, 2002). In that application, the need for variable-frequency operation by the VSC-HVDC inverter is clear: to control the speed of an induction motor. For an offshore wind farm on the other hand, variable frequency can be beneficial for the generator rotor only, but further research is required to assess whether it is more economical to achieve this locally at each individual generator than at a collector level.

7.2.4   Evaluation of >Higher (>33 kV) Collection Voltage

As wind farms and turbines have increased in power capacity, the collection voltage levels have increased from typically 11 to 33 kV. The usage of 48 or 66 kV cables to connect wind turbines to the onshore grid via 48/132 or 66/132 kV transformer onshore, instead of using a 132 kV submarine cable and a 33/132 kV transformer offshore, is being investigated (McDermott, 2009). The study shows that stepping up the voltage from 33 to 132 kV offshore is most economical for greater distance (>25 km); this is because the cost of the offshore substation is then less significant compared to the cable costs, coupled with the fact that the losses are much more reduced.

Another reason for using higher voltage collection cables is that they can bring the benefit of needing less cable strings in collection networks for large offshore wind farms (>300 MW); this is because each cable has a higher capacity. This may become a particularly attractive solution for ever-increasing wind turbine sizes. For example, with present designs of 33 kV cables, it would only be possible to connect up to four 8 MW wind turbines per cable string; however, in the London Array, up to nine 3.6 MW wind turbines are connected to one string. Higher voltages also reduce fault levels for a given MVA generation. In addition, Ohmic losses would be less. The exact overall losses, however, are design specific, which requires special consideration on a case-by-case basis. In addition, since higher voltages offer a longer transmission distance, the collection cables can be longer, so fewer offshore 'sub-transmission' platforms may be needed. These were introduced for example in the 400 MW VSC-HVDC linking the BARD Offshore wind farm to the transmission grid in north Germany, where the collection voltage is stepped up from 30 to 155 kV on two separate 'sub-transmission' platforms. These are linked via 155 kV cables to the offshore 400 MW HVDC substation (Stendius, 2007) to transmit the power via dc over a distance of 203 km to onshore grid.

The challenges of using voltages higher than 33 kV for the collection network presently are as follows:

- There is a limited supply of commercially available dry-type transformers that are both rated above 33 kV, and are capable of stepping up from a suitable generator voltage for example, 3.3 kV. They are also more expensive than the more widely available 33 kV transformers. It has been suggested that a 66 kV transformer is about double the price of a 33 kV transformer and weighs 42% more for the same MVA rating. Oil-filled transformers at the wind turbines are undesirable because of the risk of a spillage at sea, which poses an environmental threat.
- Collection cables at 33 kV and below can be of the 'wet design', which do not require metallic moisture barriers surrounding the cable as an outer sheathing layer or around the insulated core(s). The sheathing/bedding layer(s) are made from polypropylene (or jute) string/rope. Seawater fills the empty spaces inside the cable, making direct contact with the outside of the insulated core(s). Higher voltage submarine cables presently available are of the 'dry design' which use a lead sheath as a water barrier. Their drawbacks are as follows: the increased cable capital cost; the potentially higher installation cost due to the additional weight; and the possibility of lead sheath fatigue failure if movement or vibration occurs (Wald et al., 2009).

For a specific project, a thorough cost-benefit study is required to make an informed decision on whether to opt for 33 kV or higher voltage collection cables because the impact on the overall wind farm design may be profound.

7.3   Dc Transmission Protection

The protection of a multi-terminal system is a major problem, especially during a fault in the dc network. The VSC-HVDC system copes well with ac faults with appropriate control and protection methods; however, it is vulnerable to faults occurring in the dc link, and this has limited its practical application in, for example, multi-terminal networks during the last decade. The technology gap in isolation mechanisms of the dc fault and the absence of reliable dc breakers is a real problem facing multi-terminal dc operation. When a short circuit occurs between the dc link conductors, the capacitors rapidly discharge into the fault point circulating excessive dc fault current (Figure 7.7). With the absence of the zero crossing point and the excessive fault current, no dc circuit breaker has been manufactured to handle the dc link fault. Power electronic converters are very sensitive to the overload current and cannot withstand the excessive fault current (Mesut and Nikhil, 2007). A reliable protection system capable of detecting the fault location and terminating the fault is required to maintain system stability during dc faults. The dc fault could be cleared using ac circuit breakers associated with power converter blocking; however, this would require all terminals to be isolated, resulting in complete shut-down of the dc system (Hertem and Ghandhari, 2010; Adam *et al.*, 2010).

Several approaches are introduced to address the clearing of the dc fault. Ac circuit breakers can be used to eliminate the fault in the dc network. Existing VSC-HVDC systems activate the ac breakers placed on the ac networks to isolate the dc network until the fault is extinguished. Power electronic converters are blocked during the isolation to speed up the clearance time. After fault clearance, the dc system is reconnected again and the converters are unblocked gradually with a ramp function (Livermore *et al.*, 2010). For the permanent faults, the faulty dc cable is totally removed from the dc network.

IGBT circuit breakers placed between each of the converter terminal capacitors and the dc network buses are used to protect the dc network against dc network faults (Lianxiang and Boon-Teck, 2002). The limitation of this approach is the unidirectional capability of the IGBT circuit breaker. When the fault occurs on the converter side, the IGBT-CB is blocked-in and no power flows into the dc network until the fault is extinguished. If the fault occurs in the dc network side, the fault current flows into the VSCs through the anti-parallel diodes (Figure 7.8). In this case, the blocking of the IGBT-CB is associated with the blocking of the VSCs. The IGBT-CBs have the same rating of the VSC and this increases transmission system losses.

New technology is currently being introduced which is capable of addressing the dc fault issues. In the two-switch modular multilevel converter (see Appendix A), the current magnitude

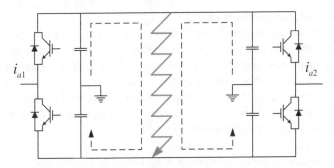

Figure 7.7   Fault in the dc link of a VSC-HVDC system.

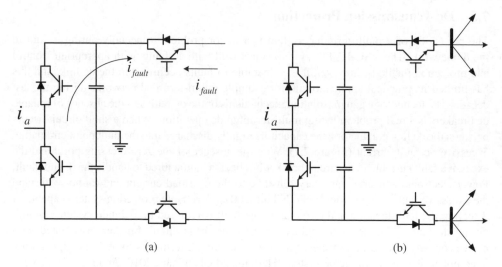

Figure 7.8   IGBT-circuit breakers.

that the converter switched may experience during a dc fault is significantly reduced due to the absence of the dc link capacitors (Adam *et al.*, 2010).

7.4   Energy Storage Systems (EESs)

Energy storage systems (ESS) can mitigate issues associated with the high penetration of wind energy such as inadequate control over generation and low power quality. In addition, ESS can potentially provide voltage and frequency regulation services. The optimal location of the ESS depends on its purpose. The main applications of energy storage systems are (Aten *et al.*, 2006; Jim E. *et al.*, 2007; Ali N. *et al.*, 2007; Makarov *et al.*, 2008; Power Beacon, 2009; Matthew and Alex, 2010; Alex, 2003):

1. **Power quality:** In such applications, the stored energy is applied for only a short period of time in the order of seconds or less to assure continuity of quality power.
2. **Bridging Power:** Here the stored energy is used to balance power demand and generation over time scales in the order of seconds to minutes. The ESS can be located anywhere in the grid where adequate network capacity exists.
3. **Energy Management:** Stored energy in these applications is used to decouple the timing of generation and consumption of electric energy over longer time scales that is, hours. A typical application is load levelling (balancing). If an ESS is located at the PCC, then the power generated from a wind farm can be stored at times when a local line in the grid would otherwise get thermally overloaded, unless the wind power is spilled. Another possible application would be to locate the ESS offshore to peak shave wind generation, thereby minimising the rating of an export cable.

Most energy storage technologies still have considerable disadvantages such as high cost, low efficiency and size, which make their practical applications limited; however, they may be gradually improved by further R&D efforts. The main energy storage technologies are discussed below (Jim E. *et al.*, 2007 Ali N. *et al.*, 2007; Makarov *et al.*, 2008).

7.4.1   Batteries

Batteries used for relatively small storage systems are nickel-cadmium (Ni-Cd), nickel-metal hydride (Ni-MH), lead acid (Pb acid), Sodium-Sulfur (Na-S) and Na-NiCl2 (Zebra) and Lithium-ion (Li-ion). All these batteries, however, require major technological advances before they can be scaled up economically for the purpose of balancing the variable power generated from large offshore wind farms. A more promising usage for them in the short term is to provide auxiliary power to the wind turbine generators and offshore substation components (switchgear, transformers, converters and control equipment).

7.4.2   Super-Capacitors

Super-capacitors, also known as ultra-capacitors, offer higher capacitance and higher energy density compared to conventional capacitors. Their energy density is still much lower than that of batteries, but their power density is much higher. Compared with batteries, super-capacitors can also be charged and discharged with much higher currents and for many more cycles, without compromising their life-time. Whereas some super-capacitor technologies use electrolytes that are safe and not harmful to the environment, others are highly flammable and toxic. One major drawback of super-capacitors is that their energy stored is proportional to the voltage squared, which means that the voltage drops as more energy is taken out, requiring an increase in current to maintain a constant power export. This means that the extractable energy and power from the energy storage system may be a compromise, unless the current rating of the power electronic interfaces (dc-dc or dc-ac) could be increased, adding to its cost. Another drawback of super-capacitors is that they are made up of many low voltage (<3 V) cells and serial connections are needed to create higher voltages. This requires voltage-balancing circuits adding to the complexity, reducing the reliability and increasing the no-load losses. In addition, self-discharge rates are considerably higher than that of an electrochemical battery.

7.4.3   Flywheel Storage System

Flywheel-based energy storage systems, unlike batteries, are sustainable technology solutions that do not use hazardous materials in their construction, nor create them during operation. They are suitable for high power applications, but their energy density is relatively low. Despite higher initial costs, it is claimed that flywheels offer a cost-effective long-term and reliable energy storage alternative (Alex, 2003; Makarov *et al.*, 2008; Beacon Power, 2009; Matthew and Alex, 2010). There are already some commercial products designed specifically for frequency regulation. The viability of employing large-scale flywheels will depend on how the capital investment costs compare with the market to provide frequency regulation in the future. More R&D work is required to achieve improved efficiency, higher energy density and lower cost.

7.4.4   Pumped-Hydro Storage

Although this is not an emerging technology, the role that pumped-hydro storage plays within future power systems will almost certainly evolve as requirements for mitigating the impact of wind variability grow. Conventional pumped-hydro uses two separate water reservoirs at different heights. During off-peak hours (when the generation exceeds the demand), water is pumped from the lower reservoir to the upper reservoir. During the peak hours where more power is required, the water flow is reversed to generate electricity. Pumped-hydro is available at almost any scale with discharge times ranging from several hours to a few days. Their efficiency is in the 70 to 85% range. Pumped-hydro storage plants are characterized by long construction times and high capital expenditure. Pumped-hydro storage is the most widespread energy storage system in use on power networks. In the context of variable energy sources such as wind power, its main applications are energy management, frequency control and provision of reserve (Toshiya and Akira, 1994; Makarov, 2008). The main drawback is that sites are limited since pumped-hydro storage requires large reservoirs in mountainous areas, which may be far from generation and demand.

7.4.5   Compressed-Air Storage Systems

Compressed-air storage technology has received significant attention recently as a means to address the variability problems with wind power. When surplus energy is available during off-peak periods, it is used to run air compressors and store compressed air in the storage tank. When electricity is needed during peak hours, the compressed air is used to generate electricity through conventional gas turbines. The first compressed air storage power station has been in operation since 1978 with a capacity of 290 MW located in Bremen, Germany, capable of delivering full output power for up to 4 hours (Lemofouet and Rufer, 2006; Swider, 2007). Compressed-air is usually stored in artificial salt caverns underground made by dissolving salt from strata of rock salt at least 100m thick and several hundred meters underground. Consequently, this method is limited to locations where there are large amounts of suitable rock salt underground.

7.4.6   Superconducting Magnetic Energy Storage (SMES)

SMES technology exploits the fundamental property of a superconductor having negligible resistance to electrical current to store energy in a superconducting coil in the form of a magnetic field that can be created by a flow of current. A typical SMES system includes three parts: a superconducting coil, a power conditioning system and a cooling system to maintain the superconducting temperature below its critical value. Once the superconducting coil is charged, the current will not decay and the magnetic energy can be stored indefinitely. The stored energy can be released back to the network by discharging the coil. The power conditioning system uses a power electronic interface such as an inverter/rectifier to allow power exchange between the superconducting storage system and the power network. The inverter/rectifier accounts for about 2–3% energy loss in each direction. Therefore, superconducting energy storage systems are highly efficient devices with round-trip efficiency greater than 95%. The use of SMES systems is limited to power quality applications due to their high cost, and their relatively low energy storage capacity (Luongo, 1996; Baev et al., 2007).

7.5   Fault Current Limiters (FCLs)

Development of medium- and high-voltage fault-current limiters can contribute to reduce fault levels at the collection network of large-scale wind farms. It may also facilitate the connection of large offshore/onshore wind farms to high-voltage networks without the need for replacement of the existing switchgear at the point of common coupling (PCC) (Martin and Claus, 2004; Brian *et al.*, 2007). Fault-current limiters may provide a cost-effective solution if there is a problem of excessive fault levels, and if they are cheaper than the cost of higher rated switchgear that would otherwise be needed due to excessive fault levels. Fault-current limiters are also an enabling technology that could allow increased load, generation or increased paralleling of feeders, which would otherwise not be possible due to limitations of the rating of available switchgear. The most common types of fault-current limiters are reactors, and high-impedance transformers. Other devices being developed are solid-state fault-current limiter circuit breakers and high-temperature superconducting fault-current limiters, but these are not commercially available yet. Superconducting fault-current limiters are designed to decrease fault levels by inserting impedance into the circuit only in the event of a fault, without adding impedance to the circuit during normal operation. This is achieved by the fact that the superconductor switches to a non-superconducting state when the current through the superconductor exceeds a critical value.

7.6   Sub-Sea Substations

Platform-based substations are used at present for the connection of offshore wind farms. As deep waters (>40 metres) are explored for the deployment of WTGs and non-fixed devices for the capture of wave and tidal energy, then subsea substations may become a competitive alternative to platform-based substations. The technologies for subsea substations have so far been developed for offshore oil and gas industrial installations. The electricity generation industry is only recently becoming involved in tidal and wave energy generation. The major obstacle to using subsea technologies so far are as follows: the high costs; health and safety risks associated with installation and maintenance; and the fact that most subsea technologies are still limited in voltage levels. The essential components required for a subsea substation are briefly reviewed:

- The largest voltage rating of a constructed and tested subsea transformer to date is 50 kV. Higher voltage levels will be needed for longer transmission links.
- Subsea connectors can be split into two categories: 'wet mate', where the physical connection can be made whilst submerged; and 'dry mate', where the connection must be made above the surface before submerging the connector (Enger and Rocke, 2006). This would require a ship, platform or similar. A 'wet mate' connection cannot be made whilst the line is energised; however, cables can be connected without regard for water ingress as the connectors contain a system for ejecting the water from the connection area.
- Wet mate connector designs are far more complicated than dry mate versions and therefore are more expensive. However, costs may be offset against the simpler cable design and installation, which does not need to include making the connection to the item of plant above the surface before lowering both the cable and plant to their subsea positions. Dry mate and wet mate connectors have been demonstrated up to 33 kV. Future designs are in

place for dry mate connectors at 145 kV; however, no such plans are in place for wet mate connectors.

- The oil and gas industries have operational subsea switchgear at 24 kV, which utilise a magnetic actuator system. The significant benefit of this system is that it is largely maintenance free. There are proposed designs for 33 kV subsea circuit breakers, but early indication is that a motorised spring charge actuator system will be used. Such a system requires periodic maintenance and is therefore not ideal.

7.7 HTSCs, GITs and GILs

7.7.1 HTSCs (High-Temperature Superconducting Cables)

Development of HTS cables may enable large-scale power transmission over long distances without the need for extra-high voltage. Therefore, it may also facilitate grid connection of remote large onshore or offshore wind farms using an ac option with reduced power losses. As high-temperature superconducting cables transmit power with essentially no electrical resistance, it may enable utilities to increase power-carrying capability of existing power corridors. This increased capacity is appealing for locations where there are space constraints for conventional lines or cables. The main benefits of HTS power cables are as follows (Masuda *et al.*, 2012):

- Current-carrying capability of 3–5 times that of conventional cables.
- Low load power losses.
- Use of environmentally friendly liquid such as nitrogen for cooling.
- No leakage of electro-magnetic field to the outside of the cable resulting in low impedance.
- As power loss is extremely low, HTS cables may eliminate the need for high voltage (400 kV) and extra high voltage for long transmission distances.

However, HTS cable technology is still at the development stage, extremely expensive and its reliability is unproven. One further aspect of concern is the requirement to have a cooling system that would have to be located approximately every 2 to 5 km, which would be very costly if not impractical for a submarine cable. Moreover, such a cooling system would add to no-load losses, which offsets the benefit of low load losses.

7.7.2 GITs (Gas-Insulated Transformers)

Gas-insulated transformers (GITs) that use SF6 as an insulator without the need for oil are a mature technology and are available up to 300 MVA. Their advantages are as follows (Toda, 2002):

- Large power rating in small size. Combined with the switchgear the size of a substation can be reduced to about a third.
- Non-flammable and non-explosive (safe).
- Higher reliability with simple internal structure.

GITs have been implemented in some niche applications where there are severe space constraints, for example in big cities. It is therefore logical to consider them for offshore substations where space is also at a premium. However, the main drawbacks of GITs are their high costs and potential environmental issues.

7.7.3 GILs (Gas-Insulated Lines)

Gas-insulated lines (GILs) are composed of pipes that house conductors with SF6 gas used as the insulator. Although the first GIL was commissioned in 1979, the technology is still at the demonstration stage and would only be better than conventional cables at powers above several GWs. Further development of GILs for long distances may improve system efficiency when used with large-scale wind farms. Since the shunt capacitance of GILs is small, the problems associated with reactive power are less than in the case of the XLPE submarine cable. Power losses are low and a high transmission capacity (several GW) can be achieved. The disadvantages of GILs are the high costs and the use of SF6 gas (Koch and Hillers, 2002).

7.8 Developments in Condition Monitoring

Condition monitoring systems can detect the potential failures of different electrical and mechanical components of an offshore wind farm; in addition, they are capable of predicting the mean time to failure and enabling a degree of remote maintenance of the wind farm. Condition monitoring systems based on Supervisory Control and Data Acquisition (SCADA) have been used extensively to monitor and to control numerous aspects of the wind farm operation, such as adjustment of the turbine blade pitch and power output, monitoring the condition of bearings, vibration of the drive-train, generator winding temperature and recording switchgear operations.

The applications of condition monitoring for offshore wind farms have increased in the last decade because of the following reasons:

- It can help to maximise the energy yield from the turbines while minimising operational and maintenance costs.
- It allows effective system management by remote control and a degree of remote maintenance that may increase the productivity of the wind farm.

In the context of the offshore electrical array system, SCADA also plays a vital role in monitoring cables, switchgears and transformers at the offshore network to facilitate remote fault detection and schedule maintenance in order to reduce downtime and increase productivity (Gardner *et al.*, 1998). With further development and improvement of the current condition monitoring techniques, it may play a major role in improving system reliability and availability which can deliver large operational savings in the future (Gardner *et al.*, 1998; Wiggelinkhuizen *et al.*, 2007; McMillan and Ault, 2008). The next section describes developments in condition monitoring in more detail for network components.

7.8.1 Partial Discharge Monitoring in HV Cables

The condition of high-voltage cables can be determined by monitoring the level of partial discharge (PD) activity in the cable. This is of interest to network operators in various countries, for instance in the UK who have a significant amount of cable approaching the 60-year design life of the cable, and need to know how to prioritise cable replacement (Renforth *et al.*, 2005). Presence of PD can be detected by attaching high-frequency current transformers to the cable, with the discharges appearing as spikes of current. The energy of the pulse can be determined by integrating the pulse current – this is more accurate than simply recording the pulse magnitude as the pulse shape will change depending on how far along the cable the detector is from the pulse location.

It is possible to estimate the remaining lifetime of XLPE cables from analysis of the PD activity, but this requires a large number of parameters to be measured for each discharge and complex analysis techniques (Yazdandoust *et al.*, 2008). XLPE cables are designed for a lifetime of at least 60 years and are tested for partial discharge due to manufacturing defects before leaving the factory (Bartnikas, 2002). PD monitoring equipment can be easily added to a cable installation later in its life-time, when the cable is more likely to suffer from age-related failure. Once the presence of PD is established further equipment can be used to locate the source of the discharge (Renforth *et al.*, 2005).

7.8.2   Transformer Condition Monitoring

For early failure detection, it is necessary to measure the transformer load current and operating voltage which will give an indication of overload currents which could damage the windings mechanically, and over-voltages which would damage the insulation (Tenbohlen *et al.*, 2002). Measuring the oil temperature and load current allows the transformer hotspot temperature to be estimated, which determines the rate of insulation degradation. The presence of various gases in the oil of an oil-filled transformer can also indicate problems with the transformer (Wang *et al.*, 2002). Gases such as hydrogen (H_2), methane (CH_4), ethane (C_2H_6), ethylene (C_2H_4), and acetylene (C_2H_2) indicate a breakdown in the oil caused by overheating, partial discharge, arcing, sparking, and so on, and the ratios of each gas can be used to determine the precise cause (Muhamad *et al.*, 2007). Presence of carbon dioxide (CO_2) and carbon monoxide (CO) indicate breakdowns in the paper insulation while presence of oxygen (O_2) and nitrogen (N_2) indicates leaks in the system. This has been shown to be the case for both mineral oils and the newer ester transformer fluids (Dai *et al.*, 2007).

At the most basic level, the health of an on-load tap-changer can be monitored by having the tap-changer in a separate compartment of the transformer tank, and monitor the temperature in the compartment. An unusually high temperature in the oil would indicate a degradation of the tap-changer contacts. Another method has been proposed where the load current of the tap-changer motor is monitored. Problems with the tap-changer would cause higher peak currents at various points in the tap-changing sequence, and this has been demonstrated in simulated faults (Tenbohlen *et al.*, 2000).

Studies on the economics of offshore wind farm condition monitoring have concentrated on monitoring the turbine systems, particularly the gearbox and generator (Tenbohlen *et al.*, 2000; Nilsson and Bertling, 2007). A study of the benefits of transformer condition monitoring used a failure rate of 1.63% a year, based on reliability of 245 and 400 kV transformers in Germany, along with estimations of the probability of a condition monitoring system detecting a fault early to estimate the benefits of installing a CM system (Tenbohlen *et al.*, 2002). If the early detection of a fault means that a transformer can be re-wound at half the cost of a new transformer, rather than installing a new transformer, a saving of 0.58% of the cost of the transformer per year could be made. For a CM system with a 10-year life, such a system would have to cost less than 5.8% of the transformer cost, and it was found that the cost was between 1 and 7.4% depending on the installation. This does not take into account the impact of lost generation while waiting to replace the transformer, which could be considerable in an offshore installation.

7.8.3 Gas-Insulated Switchgear Condition Monitoring

The insulation effectiveness and breaking capacity of switchgear insulated with sulfur hexafluoride (SF6) depends on the density of the gas. An improved gas density sensor based on micro-machined quartz tuning forks has been proposed, which offers better accuracy than conventional sensors (Zeisel et al., 2000). The density of the gas varies with temperature in a complex manner, which depends on the movement of gas and distribution of temperature inside the circuit breaker. This makes it difficult to detect small leaks of gas until a large quantity of gas has leaked, although complex models have been developed which will accurately detect small leaks of gas (Graber and Thronicker, 2008).

Presence of partial discharges in the circuit breaker indicate a degradation in the insulation effectiveness, and can also cause the SF6 to degrade into other substances, some of which are corrosive and damage components of the circuit breaker. Switching of the circuit breaker will also cause degradation of the SF6. Analysis of the gas will show the level of degradation and indicate the cumulative effect of partial discharge and switching, and simple spectrometers based on the Ion Mobility Spectrometer can be used on site, or installed permanently to monitor the gas composition (Baumbach and Pilzecker, 2005). The level of partial discharge can be checked by monitoring the RF radiation in UHF frequencies (Metwally, 2004), which will also detect partial discharge inside solid insulating materials such as spacers, which gas analysis and other PD detection methods cannot do. However, it has been suggested that PD detection cannot detect low level discharges, which can still cause degradation of the SF6 gas over time, so online PD detection would need to be combined with periodical gas analysis.

7.8.4 Power Electronics Condition Monitoring

Condition monitoring for power electronics has concentrated on monitoring two components: electrolytic capacitors, and switching devices in conventional power modules. Both of these have been found to be significant sources of converter failure (Lahyani et al., 1998), but the degradation and failure mechanism for electrolytic capacitors is much better understood. In light of this, capacitor banks can be designed to achieve a desired life-time, given expected load conditions, and any failures are usually due to unexpected operation conditions, or a capacitor bank which is deliberately undersized to give lower cost with reduced life-time.

Electrolytic capacitors degrade due to the loss of electrolyte, the rate of which depends on the temperature inside the capacitor, which depends on the ambient temperature and the capacitor loading (Harada et al., 1993). The loss of electrolyte leads to an increase in the equivalent series resistance (ESR) of the capacitor over the life-time, which can be measured. ESR can be measured using dedicated current and voltage sensors attached to the capacitor bank (Aeloiza et al., 2005), or can be estimated from the other measurements of current and voltage in the converter (Lahyani et al., 1998). In the latter case, the ESR estimation would need to be built into the control software of the converter. ESR varies with capacitor temperature and ripple current frequency, and this must be compensated for in software.

Switching devices, such as IGBTs, in conventional power modules usually fail due to one of two degradation mechanisms: lifting of the bond wires from the surface of the silicon chips (Held *et al.*, 1997), or void formation in the solder attaching the chips to the copper substrate (Ratchef *et al.*, 2004). Both of these are fatigue effects and are related to thermal cycling. There is a small thermal cycling effect from the converter input and output fundamental frequencies, for example, the generator and grid frequency, and a larger effect from variations in the total power output of the converter (Sayago *et al.*, 2008).

Condition monitoring of switching devices is relatively immature in wind farm applications, and many possible methods have been investigated. These can be divided into three main types:

- Measurements of switching device parameters, which would indicate device degradation. For instance, bond wire lift-off can be detected by measuring the on-state voltage (Lehmann *et al.*, 2003; Xiong *et al.*, 2008), and various device parameters can be used to estimate the junction temperature to calculate the thermal resistance, which indicates the level of void formation in the solder (Cova and Fantini, 1998; Barlini *et al.*, 2006).
- Embedding sensors in the device package. For instance, additional terminals can be added to measure the bond wire resistance (Lehmann *et al.*, 2003), or sensors to measure the junction temperature (Sankaran and Xu, 1996).
- Determining device degradation from changes in the system dynamic performance, for example, the reaction to certain stimuli (Judkins *et al.*, 2007; Morroni *et al.*, 2007).

The first method requires additional sensors to be installed, and often changes to the gate drive circuit and control software. The second method requires custom designed power modules, while the third method requires significant signal processing and detailed knowledge of how the changes in device parameters will affect the system response.

The trend for higher power fully-rated converters in wind turbines larger than a few MW is to move towards medium voltage converters (Badrzadeh *et al.*, 2009; Steinke, 2010; Eichler *et al.*, 2010). These often use press-pack modules which do not suffer from the same failure modes as conventional power modules (Eichler *et al.*, 2010). While the reliability of the switching device in an IGCT is claimed to be greater than that of an IGBT (Eichler *et al.*, 2010), overall reliability is reported to be significantly lower, due to the complexity of the gate drive circuit (Badrzadeh *et al.*, 2009). Monitoring the health of the gate drive circuit could be achieved by monitoring the switching time of the switching device (Sankaran and Xu, 1996), although it would be difficult to accurately calculate the remaining life.

Anecdotal reports on the reliability of fully-rated converters in wind turbines have indicated a high failure rate on the grid side converter. This is believed to be due to unexpected events leading to the switching devices operating outside their safe operating area, possibly due to problems with the converter control action, which causes chip-level damage, rather than through continuous degradation mechanisms. While this is potentially detectable using condition monitoring apparatus, failure of the device would occur fairly rapidly, giving insufficient time to replace the device before failure. At best, condition monitoring would allow the converter to be shut down, preventing the device failure from damaging other parts of the converter. Furthermore, while failures of the converter and control electronics in a wind turbine are significant in number, the resulting downtime is usually small compared with other components such as the gearbox or generator (McMillan and Ault, 2007).

7.9   Smart Grids for Large-Scale Offshore Wind Integration

The next generation of electrical grids – generation, transmission and distribution – are meant to achieve the following: operation under the Smart Grid concept; and integrate vast amounts of variable renewable energy sources. The Smart Grid conceptual model, highlighted as a vision to deeply modernise the production, delivery and use of electricity worldwide, is characterised by the bulk use of advanced enabling technologies, controls, algorithms and strategies as well as high-speed communications and cutting-edge information technology from power generation to consumption of electrical energy and from the technological and market points of view.

The Smart Grid conceptual model, endorsed by the U.S. National Institute of Standards and Technology (NIST), and afterwards by the EU, defines seven relevant domains as components: bulk generation, transmission, distribution, customers, operations, markets and service providers. The model also indicates all the communications and energy/electricity interrelations and explores the way a domain, comprised of significant Smart Grid elements, connects to each other domain through two-way high-speed communications and energy/electricity paths in order to efficiently deliver sustainable, economic and secure electricity. These domain connections are the building blocks of the next-to-come intelligent and dynamic power electricity grid. Figure 7.9 shows the domains of the Smart Grid conceptual model (IEEE Smart Grid, 2013). Offshore wind energy, as well as onshore, has a participation in the bulk generation domain of Smart Grids. Each Smart Grid domain is expanded into various layers: (1) the components (power and energy); (2) communications; (3) IT/computer/information layer; (4) function layer, and (5) the business layer. Figure 7.10 illustrates layers and domains according to EU directions (Masera, 2012).

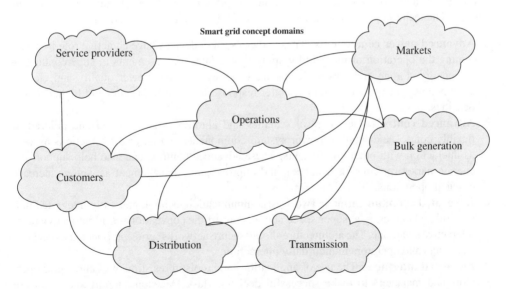

Figure 7.9   The smart grid concept domains (IEEE Smart Grid, 2013).

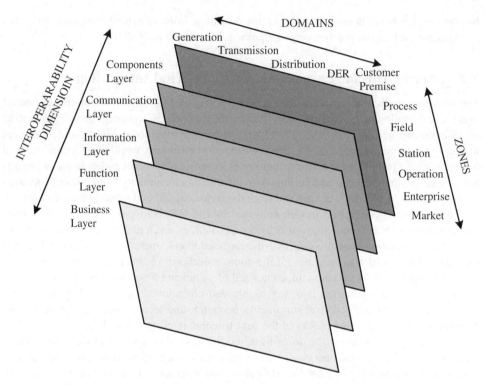

Figure 7.10   Smart grid layers and domains according EU directions (Masera, 2012).

Layers (2) and (3) are enabling infrastructure platforms of layer (1) that make the grid smarter. The technologies can be grouped into the following key areas (NETL, 2009):

a. **Advanced power components:** These power system devices have an active role in determining the operation of the grid by applying the latest achievements in superconductivity, energy storage and power electronics to produce higher power quality and electrical efficiency, increased reliability, better response to abnormal events and higher power densities.

b. **Advanced controls:** New control schemes and algorithms are adapted and tailored to flexibly control and monitor power system components and its grid interactions as well as interaction with other components. These advanced controls can also help support the energy market (pricing for instance), and improve asset management as well as increase efficient operations.

c. **Integrated communications:** Two-way communication technologies fully integrated into the grid and operating at super-speed should convert the conventional grid into a dynamic, interactive smart one. The architecture should secure networks' grid components, processes, customers and operators in their daily interaction.

d. **Enhanced interfaces:** The use of applications and tools in real-time will allow grid operators and managers to make successful decisions fast. Decision support and improved interfaces will help to perform more accurate and timely decision and data exchange at all layer levels and domains of the Smart Grid, including the consumer level.

e. **Smart sensing and measurement technologies:** These elements enhance measurements and help to convert data into information and surveillance of the network integrity and its power components, as well as supporting new protective relaying schemes.

There is a significant number of technologies ready applicable to the key areas for implementation of the Smart Grid concept (NETL, 2009). The anticipated technical and economic benefits of operational Smart Grids are numerous. For instance, for the grid side point of view, the benefits are expected to be the improvement in reliability, flexibility and efficiency of the network. In more detail, the potential benefits in these aspects are as follows:

> *Reliability*: By using smart and state-of-the-art power electronics, communication and information technologies – that enable real-time communication and the exchange of data between generation, transmission and distribution, and consumers – it may be possible to automatise a system of decisions to protect the system.

> *Flexibility*: As the flexibility in operation of the transmission and distribution grids increases, the opportunity for the grid operator to better do business also increases. On these grounds, a variety of demand side management technologies and diversified types of generation can all be provided by Smart Grid technologies, which can help achieve the reliability, environmental and economic targets.

> *Efficiency*: The transmission grid operates with a very high overall level of efficiency relative to the total overall power system losses. In contrast, the efficiency of distribution grids and the generation sector is lower. For the efficient integration of distributed generators, the structure of both transmission and distribution grids has to be changed. Next-generation of distribution grids are expected to be smart, active (with generation and similar responsibilities to current transmission systems) and with fewer dissimilarities with future transmission systems.

The Smart Grid layers and their interaction channels are under construction in the US and in Europe. In the European Union, various multinational multidimensional projects and research are under way focusing on the following: improving and accelerating grid reinforcement; make progress in developing enabling technologies; design and update standards; enlarge market balancing, augment grid flexibility; strengthen energy security; enhance reliability and maintenance; build secured data exchange between domains and layers; improve technical and economic efficiency; and integration of dispersed variable generation, mainly wind energy at bulk (Giordiano *et al.*, 2011).

For the electrical integration of large amounts of offshore wind farms into a Smart Grid, various enabling-technologies are available and others are under development (Madariaga *et al.*, 2013). Amongst these, outstanding electric topologies are the point-to-point HVDC, and the Multiterminal HVDC. The Super Grid is an ambitious project, supported by the EU, aimed to build an extra large trans-national grid which collects electrical power from multiple dispersed renewable sources of different types, mainly variable, and deliver to dispersed loads across EU country members (from northern to southern Europe). This enormous 'power interface' infrastructure is compatible with the Smart Grid concept and they can be implemented alongside although its full integration into one system, due to their different initial origin and focus, requires technical and economic conciliations and adjustments (Blarke and Jenkins, 2013).

The accommodation of numerous large wind farms has big challenges such as effective operation and maintenance of each wind power plant. To accomplish these tasks effectively, high-speed and automatic data exchange between offshore wind partners and wind power plants and operations centres are required as a way to build the correspondent layers and interaction channels for the Smart Grid. The adoption of the standard IEC 61400-25 (communications for monitoring and control of wind power plants) for data exchange and a framework for data integration can help in this pursue (IEC, 2006; T. H. Nguyen, and Prinz, 2012)

7.9.1   VPP Control Approach

A Virtual Power Plant (VPP) mainly addresses the supply side. The basic idea is to connect numerous distributed and renewable power generating facilities via modern information and communication technology (ICT). A central control entity continuously monitors the generation data and can switch individual generators in and out of the system at any time. Thus, the facilities' operation can be scheduled and optimised. By connecting multiple distributed generators, almost the same controllability as with conventional plants can be achieved (Knab and Strunz, 2010). Two effects in VPPs contribute to the achievement of this objective: First, a well-chosen mix of volatile generators can offset their inherent unreliability to a certain extent. Following basic stochastic principles, the connection of different volatile systems with different fluctuation patterns may lead to a decreased overall volatility. It is clear that this logic is applicable to power generation from renewable energy sources including both wind and solar. The aggregated control of renewable energy sources executed by the VPP concept can be centralised or decentralised supported by logic control algorithms and communication infrastructure, which then as a whole is treated as a single large power plant (Setiawan, 2007). VPPs can be divided into two categories according to the control topology in *centralised VPP* and *decentralised VPP*. Some advantages of the VPP approach and aggregated control of distributed energy sources are:

– **Scheduled power dispatch.** The power output of distributed energy resources can be dispatched quite rapidly in response to changes in generation and demand conditions, facilitating frequency control.
– **Load management.** By proper aggregation, control of DG units can be used to supply the load and reduce the demand on central generation as well as local transmission.
– **Voltage regulation.** Improved voltage regulation is possible by coordinating the operation of various DG units in a VPP approach.

An ideal VPP has the ability to solve technical barriers (e.g. by maintaining grid stability via load management), commercial barriers (e.g. by providing economic efficiency via energy trading), and regulatory barriers (e.g. by allocating cost and benefits of power consumption and production to various players involved) (Olejnczak, 2011).

7.9.2   Phasor Measurement Units

In order to keep power grids and their interconnection to wind power plants operating in secure and economic conditions in the short and long term, it is necessary to further improve protection and control systems. A phasor measurement unit (PMUs) is a useful tool for monitoring the performance of a power system. The PMUs were introduced into power systems in the 1980s, since then their values have been proved by their many applications in power system operation and planning. A PMU is essentially a digital recorder with synchronised capability that uses state-of-the-art digital signal processors that can measure 50/60 Hz ac waveforms (voltages and currents) typically at a rate of 48 samples per cycle (2880 samples per second). A phase-lock oscillator along with a Global Positioning System (GPS) reference source provides the needed high-speed synchronised sampling. Additionally, digital signal processing techniques are used to compute the voltage and current phasors. Line frequencies are also calculated by the PMU at each site. This method of phasor measurement yields a high degree of resolution and accuracy. The resultant time tagged phasors can be transmitted to a local or remote receiver at rates up to 60 samples per second.

Synchronised phasor measurements are ideal for monitoring and controlling the dynamic performance of a power system, especially during high-stress operating conditions (Roscoe *et al.*, 2013). In recent years, varieties of PMU applications have been studied, proposed and implemented with their significant benefits. The wide-area measurement system (WAMS) that gathers real-time phasor measurements by PMUs across broad geographical areas has been gradually implemented in various countries.

The 'IEEE standard for Synchrophasors for power system' defines the synchrophasor measurement and provides a method of quantifying the measurements, quality test specifications and data transmission formats for real-time data reporting. Due to the high scan rate of PMUs new communication architectures are needed to meet the requirement for wide-area monitoring, protection and control schemes. New data and information management architecture and technology are also presented to enable and enhance the applications of PMUs in wide-area protection and control. When PMUs are involved in extensive applications, an optimal strategy for PMUs placement is needed to reduce the economic burden of the utilities and maximise the performance with a limited number of PMUs. Promising potential PMU applications besides those for protection and control are (EPRI, 2007):

- Improvement on State Estimation.
- Oscillation Detection and Control.
- Voltage Stability Monitoring and Control.
- Load Modelling Validation.
- System Restoration and Event Analysis.

Acknowledgements

The material in Sections 7.2 and 7.4–7.8 of this chapter is based on research conducted by the Authors under EPSRC Supergen Wind in collaboration with Dr Max Parker, and ETI Helmwind in collaboration with Dr Martin Aten, respectively.

References

Adam, G.P., Finney, S.J., Williams, B.W. (2010) Network fault tolerant voltage-source-converters for high-voltage applications. AC and DC Power Transmission, ACDC. 9th IET International Conference on, 2010, pp. 1–5.

Aeloiza, E.C., Jang-Hwan, K., Enjeti, P.N. (2005) A Real Time Method to Estimate Electrolytic Capacitor Condition in PWM Adjustable Speed Drives and Uninterruptable Power Supplies. IEEE 36th Power Electronics Specialist Conference, PESC '05, pp. 2867–2872.

Alex, R. (2003) Integrating flywheel energy storage systems in wind power applications. Wind Power 2003, available at http://www.beaconpower.com.

Ali, N. (2007) Installation of the First Distributed Energy Storage System (DESS) at American Electric Power (AEP). Prepared by Sandia National Laboratories, http://www.ntis.gov.

Aten, M., Barton, J. and Hair, R. (2006) Benefits of an Energy Storage Device for a Wind Farm. Sixth International Workshop on Large-Scale Integration of Wind Power and Transmission Networks for Offshore Wind Farms, Delft, 26–28 October 2006, pp 333–338.

Badrzadeh, B., Smith, K.S. and Wilson, RC. (2009) Alternatives for high-power power electronic converters, switching devices and electric machines for very large wind farms: A technological and market assessment. European Wind Energy Conference (EWEC).

Baev, V.P., Buyanov, Yu.L. and Veselovs, A.S. (2007) A superconducting magnetic-energy storage system as a source of high-power current pulses with constant amplitude. *Instruments and Experimental Techniques*, **50**(4), 499–508.

Barlini, D. Ciappa, M., Castellazzi, A. (2006) New technique for the measurement of the static and of the transient junction temperature in IGBT devices under operating conditions. *Microelectronics Reliability*, **46**, 1772–1777, 9–11.

Bartnikas, R. (2002) Partial discharges, their mechanism, detection and measurement. *IEEE Transactions on Dielectrics and Electrical Insulation*, **9**, 763–808.

Baumbach, J.I. and Pilzecker (2005) Assessment of SF6 quality in gas insulated compartments of high voltage equipment using partial discharge ion mobility spectrographs. p. 2, *International Journal of Ion Mobility Spectrometry*, **8**, 5–9.

Beacon Power (2010) Article 'Smart Energy Matrix: 20 MW Frequency Regulation Plant' Power Beacon website at: http://www.beaconpower.com.

Blarke, M.B. and Jenkins, B.M. (2013) SuperGrid or SmartGrid: competing strategies for large-scale integration of intermittent renewables. *Energy Policy*, **58**, 381–390.

Bossanyi, E.A. (2000) The design of closed loop controllers for wind turbines. *Wind Energy*, **3**, 149–163. DOI:10.1002/we.34).

Bossanyi, E.A. (2005) Further load reductions with individual pitch control. *Wind Energy*, **8**, 481–485.

Brian, M., Nedeye, K. and Michael, S. (2009) An assessment of fault current limiter testing requirements, prepared for US Department of Energy, Office of Electricity Delivery and Energy Reliability, February 2009, Florida State University.

Cova, P. and Fantini, F. (1998) On the effect of power cycling stress on IGBT modules. *Microelectronics Reliability*, **38**, 1347–1352.

Dai, J. *et al.* (2007) Comparison of Hydran and Laboratory DGA Results for Electrical Faults in Ester Transformer Fluids. Annual Report Conference on Electrical Insulation and Dielectric Phenomena.

Dominguez, S. and Leithead, W. (2006) Size related performance limitations on wind turbine control performance. Proceeding of the EWEC 2006.

Eichler, M., Maibach, P. and Faulstich, A. (2008) *Full Size Voltage Converters for 5MW Offshore Wind Power Generators*, ABB Switzerland Ltd.

Enger, E. and Rocke, S. (2006) Emerging Technologies, Subsea Power Systems. Possibilities for piloting in Brazilian waters with focus on subsea to shore technologies, Rio de Janeiro, 14 September 2006.

Feltes, C. and Erlich, I. (2007) Variable frequency operation of DFIG based wind farms connected to the grid through VSC-HVDC link. IEEE Power Engineering Society general meeting, 24–28 June 2007.

Gardner, P., Craig, L.M. and Smith, G.J. (1998) *Electrical Systems for Offshore Wind Farms*, Garrad Hassan & Partners, Glasgow, UK.

Graber, L. and Thronicker, T. (2008) Thermal-Network Simulations and Computational Fluid Dynamics for Effective Gas Leakage Detection in SF6 Switchgear. Cigre session 2008.

Giordano, V. and Gangale, F. "Smart Grid projects in Europe: Wald, D., Orton, H. and Svoma, R. (2009) Requirements for different components in cables for offshore application. CIRED 2009, Prague, Paper No 0064, 8–11 June 2009.

Harada, K., Katsuki, A. and Fujiwara, M. (1993) Use of ESR for deterioration diagnosis of electrolytic capacitor. *IEEE Transactions on Power Electronics*, **8**, 355–361.

Held, M. Jacob, P., Nicoletti, G., *et al.* (1997) Fast power cycling test of IGBT modules in traction appliaction. International Conference on Power Electronics and Drive Systems.

Hertem, D.V. and Ghandhari, M. (2010) Multi-terminal VSC HVDC for the European supergrid: obstacles. *Renewable and Sustainable Energy Reviews*, **14**, 3156–3163.

Hyttinen, M. and Bentzen, K. (2002) Operating experiences with a voltage source converter HVDC-light on the gas platform troll A. ABB website.

IEC 61400-25-1 (Dec. 2006) Wind turbines – Part 25: Communications for monitoring and control of wind power plants - Overall description of principles and models. International Electrotechnical Commission.

IEEE SmartGrid (2013) http://smartgrid.ieee.org/ieec-smart grid/smart-grid-conceptual-model, Smart Grid Conceptual Model, last visit: 23 August 2013.

Jim, E., Brown, R., Norris, B. *et al.* (2007) Guide to estimating benefits and market potential for electricity storage in New York. Final Report prepared by New York State Energy Research and Development Authority, www.nyserda.org.

Johnson, S.J., Van Dam, C.P. and Berg, D.E. (2008) Active Load Control Techniques for Wind Turbines. SANDIA Report, SAND2008-4809.

Jovcic, D. (2006) Interconnecting offshore wind farms using multi-terminal VSC-based HVDC. IEEE Power Engineering Society General Meeting.

Jovcic, D. and Milanovic, J.V. (2006) Offshore wind farm based on variable frequency mini-grids with multi-terminal DC interconnection. The 8th IEE International Conference on AC-DC Power Transmission (ACDC 2006), London, UK, 28–31 March 2006.

Judkins, J., Hofmeister, J. and Vohnout, S. (2007) A prognostic sensor for voltage-regulated switch-mode power supplies. IEEE Aerospace Conference.

Knab, S. and Strunz, K. (2010) The Smart Grid – Addressing the challenges of today's power system. Report, TU Berlin.

Koch, H. and Hillers, T. (2002) Second generation gas-insulated line. *IEE Power Engineering Journal*, **16**(3), 111–116.

Lahyani, A., Venet, P., Grellet, G. *et al.* (1998) Faillure prediction of electrolytic capacitors during operation of switchmode power supply. *IEEE Transactions on Power Electronics*, **13**(6), 1199–1206.

Lehmann, J., Netzel, M., Herzer, R. *et al.* (2003) Method for electrical detection of bond wire lift-off for power semiconductor. International Sumposium on Power Semiconductor Devices & ICs (ISPSD).

Leithead, W.E. (2009) Alleviation of unbalanced rotor loads by single blade controllers. Proceeding of the EWEC.

Leithead, W.E., Dominguez, S. and Spruce, C.J. (2003) Analysis of Tower/Blade interaction in the cancellation of the tower fore-aft mode via control. *Proceeding of the EWEC*.

Lemofouet, S. and Rufer, A. (2006) A hybrid energy storage systems based on compressed air and supercapacitors with maximum efficiency point tracking (MEPT). *IEEE Transactions Industrial Electronics*, **53**(4), 1105–1115.

Lianxiang, T. and Boon-Teck, O. (2002) Protection of VSC-multi-terminal HVDC against DC faults, in Power Electronics Specialists Conference, 2002. pesc 02. 2002 IEEE 33rd Annual, vol. 2, pp. 719–724.

Livermore, L., Jun, L., Ekanayake, J. *et al.* (2010) MTDC VSC Technology and its applications for wind power, in *Universities Power Engineering Conference (UPEC), 2010 45th International*, pp. 1–6.

Lundberg, S. (2003) Performance comparison of wind park configurations. Technical Report, Chalmers University of Technology.

Luongo, C.A. (1996) Superconducting storage systems: an overview. *IEEE Transactions on Magnetics*, **32**(4), 2214–2223.

Madariaga, A., Martín, J.L. *et al.* (2013) Technological trends in electric topologies for offshore wind power plants. *Renewable and Sustainable Energy Reviews*, **24**, 32–44.

Makarov, Y.V., Nyeng, P., Yang, B. *et al.* (2008) Wide-area energy storage and management system to balance intermittent resources in the Bonneville power administration and California ISO control areas, report prepared for Bonneville Power Administration, June 2008, available online on http://www.electricitystorage.org/site/technologies.

Martin, K. and Claus, N. (2004) Fault Current Limiter in Medium and High Voltage Grids, presented at workshop on Electricity Transmission and Distribution Technology and R&D, Paris, 4–5 November 2004, available online at http://www.iea.org/Textbase/work/2004/distribution/presentations/neumann.pdf.

Masuda, T., Yumura, Watanabe *et al.* (2005) High-temperature Superconducting Cable Technology and Development Trends. Science Links Japan, http://sciencelinks.jp.

Masera, M. (Nov., 2012) Smart Grids: anticipated trends and policy directions, report, Institute for Energy and Transport, Joint Research Centre, European Union.

Mathew, L. and Alex, R. (2004) Grid Frequency Regulation by Recycling Electrical Energy in Flywheels. Power Beacon website at: http://www.beaconpower.com.

McDermott, R. (2009) Investigation of Use of Higher AC Voltages on Offshore Wind farms. EWEC 2009, Marseille, France 16–19 March 2009. www.ewec2009proceedings.info/allfiles2/283_EWEC2009presentation.pdf.

McMillan, D. and Ault, G.W. (2007) Quantification of condition monitoring benefit for offshore wind turbines. *Wind Engineering*, **31**, 267–285.

McMillan, D. and Ault, G.W. (2008) Condition monitoring benefit for onshore wind turbines: sensitivity to operational parameters. *IET Renewable Power Generation*, **2**(1), 60–72.

Meier, S. (2005) Novel Voltage Source Converter based HVDC Transmission System for offshore wind farms. PhD thesis at KTH Sweden.

Meier, S., Norrga, S. and Nee, H. (2004a) New topology for efficient AC/DC converters for future offshore wind farms. 4th Nordic Workshop on Power and Industrial Electronics, Norway, June 2004, pp. 114–119.

Meier, S., Norrga, S. and Nee, H. (2004b) New voltage source converter topology for HVDC connection of offshore wind farms. 11th International Power Electronics and Motion Control Conference, EPE-PENorway, June 2004, pp. 114–119.

Mesut, E.B. and Nikhil, R.M. (2007) Overcurrent protection on voltage-source-converter-based multiterminal DC distribution systems. *IEEE Transactions on Power Delivery*, **22**, 406–412.

Metwally, I.A. (2004) Status review on partial discharge measurement techniques in gas insulated switchgear/lines. *Electric Power Systems Research*, **69**, 25–36.

Meyer, C., Hoing, M., Peterson, A. *et al.* (2007) Control and design of DC grids for offshore wind farms. *IEEE Transactions Industry Applications*, **43**(6), 1475–1482.

Morroni, J., Dolgov, A., Shirazi, M. *et al.* (2007) Online health monitoring in digitally controlled power converters. IEEE Power Electronics Specialist Conference (PESC), pp. 112–118.

Muhamad, N.A. *et al.* (2007) Comparative Study and Analysis of DGA Methods for Transformer Mineral Oil. Proceedings from Power Tech, pp. 45–50.

National Energy Technology Laboratory (July 2009) A Compendium of Smart Grid Technologies. Report, Office of Electricity Delivery and Energy Reliability – the U.S. Department of Energy.

Nguyen, T.H., Prinz, A. *et al.* (2012) Smart Grid for Offshore Wind Farms: Towards an Information Model based on the IEC 61400-25 Standard. Proceedings of 2012 Innovative Smart Grid Technologies (ISGT), pp. 1–6.

Nilsson, J. and Bertling, L. (2007) Maintenance management of wind power systems using condition monitoring systems - lifecycle cost analysis for two case studies. *IEEE Transactions on Energy Conversion*, **22**, 223–229.

Olejnczak, T. (2011) Distributed generation and virtual power plants: barriers and solutions. MSc thesis, Utrecht University, 2011.

Parker, M. and Anaya-Lara, O. (2011) An evaluation of collection network designs which eliminate the turbine converter. Supergen Wind Report, 2011.

Parker, M. and Anaya-Lara, O. (2012) An evaluation of collection network designs which eliminate the turbine converter. European Wind Energy Association Conference, EWEC, 2012.

Parker, M. and Anaya-Lara, O. (2013) The Cost and Losses Associated with Offshore Windfarm Collection Networks which Centralise the Turbine Power Electronic Converters. IET-RPG.

Phasor Measurement Unit (PMU) (2007) *Implementation and Applications*, EPRI, Palo Alto, CA, 1015511.

Ratchef, P., Vandevelde, B., De Wolf, I. *et al.* (2004) Reliability and failure analysis of Sn-Ag-Cu solder interconnections for PSGA packages on Ni/Au surface finish. *IEEE Transactions on Device and Materials Reliability*, **4**, 5–10.

Renforth, L., Mackinlay, R. and Michel, M. (2005) MV Cable Diagnostics – Applying Online PD Testing and Monitoring. *Asia Pacific Conference on MP Power Cable Technologies*.

Roscoe, A., Abdulhadi, I.F. and Burt, G. (2013) P and M class phasor measurement unit algorithms using adaptive cascaded filters. *IEEE Transactions on Power Delivery*, **28**, 1447–1459.

Sankaran, V. and Xu, X. (1996) Integrated Power Module Diagnostic Unit. US Patent 5,528,446 1996. to Ford Motor Company.

Sayago, J.A., Bruckner, T. and Bernet, S. (2008) How to select the system voltage of MV drives – a comparison of semiconductor expenses. *IEEE Transactions on Industrial Applications*, **55**, 3381–3390.

Setiawan, E.A. (2007) Concept and controllability of Virtual Power Plant. PhD Thesis, Kassel University.

Steinke, J.K. and Apeldoorn, O. (2002) Applying the experience of industrial high power converter design to windpower conversion. ABB Schweiz AG, Automation Technology Products.

Stendius, L. (2007) Nord E.ON 1 400 MW HVDC Light Offshore wind project. EOW 2007, Berlin, 5 December 2007, http://www.eow2007proceedings.info/allfiles/290_Eow2007presentation.ppt.

Swider, D.J. (2007) Compressed air energy storage in an electricity system with significant wind power generation. *IEEE Transactions Energy Conversion*, **22**(1), 95–102.

Tenbohlen, S., Uhde, D., Poittevin, J. *et al.* (2000) Enhanced Diagnosis of Power Transformers using On- and Off-line Methods: Results, Examples and Future Trends. Cigre session 2000.

Tenbohlen, S., Stirl, T., Bastos, G., *et al.* (2002) Experience-based Evaluation of Economic Benefits of On-line Monitoring Systems for Power Transformers. Cigre session.

Toda, K. (2002) Structural features of gas insulated transformers. IEEE Power Engineering Society, Transmission and Distribution Conference and Exhibition 2002: Asia Pacific. IEEE/PES, Vol. 1, 6–10 October 2002, pp. 508–510.

Toshiya, N. and Akira, T. (1994) A study on required size for pumped hydro storage. *IEEE Transactions Power Systems*, **9**(1), 318–323.

Wang, M., Vandermaar, A.J. and Srivastava, K.D. (2002) Review of condition assessment of power transformers in service. *IEEE Electrical Insulation Magazine*, **18**(6), 12–25.

Wiggelinkhuizen, E., Verbruggen, T., Braam, H. *et al.* (2007) CONMOW: Condition monitoring for offshore wind farms (Available online) http://www.supergen-wind.org.uk.

Wright, A.D. and Fingersh, L.J. (2008) Advanced Control Design for Wind Turbines - Part I: Control Design, Implementation, and Initial Tests. NREL Technical Report, NREL/TP-500-42437.

Xiong, Y., Cheng, X., Shen, Z. *et al.* (2008) A prognostic and warning system for power electronic modules in electric, hybrid and fuel cell vehicles. *IEEE Transactions on Industrial Electronics*, **55**, 2268–2276.

Yazdandoust, A.R., Haghjoo, F. and Shahrtash, S.M. (2008) Insulation status assessment in high voltage cables based on decision tree algorithm. IEEE Power & Energy Conference.

Zeisel, D., Menzi, H. and Ullrich, L. (2000) A precise and robust quartz sensor based on tuning fork technology for SF6-gas density control. *Sensors and Actuators*, **80**, 233–236.

A

Voltage Source Converter Topologies

This appendix describes the operating principles and modulation methods of the most commonly used voltage source converter (VSC) topologies. In addition, a number of illustrative examples are provided to aid in the understanding of the relevance of each converter and its power system application(s). The level of detail presented in this appendix is intended to be adequate to allow a wide range of readers to appreciate most of the power electronic concepts and devices presently used in power networks.

A.1 Two-Level Converter

A.1.1 Operation

Figure A.1 shows the schematic diagram of a three-phase two-level VSC and its output phase voltage relative to the supply mid-point. The converter in Figure A.1 is called a two-level converter because it can generate only two voltage levels, $^1/_2V_{dc}$ and $-^1/_2V_{dc}$, at the output phases 'a', 'b' and 'c' relative to the supply mid-point '0' (Holmes and McGrath, 1999; Bowes et al., 2000a; Flourentzou and Agelidis, 2007; Dahidah et al., 2010; Zhang et al., 2010; Eckel and Runge, 2011). For example, for phase 'a', the voltage level of $^1/_2V_{dc}$ is achieved by turning the upper switch 'S$_1$' on and lower switch 'S$_2$' off, while the voltage level of $-^1/_2V_{dc}$ is generated by turning the upper switch 'S$_1$' off and the lower switch 'S$_2$' on. Turning on the upper and lower switches 'S$_1$' and 'S$_2$' simultaneously creates a short circuit across the dc link and exposes the converter switching devices to the risk of damage from over-current, therefore this must avoided. Thus, safe operation of a two-level VSC requires the upper and lower switches of the same phase-leg to operate in a complementary manner. This means, turning on for example upper switch 'S$_1$' precludes the lower switch S$_2$ from being on and vice versa. To ensure such complementary operation in a real application, a dead time is introduced to the gate signals of the upper and lower switches to prevent both switches from

Offshore Wind Energy Generation: Control, Protection, and Integration to Electrical Systems, First Edition.
Olimpo Anaya-Lara, David Campos-Gaona, Edgar Moreno-Goytia and Grain Adam.
© 2014 John Wiley & Sons, Ltd. Published 2014 by John Wiley & Sons, Ltd.
Companion Website: www.wiley.com/go/offshore_wind_energy_generation

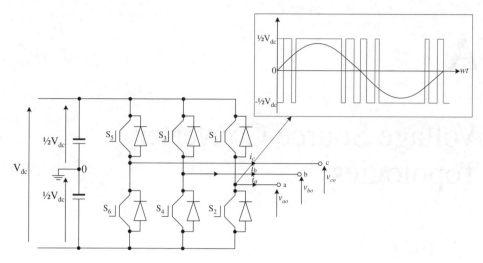

Figure A.1   Two-level voltage source converter.

being simultaneously on. Observe that when the upper switch 'S$_1$' is on, switch 'S$_2$' blocks the full dc voltage 'V$_{dc}$', and switch 'S$_1$' blocks full dc link voltage 'V$_{dc}$' when switch 'S$_2$' is on. This means that every switching device in the two-level VSC must be rated to tolerate full dc link voltage, so series connected devices are required to enable operation in medium and high voltage applications. A VSC uses composite self-commutated switching devices such as the insulated gate bipolar transistor (IGBT) and anti-parallel diodes as shown in Figure A.1 to enable bi-directional current flow in each phase-leg, hence bidirectional power flow between ac and dc sides.

A.1.2   Voltage Source Converter Square-Mode Operation

In the past there were two known methods to operate a three-phase VSC in square mode: the 180° conduction method, and the 120° conduction method. Figure A.3 shows the gate signal mapping, and corresponding converter phase and line voltage for the 180° conduction method with the converter switches relabelled as in Figure A.2. Observe that in the 180° conduction method, each switching device is turned on for 180° (or half of the fundamental period), with 60° phase shift between successive gate signals of the switching devices.

Figure A.4 shows the gate signal mapping and corresponding phase and line voltages for the 120° conduction method. Observe that with the 120° conduction method, the phase voltage has three voltage levels ($\frac{1}{2}V_{dc}$, 0 and $-\frac{1}{2}V_{dc}$), rather than only two voltage levels as previously discussed in Section A.1.1. Additionally, with the 120° conduction method, the gate signals for the upper and lower switching devices of the same phase-leg are not complementary, and there are instances where both switches are off. This is undesirable for many applications for the following reasons:

- Converter output phase currents tend to be discontinuous as the 120° conduction method uses tri-state operation that simultaneously turns converter upper and lower switches of the same phase-leg.
- Exposes converter switching devices to over-voltage when supplying inductive loads.

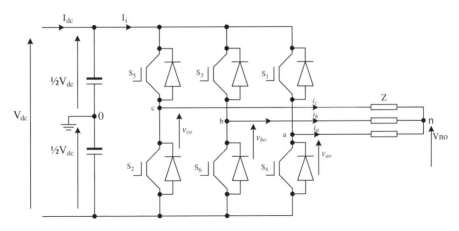

Figure A.2 Two-level VSC with switching device renumbered in specific order to assist with the explanation of the 180° and 120° conduction methods.

Additionally, when the converter operates with the 180° conduction method it generates a higher fundamental component of the output voltage compared to that when operating with the 120° conduction method, provided they have the same dc link voltage. From the Fourier series of the output phase voltage v_{ao} for 180° and 120° conduction methods in Figures A.3 and A.4, the fundamental peak phase voltages are $V_{m1} = \frac{2}{\pi} V_{dc}$ and $V_{m1} = \frac{3}{2\pi} V_{dc}$, respectively; where V_{dc} represents the converter dc voltage and V_{m1} is the fundamental peak phase voltage. This indicates that the converter ac gain (also known as the amplitude modulation index, m) for the 180° and 120° methods is $\frac{4}{\pi}$ and $\frac{3}{2\pi}$, respectively; where the modulation index is defined as: $m = \frac{V_{m1}}{\frac{1}{2} V_{dc}}$. Both methods are no longer in use because they cannot vary the magnitude of the fundamental voltage, and they generate low-order harmonics, which are expensive to filter.

A.1.3   Pulse Width Modulation

Several pulse width modulation (PWM) strategies have been developed to control the magnitude of the fundamental component of the output voltage and to reduce the harmonic content in voltage and current waveforms. Some well-known modulation strategies in use today are selective harmonic elimination (SHE), sinusoidal pulse width modulation (SPWM) and space vector modulation (SVM) (Bowes, 1975; Bowes and Bird, 1975; Bowes, 1988; Bowes and Clare, 1988; Bowes and Clark, 1988; Bowes, 1990; Bowes and Clark, 1990; Jiang *et al.*, 1991; Bowes and Clark, 1992; Bowes, 1993a; Bowes, 1993b; Bowes, 1993c; Bowes, 1994; Bowes, 1995; Holmes, 1996; Bowes and Grewal, 1998a; Bowes and Grewal, 1998b; Holmes and McGrath, 1999; Bode and Holmes, 2001; Holmes and McGrath, 2001; Adam *et al.*, 2012b).

A.1.3.1   Selective Harmonic Elimination

Selective harmonic elimination explicitly defines the switching angles on the output phase voltage that are needed to set the magnitude of the fundamental component of the phase voltage and to eliminate specific harmonics. Figure A.5 shows an example where selective

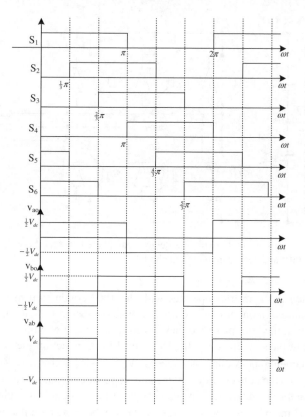

Figure A.3 Gate signal mapping and converter output voltage for the 180° conduction method.

harmonic elimination defines three angles (α_1, α_2 and α_3) of the phase voltage, and the gate signals for the upper and lower switches of phase 'a' for a three-phase voltage source converter. One of these angles will be used to set the magnitude of the voltage fundamental component and the remaining two will be used to eliminate specific nontriplen harmonics, normally the dominant lower-order harmonics such as the 5th and 7th. From the Fourier series of the phase voltage v_{ao} (Azmi *et al.*, 2013; Baker *et al.*, 1988; Ben-Sheng and Hsu, 2007; Bierk *et al.*, 2011; Bierk *et al.*, 2009):

$$a_n = 0$$

$$b_n = \begin{cases} \dfrac{2V_{dc}}{n\pi}\left[2\cos n\alpha_1 - 2\cos n\alpha_2 + 2\cos n\alpha_3 - 1\right] & \text{for } \forall n = odd \\ 0 & \text{for } \forall n = even \end{cases} \qquad \text{(A.1)}$$

For adjustment of the voltage fundamental component, setting n = 1 and $b_1 = V_{m1}$ in Eq. (A.1), the following equation is obtained:

$$2\cos\alpha_1 - 2\cos\alpha_2 + 2\cos\alpha_3 - 1 = \frac{\pi}{4}m \qquad \text{(A.2)}$$

recalling that $m = \dfrac{V_{m1}}{\frac{1}{2}V_{dc}}$.

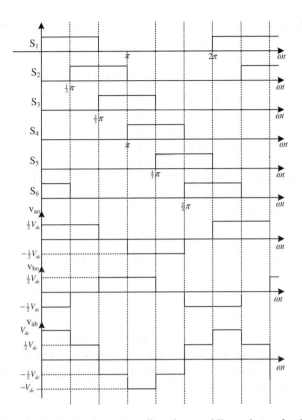

Figure A.4 Gate signal mapping, and corresponding phase and line voltages for the 120° conduction method.

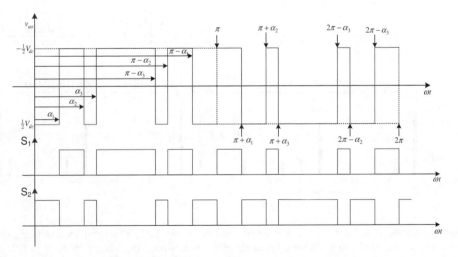

Figure A.5 Output phase voltage and corresponding gate signals of the upper and lower switches of phase 'a'.

To eliminate the fifth and seventh harmonics from the phase voltage, set n = 5 and 7, and $b_5 = 0$ and $b_7 = 0$, to obtain the following equations:

$$2\cos 5\alpha_1 - 2\cos 5\alpha_2 + 2\cos 5\alpha_3 - 1 = 0 \tag{A.3}$$

$$2\cos 7\alpha_1 - 2\cos 7\alpha_2 + 2\cos 7\alpha_3 - 1 = 0 \tag{A.4}$$

The switching angles (α_1, α_2 and α_3) are obtained by solving Eqs. (A.2), (A.3) and (A.4). Normally, such nonlinear transcendental equations have more than one solution. However, the only correct solution that ensures quarter symmetry in the phase voltage (shown in Figure A.5) must satisfy the following set of inequalities: $\alpha_1 < \alpha_2 < \alpha_3 < \frac{1}{2}\pi$. In this manner, the quality of the converter output voltage can be improved as the number of eliminated harmonics increases. In contrast, the maximum attainable fundamental voltage reduces as the number of harmonics eliminated increases.

Example A.1   Solve Eqs. (A.2), (A.3) and (A.4) with a modulation index m = 0.9, and plot the phase and line voltages along with their spectra when the converter dc link voltage is $V_{dc} = 400$ V.

Solution. Solving Eqs. (A.2), (A.3) and (A.4) for m = 0.9 provides the following values: $\alpha_1 = 16.66°$, $\alpha_2 = 37.57°$ and $\alpha_3 = 46.52°$. Figure A.6(a) shows the output phase voltage measured at the pole point relative to the supply mid-point and its fundamental component. Figure A.6(c) shows the spectrum of the phase voltage, it can be seen that the fifth and seventh harmonics have been eliminated. Also, it should be noted that the fundamental component of

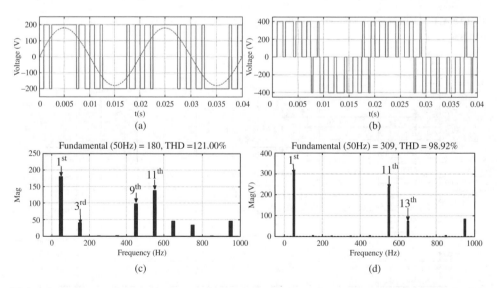

Figure A.6   Phase and line voltage waveforms obtained when a three-phase two-level voltage source converter is controlled using selective harmonic elimination with a modulation index m = 0.9, and elimination of the 5th and 7th harmonics. (a) Phase voltage relative to the supply mid-point; (b) Line voltage; (c) Phase voltage spectrum; (d) Line voltage spectrum.

the output phase voltage is 180 V (which is $V_{m1} = \frac{1}{2}mV_{dc} = \frac{1}{2} \times 0.9 \times 400$ V = 180 V). Figure A.6(b) and (d) show the output line voltage of the converter and its harmonic spectrum; observe that the triplen harmonics (such as the 3rd and 9th) which appear in the phase voltage spectrum have been cancelled in the line voltage. For this reason, only non-triplen dominant odd harmonics need to be eliminated in the three-phase converter.

Example A.2 Repeat Example A1 for a modulation index m = 1.16, and assume the converter is connected to an RL load of 10 Ω and 23.8 mH per phase.

Solution. With modulation index m = 1.16, switching angles that ensure quarter symmetry are: $\alpha_1 = 10.69°$, $\alpha_2 = 30.18°$ and $\alpha_3 = 33.14°$. Figure A.7(a) shows that when m = 1.16 the phase voltage fundamental component exceeds $\frac{1}{2}V_{dc}$. Figure A.7(c) shows that as the modulation index increases the magnitudes of the remaining harmonics in the phase voltage decrease. This shows that selective harmonic elimination can be used to maximise the output voltage fundamental component and to improve the quality of the output voltage waveform.

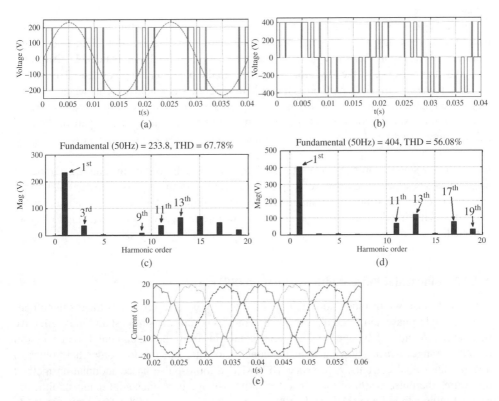

Figure A.7 Phase and line voltages, and current waveforms obtained when a three-phase two-level voltage source converter is controlled using selective harmonic elimination. With modulation index m = 1.16, and elimination of the 5th and 7th harmonics. (a) Phase voltage relative to the supply mid-point; (b) Line voltage at converter terminal; (c) Phase voltage spectrum; (d) Line voltage spectrum; (e) Three-phase output currents.

Total harmonic distortion is the figure of merit that expresses the harmonics content in the waveform relative to fundamental, and it can be defined as (Chang *et al.*, 2006; Hoevenaars; *et al.*):

$$THD = \frac{\sqrt{V_{rms}^2 - V_{rms1}^2}}{V_{rms1}} \tag{A.5}$$

or

$$THD = \frac{\sqrt{V_{rms2}^2 + V_{rms3}^2 + V_{rms5}^2 + V_{rms6}^2 + \ldots\ldots V_{rmsn}^2}}{V_{rms1}} = \frac{\sqrt{\sum_{j=2}^{n} V_{rmsj}^2}}{V_{rms1}} = \frac{\sqrt{\sum_{j=2}^{n} V_{mj}^2}}{V_{m1}} \tag{A.6}$$

where, V_{rmsj} and V_{mj} represent the Root-Mean Square (RMS) and peak voltages of the jth harmonic respectively. The same expressions are applicable to calculate the THD in the current signal.

Figure A.7(d) shows the harmonic spectrum of the line voltage, illustrating that the eleventh, thirteenth, seventeenth and nineteenth are the only harmonics present in the frequency range from 0 to 1 kHz. Figure A.7 shows that the load current is sinusoidal with noticeable distortions caused by the presence of the low-order harmonics in the phase and line voltages. This shows the effectiveness of the SHE method as it is able to generate sinusoidal currents when only two non-triplen harmonics near to the voltage fundamental component are eliminated. Generally, when a voltage source converter is supplying inductive loads, the inductive part of the load acts as a low-pass filter, hence resulting in improved current waveforms.

In practical systems, improved current waveforms can be achieved increasing the number of harmonics to be eliminated (usually, up to 29th harmonics for three-phase systems). Additionally, when selective harmonic elimination is used in medium and high voltage applications the converter switching devices operate at relatively low frequency; thus, low switching losses are expected.

A.1.3.2   Sinusoidal Pulse Width Modulation (SPWM)

Sinusoidal pulse width modulation is a method widely used to control voltage source converters (single-phase and three-phase). This method is one of the simplest and most effective modulation methods that suppress all the harmonics far from the fundamental frequency (50 or 60Hz) range, hence allowing a small inductor to filter out the low-order harmonics to achieve sinusoidal currents. In essence, SPWM is a multi-pulse based modulation method that varies the pulse width of the converter output voltage in a sinusoidal manner following a target reference voltage (Dai *et al.*, 1992; Mwinyiwiwa *et al.*, 1998a; Mwinyiwiwa *et al.*, 1998b; Mwinyiwiwa *et al.*, 1999; Patel and Agarwal, 2008; Hava and Ün, 2011; Pongiannan *et al.*, 2011). In an analogue implementation, this is achieved by comparing a low-frequency reference signal with a high-frequency carrier signal. The frequency of the reference signal (f) must be equal to the desired fundamental frequency of the output voltage. The frequency of the

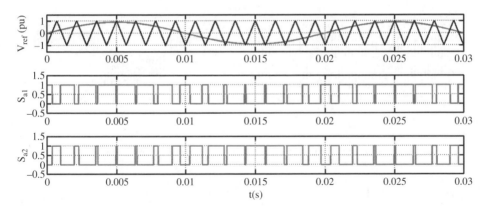

Figure A.8 Gate signal generation using sinusoidal pulse width modulation.

carrier signal (f_c) must be much higher than that of the reference signal. A higher frequency ratio ($m_f = \frac{f_c}{f}$) produces higher quality output voltage and current (Carter and Johnson, 1988; Carrara *et al.*, 1990; Carrara *et al.*, 1992; Bowes *et al.*, 2000a; Bowes *et al.*, 2000b; Bruckner and Holmes, 2003; Burgos *et al.*, 2007; Cataliotti *et al.*, 2007; Burgos *et al.*, 2008; Casadei *et al.*, 2008; Bradshaw *et al.*, 2012; Carnielutti *et al.*, 2012; Azmi *et al.*, 2013). Figure A.8 shows the gate signal generation for one phase leg of the two-level VSC, where S_{a1} and S_{a2} stand for the signals corresponding to upper and lower switches respectively. When the reference signal exceeds the carrier signal, the upper switch (S_{a1}) is turned on. When the reference signal is less than the carrier signal the upper switch (S_{a1}) is turned off. The gate signal for the lower switch (S_{a2}) is the complement of S_{a1} ($S_{a2} = 1 - S_{a1}$). For a three-phase two-level converter, three reference signals are required (one per phase) and one high-frequency carrier. The gate signals for the switching devices of each phase-leg are generated in a similar manner as shown in Figure A.8.

For a reference signal defined as: $v_{aref} = m\sin(\omega t + \delta)$, the converter output voltage at the fundamental frequency measured relative to the supply mid-point is given by: $v_a = \frac{1}{2}mV_{dc}$ $\sin(\omega t + \delta)$; where m is the modulation index. Notice that for SPWM when m>1, the peak of the reference signal exceeds the peak of the high-frequency triangular carrier, and this leads to over-modulation. Operation in over-modulation is not desirable because the linear relationship of the converter output voltage and reference signal described above is no longer applicable, and suppression of the harmonics toward the carrier frequencies cannot be guaranteed.

Example A.3 Using the same parameters as in Example A.2 for a two-level converter, plot the output phase and line voltage and the current when the converter is controlled using SPWM with 2kHz switching frequency to generate a 50-Hz output voltage at 0.9 modulation index.

Solution. Figure A.9(a) and (c) show the phase voltage and its harmonic spectrum for the system described; observe that the peak of the fundamental component of the voltage of phase 'a' is equal to $\frac{1}{2}\,mV_{dc} = \frac{1}{2} \times 0.9 \times 400 = 180$ V and as described above, all harmonics are suppressed toward the first and second carrier frequencies. The closest harmonics to the fundamental component of the voltage are the 38th, 40th and 42nd. These harmonics have small

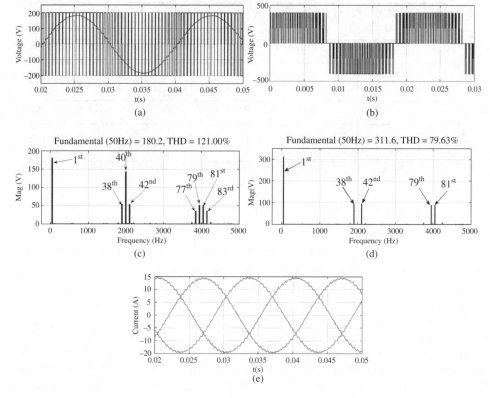

Figure A.9   Waveforms obtained when a three-phase two-level voltage source converter uses SPWM with 2 kHz carrier frequency and 0.9 modulation index. (a) Converter output phase voltage relative to the supply mid-point and its fundamental component (phase a); (b) Line voltage; (c) Phase voltage spectrum; (d) Line voltage spectrum; (e) Three-phase output currents.

contribution to current distortion because their corresponding harmonic impedances are sufficiently large. Figure A.9(b) and (d) display the line voltage and its harmonic spectrum. Observe that the line voltage has a much lower harmonic content than that of the phase voltage. This is because some of the harmonics in all three phases are located at the same position (in other words, they have the same phase and magnitude); therefore, they cancel in the line voltage. Figure A.9 shows that a two-level voltage source converter with SPWM produces an output current that is sinusoidal. The limited modulation index range (up to 1) of SPWM can be overcome in three-phase three-wire systems by adding the 3rd harmonic to the reference signals.

With the addition of the third harmonic (zero sequence) to the reference signal, the linear modulation range of SPWM can be extended to 1.154, meaning that a larger fundamental component of voltage can be synthesised. In reference (Holmes and Lipo, 2003) it has been shown that the optimal performance is achieved when the peak magnitude of the third harmonic added to the reference signals is maintained at one-sixth of the fundamental. Therefore, the reference signals for SPWM plus 3rd harmonic are defined as: $v_a^* = m \sin \omega t + \frac{1}{6} m \sin \omega t$, $v_b^* = m \sin(\omega t + \frac{4}{3}\pi) + \frac{1}{6} m \sin \omega t$ and $v_c^* = m \sin(\omega t + \frac{2}{3}\pi) + \frac{1}{6} m \sin \omega t$ (Azmi $et\ al.$, 2013).

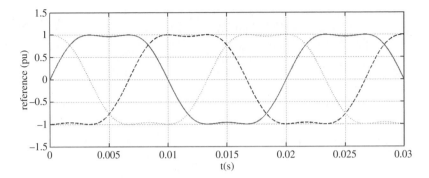

Figure A.10   Reference signals for SPWM plus 3rd harmonic injection.

Figure A.10 shows the reference signals for a modulation index m = 1.154 with the 3rd harmonic added. Observe that the positive and negative peaks of the reference signals are contained within 1 and −1 even when the modulation index exceeds 1, that is, the modulation index linear range of SPWM is extended.

Example A.4   Repeat Example A.3 with a modulation index of 1.1.

Solution. Figure A.11(a) to (d) show the phase and line voltages, with their spectra. Observe that the phase voltage contains the 3rd harmonic, but it is cancelled out in the line voltage. Additionally, it can be seen that injection of the 3rd harmonic increases the voltage fundamental component but also a number of sidebands (harmonics) around the first and second carrier frequency. However, the magnitude of these harmonics are much smaller than that in SPWM, and for this reason it exhibits lower THD in phase and line voltages. Figure A.11(e) shows that the output currents are sinusoidal.

An alternative method to extend the linear range of the modulation index is proposed by Steinke in (Steinke, 1992). This method calculates the value of the third harmonic (zero sequence) to be added to the sinusoidal reference signal as: $v_0 = \frac{1}{2}(\min(v_a, v_b, v_c)+\max(v_a, v_b, v_c))$. This zero sequence component is subtracted from the reference of each phase as: $v_a^* = v_a - v_0, v_b^* = v_b - v_0$ and $v_c^* = v_c - v_0$, where $v_a = m\sin\omega t, v_b = m\sin(\omega t + \frac{4}{3}\pi)$ and $v_c = m\sin(\omega t + \frac{2}{3}\pi)$. Figure A.12 shows the reference signals when a zero sequence calculated based on Steinke method is added to the sinusoidal references for a modulation index m = 1.154. It can be seen that all the positive and negative peaks of the signal are contained within 1 and −1 limits, despite m exceeding 1.

A.1.3.3   Space Vector Modulation

Space vector modulation maps all possible switch combinations of the two-level VSC into the $\alpha\beta$ plane. This mapping is achieved by transforming the converter output voltages (v_{a0}, v_{b0} and v_{c0}) corresponding to each switching combination into equivalent v_α and v_β. Changing the ground position of the dc side from the supply mid-point to the negative dc link as shown

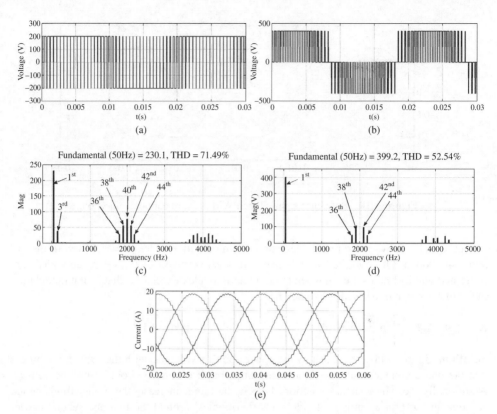

Figure A.11 Output phase and line voltages and current waveforms when a three-phase two-level converter is controlled with SPWM plus 3rd harmonic injection. (a) Output phase voltage measured relative to the supply mid-point; (b) Line voltage (c) Phase voltage spectrum; (d) Line voltage spectrum; (e) Three-phase output currents.

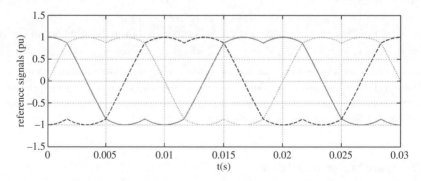

Figure A.12 Reference signals when the 3rd harmonic added to the SPWM is calculated based on Steinke method (with modulation index m = 1.154).

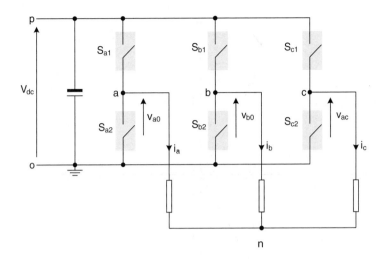

Figure A.13 Simple representation of a two-level converter.

in Figure A.13, the two voltage levels at the output of the converter will be V_{dc} and 0 instead of $+^1/_2 V_{dc}$ and $-^1/_2 V_{dc}$ (this will not affect the final outcomes from a conceptual point of view). Since the switching devices of the same phase-leg operate in complementary manner, knowing the state of one is sufficient to determine the converter output voltage. For phase a, $S_{a1} = 1$ and $S_{a1} = 0$ denote on and off states of switch S_{a1}, respectively. The same convention is applicable for phases b and c. In this manner, all possible switch combinations of the two-level voltage source converter with their corresponding output phase voltages (v_{a0}, v_{b0} and v_{c0}) and resultant voltage vectors in the $\alpha\beta$ plane are summarised in Table A.1. The converter output voltages (v_{a0}, v_{b0} and v_{c0}) are transformed into the $\alpha\beta$ plane using Eq. (A.7). This transformation equation is derived based on the voltage vector alignments shown in Figure A.14 (Buja and

Table A.1 Summary of the relationships between switching combinations, output phase voltages and voltage vectors in $\alpha\beta$ plane in a two-level VSC.

S_{a1}	S_{b1}	S_{c1}	V_{a0}	V_{b0}	V_{c0}	v_α	v_β	$\left|\vec{V}\right| = \sqrt{v_\alpha^2 + v_\beta^2}$	$\theta = \tan^{-1}\frac{v_\beta}{v_\alpha}$
0	0	0	0	0	0	0	0	0	0°
1	0	0	V_{dc}	0	0	V_{dc}		V_{dc}	0°
1	1	0	V_{dc}	V_{dc}	0	$\frac{1}{2}V_{dc}$	$\frac{\sqrt{3}}{2}V_{dc}$	V_{dc}	60°
0	1	0	0	V_{dc}	0	$-\frac{1}{2}V_{dc}$	$\frac{\sqrt{3}}{2}V_{dc}$	V_{dc}	120°
0	1	1	0	V_{dc}	V_{dc}	$-V_{dc}$	0	V_{dc}	180°
0	0	1	0	0	V_{dc}	$-\frac{1}{2}V_{dc}$	$-\frac{\sqrt{3}}{2}V_{dc}$	V_{dc}	240°
1	0	1	V_{dc}	0	V_{dc}	$\frac{1}{2}V_{dc}$	$-\frac{\sqrt{3}}{2}V_{dc}$	V_{dc}	300°
1	1	1	V_{dc}	V_{dc}	V_{dc}	0	0	0	0°

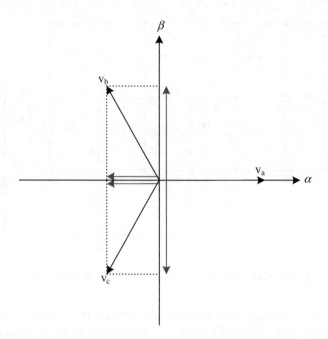

Figure A.14   Resolution of the three-phase *abc* variables into the $\alpha\beta$ plane.

Indri, 1977; Green and Boys, 1982; Fukuda *et al.*, 1990; Dunford and Van Wyk, 1992; Handley and Boys, 1992; Holmes, 1995; Holmes, 1996; Bowes and Yen-Shin, 1997; Hava *et al.*, 1997a; Hava *et al.*, 1997b; Hava *et al.*, 1998; Holmes, 1998; Bowes and Grewal, 1999; Ching-Tsai and Jenn-Jong, 1999; Bech *et al.*, 2000; Changjiang *et al.*, 2001; Changjiang *et al.*, 2003; Bowes and Holliday, 2006; Hadiouche *et al.*, 2006; Adam, 2007; Do-Hyun, 2007; Dyck *et al.*, 2010; Beig, 2012; Bradshaw *et al.*, 2012; Azmi *et al.*, 2013; Dordevic *et al.*, 2013b; Duran *et al.*, 2013; Gruson *et al.*, 2013).

$$\begin{bmatrix} v_\alpha \\ v_\beta \end{bmatrix} = \begin{bmatrix} 1 & -\dfrac{1}{2} & -\dfrac{1}{2} \\ 0 & \dfrac{\sqrt{3}}{2} & -\dfrac{\sqrt{3}}{2} \end{bmatrix} \begin{bmatrix} v_a \\ v_b \\ v_c \end{bmatrix} \qquad (A.7)$$

Based on the last two columns of Table A.1, the space vector diagram of the two-level converter can be drawn as shown in Figure A.15. Observe that the space vector diagram comprises of six sectors, separated by six active voltage vectors 60° apart, and two zero vectors in the centre. The vertex of each voltage vector in Figure A.15 represents the state of the upper three switching devices of the two-level converter shown in Figure A.13. For example, the vertex of active voltage vector $\vec{V}_1(100)$, means phase *a* upper switch 'S_{a1}' is on ($S_{a1} = 1$), while upper switches of phases *b* and *c* are off ($S_{b1} = 0$ and $S_{c1} = 0$). Vertices of the two null or zero vectors $\vec{V}_0$ (000) and $\vec{V}_7$ (111) mean that all upper switches (S_{a1}, S_{b1} and S_{c1}) are off and on, respectively. Note that zero vectors $\vec{V}_0$ (000) and $\vec{V}_7$ (111) clamp the converter output phases *a*, *b* and *c* to the negative bus (0 voltage level) and the positive dc bus (V_{dc} voltage level), hence all the line voltages are zero.

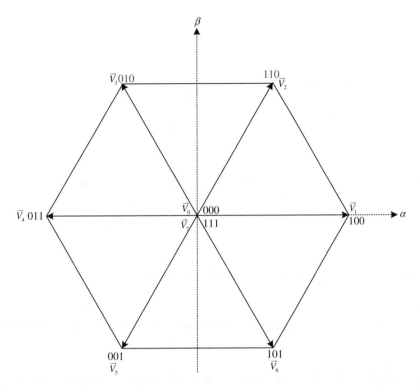

Figure A.15 Space vector diagram of a two-level VSC.

The dwell times of the switch states are calculated assuming that the voltage-second average of the vectors over one switching period is equal to that of the target reference signal. For example, when the reference voltage vector is located in the first sector, see (Figure A.16), it can be synthesised as follows:

$$T_1 \vec{V}_1 + T_2 \vec{V}_2 + T_0 \vec{V}_0 = T_s \vec{V}_{ref} \tag{A.8}$$

where T_0, T_1 and T_2 represent the dwell times, and assuming that the sum of the dwell times of all vectors within one switching period is equal to T_s. This can be expressed as:

$$T_0 + T_1 + T_2 = T_s \tag{A.9}$$

From Table A.1 or Figure A.15 voltage vectors $\vec{V}_0$, $\vec{V}_1$ and $\vec{V}_2$ can be expressed as: $\vec{V}_0 = 0$, $\vec{V}_1 = V_{dc} \angle 0^0 = V_{dc}$, $\vec{V}_2 = V_{dc} \angle 60^0 = V_{dc}(\cos 60^0 + j \sin 60^0) = V_{dc}(\frac{1}{2} + j\frac{\sqrt{3}}{2})$ and $\vec{V}_{ref} = V_{ref} \angle \theta = V_{ref}(\cos \theta + j \sin \theta)$. Substituting these expressions in Eq. (A.8) and equating the real and imaginary parts of both sides, results in the following equations:

$$V_{dc} T_1 + \frac{1}{2} V_{dc} T_2 = T_s V_{ref} \cos \theta \tag{A.10}$$

$$T_2 = \frac{2}{\sqrt{3}} T_s \frac{V_{ref}}{V_{dc}} \sin \theta = \frac{2}{\sqrt{3}} m T_s \sin \theta \tag{A.11}$$

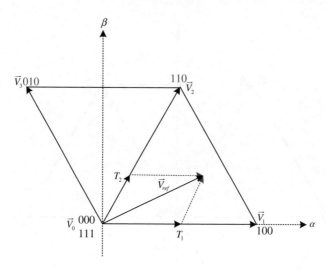

Figure A.16   Illustration of switching devices dwell times calculation in the first sector.

Where $m = \dfrac{V_{ref}}{V_{dc}}$ is the modulation index. By solving Eqs. (A.10) and (A.11), T_1 is obtained as:

$$T_1 = T_s \frac{V_{ref}}{V_{dc}} \cos\theta - \frac{1}{2}T_2 = mT_s \cos\theta - \frac{1}{2} \times \frac{2}{\sqrt{3}}mT_s \sin\theta = \frac{2}{\sqrt{3}}mT_s \cos(\theta + 30°)$$

$$T_1 = \frac{2}{\sqrt{3}}mT_s \cos(\theta + 30°)$$

(A.12)

The dwell time of zero vectors is calculated from Eq. (A.9) as: $T_0 = T_s - T_1 - T_2$. After calculating the dwell times, space vector modulation offers more flexibility than SPWM in terms of gate signal generation. For example, it offers a number of possible switching sequences in each sector that can be used to generate the gate signals for the three-phase converter switches. Table A.2 summarises the switching sequences in the first sector that ensure switching of no more than one voltage level at the same time. It also provides a pictorial representation of the state of the converter upper switches, provided the lower switches operate complementary.

Table A.3 presents similar switching sequences for the second sector. To ensure an integer number of sampling (switching period) within each sector to prevent switching simultaneously between more than one phase during the transition between successive sectors, the frequency ratio $m_f = f_s/f$ must be an integer multiple of six.

Example A.5   Repeat Example A.3 when space vector modulation is used with 2.1 kHz switching frequency and 1.1 modulation index.

Solution. In this case, switching sequence I uses both zero vectors $\vec{V}_0(000)$ and $\vec{V}_7(111)$ for gate signal generation. Figure A.17(a), (b), (c) and (d) show the phase and line voltages of the two-level VSC and their spectra when space vector modulation is used. Observe that the phase

Table A.2 Summary of the switching sequences in the first sector that ensure switching of one voltage level, including the transition between successive sectors.

Switching sequences that exploit both zero vectors $\vec{V}_0(000)$ and $\vec{V}_7(111)$

I	II
$000 \to 100 \to 110 \to 111 \to 110 \to 100 \to 000$	$111 \to 110 \to 100 \to 000 \to 100 \to 110 \to 111$

T_s

$\frac{1}{4}T_0$	$\frac{1}{2}T_1$	$\frac{1}{2}T_2$	$\frac{1}{2}T_0$	$\frac{1}{2}T_2$	$\frac{1}{2}T_1$	$\frac{1}{4}T_0$
000	100	110	111	110	100	000

S_{a1}, S_{b1}, S_{c1}

T_s

$\frac{1}{4}T_0$	$\frac{1}{2}T_1$	$\frac{1}{2}T_2$	$\frac{1}{2}T_0$	$\frac{1}{2}T_2$	$\frac{1}{2}T_1$	$\frac{1}{4}T_0$
111	110	100	000	100	110	111

S_{a1}, S_{b1}, S_{c1}

Switching sequences that exploit only one of the zero vectors $\vec{V}_0(000)$ and $\vec{V}_7(111)$

III	IV
$000 \to 100 \to 110 \to 100 \to 000$	$111 \to 110 \to 100 \to 110 \to 111$

T_s

$\frac{1}{2}T_0$	$\frac{1}{2}T_1$	$\frac{1}{2}T_2$	$\frac{1}{2}T_2$	$\frac{1}{2}T_1$	$\frac{1}{2}T_0$
000	100	110	110	100	000

S_{a1}, S_{b1}, S_{c1}

T_s

$\frac{1}{2}T_0$	$\frac{1}{2}T_1$	$\frac{1}{2}T_2$	$\frac{1}{2}T_2$	$\frac{1}{2}T_1$	$\frac{1}{2}T_0$
111	110	100	100	110	111

S_{a1}, S_{b1}, S_{c1}

voltage spectrum contains the 3rd harmonic and its multiples; however, they cancel in the line voltage as before, resulting in low THD. Figure A.17(c) shows that the converter output currents are sinusoidal, with some measure of distortion. This example shows the ability of space vector modulation to extend the modulation index beyond 1, with increased flexibility of switching pattern selections.

Table A.3 Summary of the switching sequences in the second sector that ensure switching of one voltage level, including the transition between successive sectors.

Switching sequences that exploit both zero vectors $\vec{V}_0(000)$ and $\vec{V}_7(111)$

I	II
$000 \to 010 \to 110 \to 111 \to 110 \to 010 \to 000$	$111 \to 110 \to 010 \to 000 \to 010 \to 110 \to 111$

Switching sequences that exploit only one of the zero vectors $\vec{V}_0(000)$ and $\vec{V}_7(111)$

III	IV
$000 \to 010 \to 110 \to 010 \to 000$	$111 \to 110 \to 010 \to 110 \to 111$

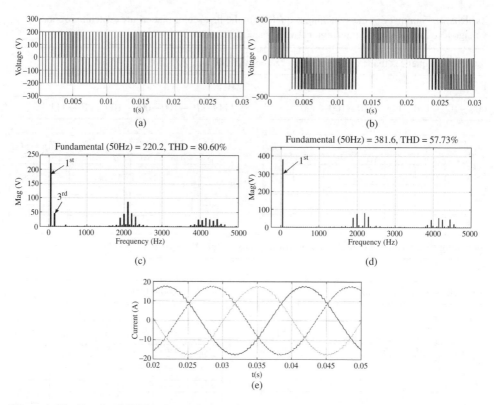

Figure A.17 Two-level VSC output voltage and current waveforms when space vector modulation is used, with 2.1 kHz switching frequency and 1.1 modulation index. (a) Phase voltage relative to supply mid-point; (b) Line voltage; (c) Phase voltage spectrum; (d) Line voltage spectrum; (e) Three-phase output phase currents.

A.2 Neutral-Point Clamped Converter

Figure A.18 shows the configuration of a three-phase three-level diode-clamped converter also known as neutral-point clamped (NPC) converter. It can generate three voltage levels ($\frac{1}{2}V_{dc}$, 0 and $-\frac{1}{2}V_{dc}$) at the output of each phase (a, b and c) relative to supply mid-point or neutral-point 'o'. Its proper operation requires the dc voltage across the dc link capacitors C_1 and C_2 at $\frac{1}{2}V_{dc}$ to be maintained. The voltage stress across each switching device and clamping diode is equal to the voltage across the dc link capacitors, which is equal to $\frac{1}{2}V_{dc}$. Switches (S_{a1}, S_{a3}) and (S_{a2}, S_{a4}) represent two complementary switch pairs. This means that turning switch S_{a1} excludes switch S_{a3} from being on, and the same is true for complementary pair S_{a2} and S_{a4}.

Considering phase a as an example, the neutral-point clamped converter in Figure A.18 generates voltage level $\frac{1}{2}V_{dc}$ at the output of phase a relative to neutral-point 'o' by turning on switches S_{a1} and S_{a2}, while S_{a3} and S_{a4} are turned off. Voltage level 0 is achieved by turning switches S_{a2} and S_{a3} on, with switches S_{a1} and S_{a4} off. Voltage level of $-\frac{1}{2}V_{dc}$ is generated by turning on switches S_{a3} and S_{a4}, while S_{a1} and S_{a2} are turned off. Table A.4 summarises the switch states that generate different voltage levels for phase a and corresponding conduction

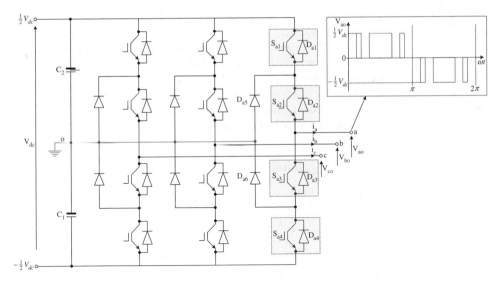

Figure A.18   Three-phase NPC converter.

paths, assuming the current polarities shown in Figure A.18 are positive. Observe that when the converter output phases are connected to voltage levels $^1/_2 V_{dc}$ and $-^1/_2 V_{dc}$ the state-of-charge of the dc link capacitors C_1 and C_2 remains unaffected since the output phases are directly connected to positive and negative dc buses. The state-of-charge of the dc link capacitors C_1 and C_2 is affected when the converter output phases are connected to neutral-point 'o'. Capacitor C_1 discharges for positive output phase currents ($i_a > 0$ for phase a) and charges for negative output phase currents ($i_a < 0$ for phase a). Since the circuit structure of the neutral-point clamped converter imposes the sum of the voltages across the upper and lower capacitors to be equal to full dc link voltage (V_{dc}), discharge of the lower capacitor C_1 means charge of the upper capacitor C_2, and vice versa (Fukuda *et al.*, 1999; Dieckerhoff *et al.*, 2003; Bor-Ren and Ta-Chang, 2004; Barbosa *et al.*, 2005; Dieckerhoff *et al.*, 2005; Ben-Brahim and Tadakuma, 2006; Ceballos *et al.*, 2006a; Ceballos *et al.*, 2006b; Meili *et al.*, 2006; Carpaneto *et al.*, 2007; Ceballos *et al.*, 2007; Ben-Brahim, 2008; Ceballos *et al.*, 2008a; Ceballos *et al.*, 2008b; D'Errico *et al.*, 2009; Beza and Norrga, 2010; Ceballos *et al.*, 2010; Chaudhuri *et al.*, 2010; Floricau *et al.*, 2010a; Floricau *et al.*, 2010b; Gonza *et al.*, 2010; Acuna *et al.*, 2011; Ceballos *et al.*, 2011b; da Silva *et al.*, 2011; Kouzou *et al.*, 2011; Mohzani *et al.*, 2011; Nademi

Table A.4   Summary of the switch states that generate different voltage levels in the NPC converter, including conduction paths.

Voltage level	State of switching devices				Conduction path	
	S_{a1}	S_{a2}	S_{a3}	S_{a4}	$i_a > 0$	$i_a < 0$
$^1/_2 V_{dc}$	ON	ON	OFF	OFF	S_{a1} and S_{a2}	D_{a1} and D_{a2}
0	OFF	ON	ON	OFF	S_{a2} and D_{a5}	S_{a3} and D_{a6}
$-^1/_2 V_{dc}$	OFF	OFF	ON	ON	D_{a3} and D_{a4}	S_{a3} and S_{a4}

et al., 2011; Das and Narayanan, 2012; Mohzani *et al.*, 2012; Barros *et al.*, 2013; Betanzos *et al.*, 2013; Dordevic *et al.*, 2013a).

Neutral-point clamped converters can be controlled using SHE, SPWM and space vector modulation. Standard space vector modulation is normally used for medium-voltage drives system where the three phases are balanced. To be applicable to unbalanced three-phase systems, substantial modifications to the space vector diagram are needed. Selective harmonic elimination and SPWM are discussed, briefly, benefiting from the basic concepts previously discussed when explaining two-level converter modulation methods.

A.2.1 Selective Harmonic Elimination

Figure A.19 shows one-phase of the neutral-point clamped converter and the phase output voltage when selective harmonic elimination is used. From the phase voltage in Figure A.19, the Fourier series coefficients are $A_n = 0$, and

$$
\begin{aligned}
B_n = \frac{1}{\pi} &\left[\int_{\alpha_1}^{\alpha_2} \left(\tfrac{1}{2} V_{dc} \sin n\omega t \right) d\omega t + \int_{\alpha_3}^{\pi-\alpha_3} \left(\tfrac{1}{2} V_{dc} \sin n\omega t \right) d\omega t + \int_{\pi-\alpha_2}^{\pi-\alpha_1} \left(\tfrac{1}{2} V_{dc} \sin n\omega t \right) d\omega t \right. \\
&\left. - \int_{\pi+\alpha_1}^{\pi+\alpha_2} \left(\tfrac{1}{2} V_{dc} \sin n\omega t \right) d\omega t - \int_{\pi+\alpha_3}^{2\pi-\alpha_3} \left(\tfrac{1}{2} V_{dc} \sin n\omega t \right) d\omega t - \int_{2\pi-\alpha_2}^{2\pi-\alpha_1} \left(\tfrac{1}{2} V_{dc} \sin n\omega t \right) d\omega t \right] \\
&= \frac{2V_{dc}}{n\pi} \left[\cos n\alpha_1 - \cos n\alpha_2 + \cos n\alpha_3 \right]
\end{aligned}
\tag{A.13}
$$

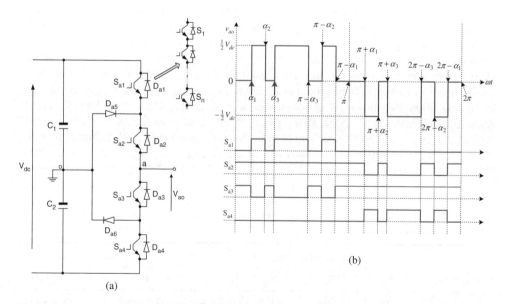

(a)   (b)

Figure A.19 (a) One-phase of the NPC converter; (b) Sample of the output phase voltage with corresponding switching signals when SHE is used.

Equation (A.13) can be written in general form as:

$$B_n = \frac{2V_{dc}}{n\pi} \sum_{j=1}^{\infty} \left[(-1)^{j+1} \cos n\alpha_j\right] \tag{A.14}$$

Example A.6 Plot output voltage and current waveforms, when the NPC converter is connected to 10 Ω and 23 mH load, and controlled using selective harmonic elimination. Assume 5th, 7th, 11th and 13th harmonics are eliminated. Use a modulation index of $m = 1.1$, dc link voltage $V_{dc} = 800$ V, and dc link capacitance $C_1 = C_2 = 2$ mF.

Solution. By setting $m = 1.1$ and B_5, B_7, B_{11} and B_{13} to zero, the following switching angles are obtained: $\alpha_1 = 15.47°$, $\alpha_2 = 23.25°$, $\alpha_3 = 30.69°$, $\alpha_4 = 45.88°$ and $\alpha_5 = 49.06°$. Notice that this solution is valid because it obeys the quarter symmetry criteria previously discussed. Figure A.20(a), (b), (c) and (d) show phase and line voltage of the neutral-point clamped converter and their spectra. Observe that the first non-triplen harmonic present in the phase voltage is the 17th, and this is the first harmonic that can be observed in the line voltage as all the triplen harmonics (3rd, 9th, 13th and 15th) in the phase voltage cancel in the line voltage by common mode.

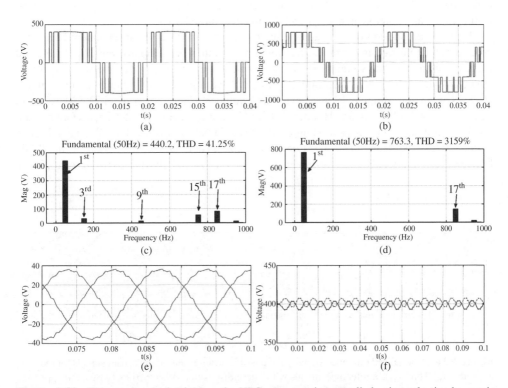

Figure A.20 Key results obtained when the NPC converter is controlled using selective harmonic elimination. (a) Phase voltage; (b) Line voltage; (c) Phase voltage spectrum; (d) Lime voltage spectrum; (e) Three-phase load currents (f) Voltages across upper and lower dc link capacitors (C_1 and C_2).

Figure A.20(e) shows that the NPC converter generates sinusoidal output currents when each switching device operates at approximately 250 Hz, thus low switching losses are expected (this feature made SHE attractive in medium and high-voltage applications). Figure A.20(f) shows the voltages across the upper and lower dc link capacitors. Observe that these voltages remain stable around $\frac{1}{2}V_{dc}$, but oscillate with three times the fundamental frequency. The peak of these oscillations must be kept tightly to avoid exposing converter switches to high voltage stress.

A.2.2   Sinusoidal Pulse Width Modulation

Neutral-point-clamped converters can be controlled using sinusoidal pulse width modulation, with two carriers that must be arranged to fully occupy the contiguous band between 1 and −1 as shown Figure A.21. The gate signal generation is similar to that of the two-level converter. Considering phase a as an example, when $V_a \geq Y_{c1}$ converter output phase a must be connected to the positive bus (or voltage level $\frac{1}{2}V_{dc}$); this means switches S_{a1} and S_{a2} must be turned on, and S_{a3} and S_{a4} must be turned off. When $V_a \leq Y_{c2}$ converter output phase a is connected to the negative bus (or voltage level $-\frac{1}{2}V_{dc}$) which means S_{a1} and S_{a2} are turned off, and S_{a3} and S_{a4} are turned on. Converter output phase a is connected to the neutral-point (or voltage level 0) when $V_a < Y_{c1}$ and $V_a > Y_{c2}$; this means the two middle switches S_{a2} and S_{a3} must be turned on, and outer switches S_{a1} and S_{a4} must be turned off.

The phase disposition carriers are the predominantly used carrier arrangement because it produces an output voltage with less harmonic content than phase opposition disposition, and ensures switching of one voltage level at the output; hence low dv/dt is expected (Carrara *et al.*, 1990; Carrara *et al.*, 1992; Carrara *et al.*, 1993; Newton *et al.*, 1997; Newton and Summer, 1998; Newton and Sumner, 1998; Newton and Sumner, 1999; Ooi *et al.*, 1999; Marchesoni and Tenca, 2002; McGrath and Holmes, 2002a; McGrath and Holmes, 2002b; McGrath *et al.*, 2003; Leon *et al.*, 2005; Khajehoddin *et al.*, 2006; Ceballos *et al.*, 2006a; Busquets-Monge *et al.*, 2007; Franquelo *et al.*, 2007; Leon *et al.*, 2007; Nami *et al.*, 2007; Akagi *et al.*, 2008; Franquelo *et al.*, 2008; Busquets-Monge *et al.*, 2009; Beza and Norrga, 2010; Ito *et al.*, 2010;

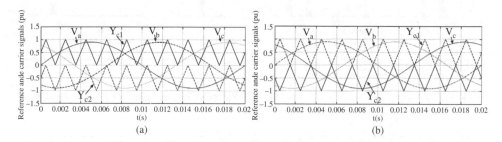

Figure A.21   Normalised reference signals, and possible arrangement of the carrier signals for the NPC converter. (a) Three-phase reference signals with phase disposition (PD) carriers (where the upper and lower carriers are in phase); (b) Three-phase reference signals with phase opposition disposition (POD) carriers (where the carriers above time axis are 180° out of phase from those below).

Minshull *et al.*, 2010; Peng *et al.*, 2010; Adam, Adam *et al.*, 2010b; Grigoletto and Pinheiro, 2011; Hosseini and Sadeghi, 2011; Akagi, 2011b; Adam *et al.*, 2012c; Congzhe *et al.*, 2013).

Example A.7   Repeat Example A.6 when the NPC converter is controlled using SPWM with 3rd harmonic injection based on Steinke's approach, assuming 1.35kHz carrier frequency.

Solution. Figure A.22(a) through (d) show the phase and line output voltage of the NPC converter and their spectra when optimal switching frequency (or SPWM plus 3rd harmonic based on Steinke approach) is used. Observe that the fundamental component of phase and line voltages have lower THD, with similar magnitudes to that of SHE in Example 6. Additionally, it produces high quality output currents with comparable switching frequency, see Figure A.22(e). Figure A.22(f) shows the voltage across the dc link capacitors remaining balanced, without any external intervention (this confirms the natural balancing of the dc link capacitors of the NPC converter). Nevertheless, in practical applications, some balancing strategy will be required to ensure the capacitor voltage remains around half the dc voltage,

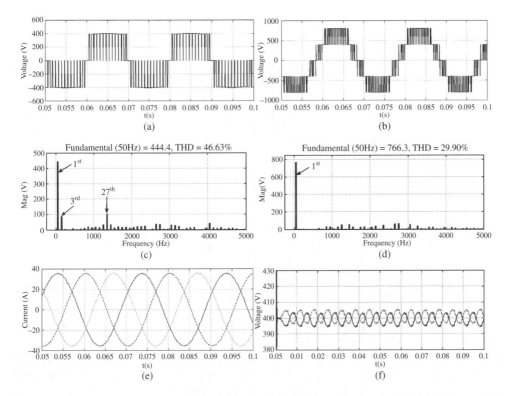

Figure A.22   Key results obtained when the NPC converter is controlled using SPWM plus 3rd harmonic based on Steinke's approach. (a) Phase voltage at converter terminal relative to the neutral-point; (b) Line voltage at converter terminal; (c) Phase voltage spectrum; (d) Line voltage spectrum; (e) Three-phase load currents; (f) Voltage across upper and lower dc link capacitors.

independent of operating conditions and network disturbances. It is worth mentioning that the performance of all carrier-based pulse width modulation deteriorates rapidly at low modulation indices, while that of the selective harmonic elimination deteriorates at a much slower rate compared to carrier-based modulation. Note that in this example, the carrier arrangement is assumed to be PD.

Figure A.23 shows a five-level version of the diode-clamped converter, which was developed to further improve the quality of the output voltage (low THD), to reduce dv/dt, and to

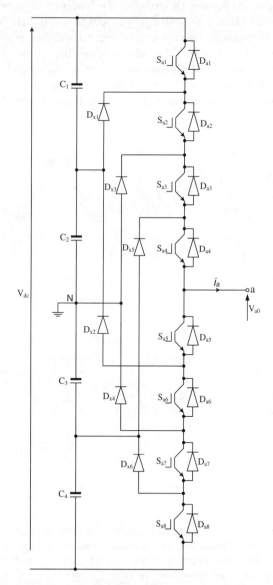

Figure A.23 One phase-leg of a five-level NPC converter.

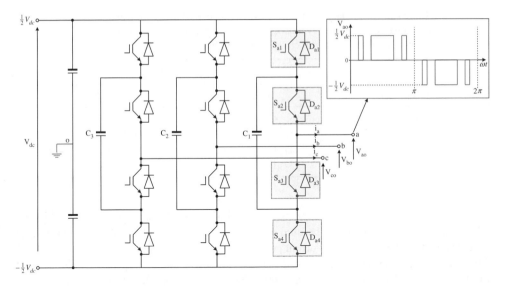

Figure A.24   Three-phase three-level flying capacitor converter.

increase converter dc and ac operating voltage. This configuration was designed with the aim of providing a generic multilevel converter for high-voltage applications. However, this attempt suffers from major setbacks as the balancing of the dc link capacitors turns out to be more challenging than expected.

A.3   Flying Capacitor (FC) Multilevel Converter

Figure A.24 shows the configuration of a three-phase three-level flying-capacitor multilevel converter. It can generate three voltage levels $\frac{1}{2}V_{dc}$, 0 and $-\frac{1}{2}V_{dc}$ at each output phase relative to supply mid-point. The voltage across each clamping capacitor must be maintained at $\frac{1}{2}V_{dc}$. Considering phase a as an example, voltage level $\frac{1}{2}V_{dc}$ is generated by turning on switch S_{a1} and turning off S_{a2}, S_{a3} and S_{a4}. A zero voltage level can be generated using two distinct switch combinations:

(a) Turning on switches S_{a1} and S_{a3}, while turning off switches S_{a2} and S_{a4}. Notice that this switch combination achieves zero voltage level as: $\frac{1}{2}V_{dc} - V_{c1}$, where V_{c1} represent voltage across capacitor C_1.
(b) Turning on switches S_{a2} and S_{a4}, while turning off switches S_{a1} and S_{a3}. This switch combination produces zero voltage as: $-\frac{1}{2}V_{dc} + V_{c1}$.

Voltage level $-\frac{1}{2}V_{dc}$ is achieved by turning off switches S_{a1} and S_{a2}, and turning on switches S_{a3} and S_{a4}. Observe that the switch combinations that generate voltage levels $\frac{1}{2}V_{dc}$ and $-\frac{1}{2}V_{dc}$ connect output phases directly to the positive and negative dc buses, and they do not affect the state-of-charge of the converter clamping capacitors. Only switch combinations that generate a zero voltage level will influence the state-of-charge of the flying capacitor,

Table A.5 Summary of the three-level flying capacitor switch combinations and power path, and their influence on the clamping capacitor state-of-charge (↑ and ↓ represent capacitor charging and discharging, respectively).

Voltage level	Power paths	
	$i_a \geq 0$	$i_a < 0$
$\frac{1}{2}V_{dc}$	$S_{a1}S_{a2}$	$D_{a1}D_{a2}$
0	$S_{a1}C_1D_{a3}$ (C_1↑)	$D_{a1}C_1S_{a3}$ (C_1↓)
	$D_{a4}C_1S_{a3}$(C_1↓)	$S_{a4}C_1D_{a3}$(C_1↑)
$-\frac{1}{2}V_{dc}$	$D_{a4}D_{a3}$	$S_{a4}S_{a3}$

and it should be noted that with these switch combinations one capacitor charges whilst the other discharges. Thus, they can be used alternatively to maintain the voltage balance of the flying capacitor when the output current polarity is known (Jih-Sheng and Fang Zheng, 1995; Chunmei et al., 2003; Deschamps et al., 2003; Fan et al., 2004; Feng and Agelidis, 2004; In-Dong et al., 2004; Dae-Wook et al., 2005; Dieckerhoff et al., 2005; McGrat et al., 2005; Chunmei et al., 2006; Chunmei et al., 2007; Fazel et al., 2007; McGrath and Holmes, 2007a McGrath and Holmes, 2007b; McGrath and Holmes, 2007c; Ceballos et al., 2008a; Franquelo et al., 2008; Adam et al., 2009; McGrath and Holmes, 2009a; McGrath and Holmes, 2009b; Khazraei et al., 2010; Gonzalez et al., 2010; Adam et al., 2011b; Dahidah et al., 2011; Khazraei et al., 2012; Davoodnezhad et al., 2012; Escalante, 2012; Ghias et al., 2012a; Ghias et al., 2012b; Mathew et al., 2013).

Table A.5 summarises the switch combinations of the three-level flying capacitor converter, and their effect on the converter clamping capacitor voltages. Three-level flying capacitor converters can be controlled using the same modulation strategies as with the NPC converter.

Example A.8 Plot the output voltage and current waveforms of the three-phase three-level flying capacitor converter assuming the following parameters: dc link voltage $V_{dc} = 300$ V, clamping capacitance $C = 2$ mF, load of 10 Ω resistance and 23 mH inductance, 1.35 kHz switching frequency, and unity modulation index.

Solution. In this example, two carriers arranged in PD are used with SPWM. Figure A.25(a) to (d) show the output phase and line voltages, and their spectra. Observe that the output voltages and their harmonic distribution are similar to that of the NPC. Figure A.25(e) and (f) show the three-phase output currents and the balanced clamping capacitor voltages. This example shows that the three-level flying capacitor converter generates similar results as the NPC converter.

A.4   Cascaded Multilevel Converter

Figure A.26 shows the configuration of the three-phase cascaded multilevel converter with electrically isolated passive dc sources. Each phase consists of two H-bridge cells, and each H-bridge with a cell capacitor voltage of $\frac{1}{2}V_{dc}$ can generate three-voltage levels $\frac{1}{2}V_{dc}$, 0 and

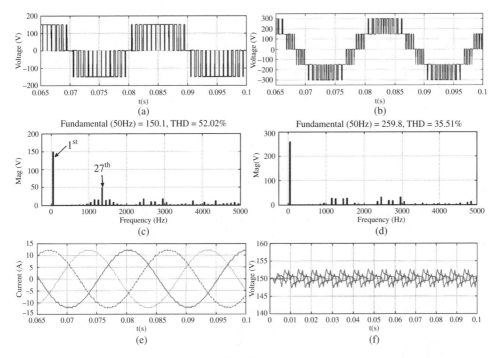

Figure A.25 Key results obtained when the three-phase three-level flying capacitor converter is controlled using SPWM with PD carriers and 1.35 kHz switching frequency. (a) Phase voltage; (b) Line voltage; (c) Phase voltage spectrum; (d) Line voltage spectrum; (e) Load currents; (f) Voltage across the clamping capacitors of the three phases.

$-\frac{1}{2}V_{dc}$. Consider cell 1 of phase a as an example. It achieves voltage level $\frac{1}{2}V_{dc}$ by turning on switches S_{a11} and S_{a14}. Voltage level 0 is achieved by either turning on two upper switches S_{a11} and S_{a13}, or two lower switches S_{a12} and S_{a14}. Voltage level $-\frac{1}{2}V_{dc}$ is generated by turning on switches S_{a13} and S_{a12}. Therefore each phase of the cascaded multilevel converter with two cells per leg, as shown in Figure A.26, can generate five voltage levels at output phase a relative to neutral-point n: V_{dc}, $\frac{1}{2}V_{dc}$, 0, $-\frac{1}{2}V_{dc}$ and $-V_{dc}$ (Dae-Wook *et al.*, 2003; Akagi *et al.*, 2007; Alizadeh and Farhangi, 2007; Carpaneto *et al.*, 2007; Chong *et al.*, 2007; de Leon Morales *et al.*, 2007; Dahidah and Agelidis, 2008; Abu-Rub *et al.*, 2010; Acuna *et al.*, 2011; Baier *et al.*, 2011; de Alvarenga and Pomilio, 2011; Akagi, 2011b; Aceiton *et al.*, 2012; Dahidah *et al.*, 2012; Fang *et al.*, 2012; Adam *et al.*, 2013a).

A.5 Modular Multilevel Converter

For ease of illustration, this section uses Figure A.27(a) and Figure A.28 which show one-phase and three-phase versions of the modular converter, with two half-bridge cells per arm to illustrate its operating principle and modulation. For a dc link voltage of V_{dc}, the voltage across each cell capacitor must be maintained at $\frac{1}{2}V_{dc}$, hence the voltage stress in each

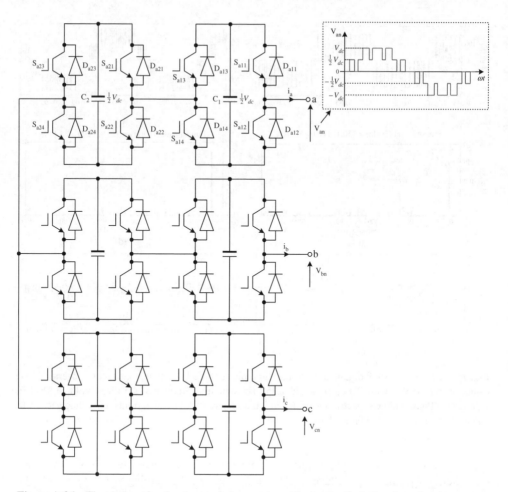

Figure A.26   Three-phase five-level cascaded converter with electrically isolated passive dc sources.

switching device is limited to $^1/_2V_{dc}$ (for one cell capacitor). In this case, each half-bridge cell generates two voltage levels $^1/_2V_{dc}$ and 0 ($\frac{1}{N}V_{dc}$ and 0 for a converter with N cells per arm as shown in Figure A.29). Consider the half-bridge cell in Figure A.27(b) as an example. Voltage level $^1/_2V_{dc}$ is achieved between a_2 and a_1 by turning on switch S_x and turning off switch S_a; and voltage level 0 is achieved by turning off switch S_x and turning on switch S_a. When S_x is on and $i_x > 0$ the cell capacitor is charging, while when $i_x < 0$ it is discharging. To prevent a short circuit across half-bridge cells switch S_a must be operated in a complementary manner with S_x. This means switching S_a on prevents S_x from being on and vice versa. Each phase of the modular converter in Figure A.27 (with two cells per arm) generates three-voltage levels $^1/_2V_{dc}$, 0 and $-^1/_2V_{dc}$. Voltage level $^1/_2V_{dc}$ is generated at output phase *a* relative to the supply mid-point by bypassing the capacitors of the two half-bridge cells in the upper arm, and this is achieved by turning on switches S_{a1} and S_{a2}. Thus, in this voltage level, the

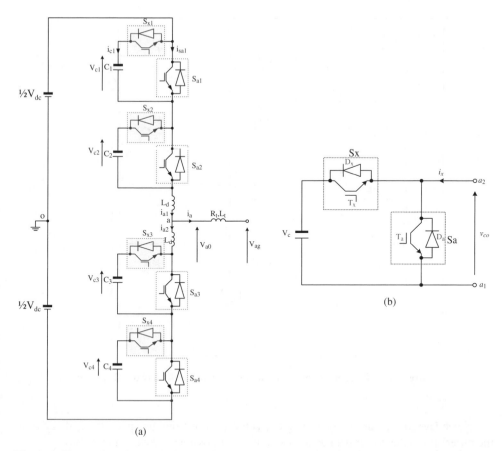

(a)

(b)

Figure A.27   (a) One phase-leg of the modular converter with two cells per arm, and (b) one half-bridge cell.

state-of-charge of the cell capacitors remains unaffected. Notice that in this voltage level, both cell capacitors of the lower arm block the full dc link voltage (Balzani *et al.*, 2006; Carpento *et al.*, 2006; Allebrod *et al.*, 2008; Adam *et al.*, 2009; Antonopoulos *et al.*, 2009; Adam *et al.*, 2010a; Adam *et al.*, 2010b; Akagi and Kitada, 2010; Angquist *et al.*, 2010; Antonopoulos *et al.*, 2010; Adam *et al.*, 2011a; Ahmed *et al.*, 2011; Akagi, 2011a; Akagi, 2011b; Akagi and Kitada, 2011; Angquist *et al.*, 2011a; Angquist *et al.*, 2011b; Barker and Kirby, 2011; Baruschka and Mertens, 2011b; Baruschka and Mertens, 2011a; Bergna *et al.*, 2011; Ceballos *et al.*, 2011a; Adam *et al.*, 2012a; Adam *et al.*, 2012b; Adamowicz *et al.*, 2012; Ahmed *et al.*, 2012; Amankwah *et al.*, 2012a; Amankwah *et al.*, 2012b; Antonopoulos *et al.*, 2012; Araque *et al.*, 2012; Barnklau *et al.*, 2012; Bergna *et al.*, 2012b; Bergna *et al.*, 2012a; Bergna *et al.*, 2012c; Bergna *et al.*, 2012d; Buschendorf *et al.*, 2012; Casadei *et al.*, 2012; Chao *et al.*, 2012; Cherix *et al.*, 2012; Antonopoulos *et al.*, 2013; Adam *et al.*, 2013a; Adam *et al.*, 2013b; Alam and Khan, 2013; Barnklau *et al.*, 2013; Bergna *et al.*, 2013; Brando *et al.*, 2013).

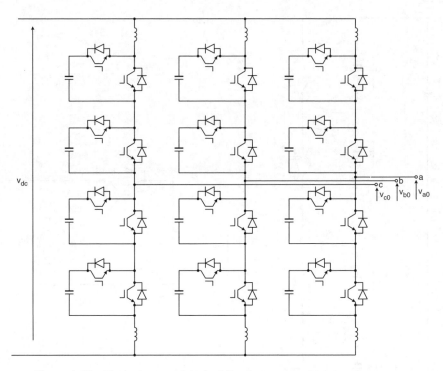

Figure A.28   Three-phase modular multilevel converter with two cells per arm.

Voltage level 0 is achieved by placing one cell capacitor from the upper arm and one from the lower arm in the power path using one of the four possible switch combinations:

(a) Turn on switches S_{a1} and S_{a3} with switches S_{a2} and S_{a4} turned off. Assume the polarities of arm currents i_{a1} and i_{a2} in Figure A.27 are positive. This switch combination charges capacitors C_2 and C_4 for $i_{a1} > 0$ and $i_{a2} > 0$, and discharges C_2 and C_4 for $i_{a1} < 0$ and $i_{a2} < 0$.

(b) Turn on switches S_{a1} and S_{a4}, with switches S_{a2} and S_{a3} turned off. This switch combination charges capacitors C_2 and C_3 for $i_{a1} > 0$ and $i_{a2} > 0$, and discharges C_2 and C_3 for $i_{a1} < 0$ and $i_{a2} < 0$.

(c) Turn on switches S_{a2} and S_{a3}, with switches S_{a1} and S_{a4} turned off. This switch combination charges capacitors C_1 and C_4 for $i_{a1} > 0$ and $i_{a2} > 0$, and discharges C_1 and C_4 for $i_{a1} < 0$ and $i_{a2} < 0$.

(d) Turn on switches S_{a2} and S_{a4}, with switches S_{a1} and S_{a3} turned off. This switch combination charges capacitors C_1 and C_3 for $i_{a1} > 0$ and $i_{a2} > 0$, and discharges C_1 and C_3 for $i_{a1} < 0$ and $i_{a2} < 0$.

From the above discussions, it is clear that switch combinations that generate voltage 0 can be exploited to maintain the voltage balance of the cell capacitors in the modular converter with two cells per arm in Figure A.27.

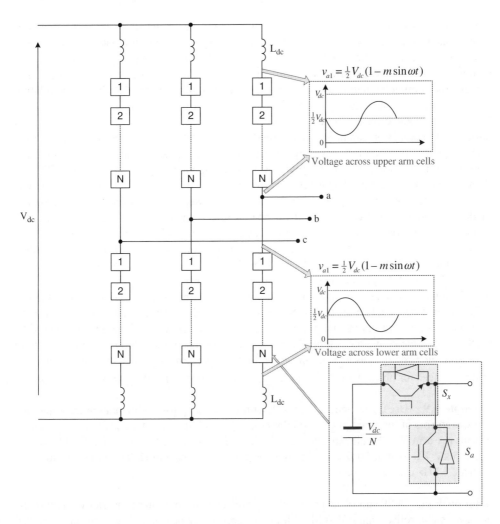

Figure A.29 Three-phase modular multilevel converter with N cells per arm.

Voltage level $-\frac{1}{2}V_{dc}$ is generated at output phase *a* relative to the supply mid-point by turning on switches S_{a3} and S_{a4}, while switches S_{a1} and S_{a2} remain off (which bypasses the lower arm cell capacitors). Thus, the cell capacitor voltages remain unaffected. At this voltage level, both upper arm cells block the full dc link voltage. This indicates that each arm of the modular converter must be designed to be able to block the full dc link voltage.

The arm inductance of the modular converter must be sized sufficiently large to limit the inrush and circulating currents between phases when there is a mismatch between the sum of the voltage across the cell capacitors of each phase and the dc link voltage. Since the cell capacitors are used to synthesise different voltage levels at the converter output, it must be sized to limit the voltage ripple associated with the flow of fundamental current and some of the dominant low-order harmonics, such the 2nd harmonic, to less than 10%.

In general, modular multilevel converters can be controlled using all the mainstream modulation strategies such as space vector modulation, selective harmonic elimination, carrier-based sinusoidal pulse modulation, and staircase (or amplitude) modulation. However, space vector modulation and selective harmonic elimination are not preferred because of the increased complexity of implementation as the number of cells per arm increases. Staircase modulation is the easiest to implement for modular converters with a large number of cells per arm. Pulse width modulation is applicable regardless of number of cells per arm. Staircase and pulse width modulation can be applied to modular converter on a per-phase or per-arm basis. The per-phase approach uses one reference signal to synthesise a sinusoidal voltage at the converter output as the conventional multilevel modulation techniques used with the NPC and flying-capacitor converters. The per arm approach uses two complementary reference signals for each phase (one for the upper arm and one for the lower arm (see Figure A.29). Staircase modulation does not use carrier signals, instead it selects the number of cells to be switched in and out of the power path in an attempt to track the sinusoidal references on a per-phase or per-arm basis. However, in both staircase and pulse width modulation, the decision of which cells should be switched in or out of the power path take into account the voltage magnitude (or state-of-charge) of the cell capacitors and arm or phase currents polarities. The cell capacitors with the lowest voltages must be switched in for positive arm currents, and those with highest voltages must be switched in for negative arm currents. However, the actual number of cells to be switched in and out in each sampling period varies with the voltage level needed to be synthesised at the converter output. For a per-phase approach, the reference signal for phase a for example is defined as $v_a = m \sin \omega t$; while in the per-arm approach, the upper and lower arm references for phase a are defined as: $v_{a1} = \frac{1}{2}(1 - m \sin \omega t)$ and $v_{a2} = \frac{1}{2}(1 - m \sin \omega t)$.

Example A.9 Use the per-phase approach with sinusoidal pulse width modulation to plot the key waveforms of the three-phase modular converter with the following parameters: Number of cells per arm N = 2, cell capacitance = 2 mF, arm inductance L_{dc} = 3 mH, dc link voltage V_{dc} = 5 kV, RL load of 10 Ω and 23.5 mH per phase. Assume a modulation index of m = 0.9 with a 2.1 kHz carrier frequency.

Solution. Figure A.30 shows the key results obtained when a modular converter with two cells per arm is controlled using SPWM with one reference per phase, and two carriers arranged in phase disposition (PD) with 2.1 kHz frequency. Figure A.30(a) to (d) show the phase and line voltage waveforms, and their spectra. Observe that with SPWM and PD carriers these voltages and their harmonic distribution are identical to that of the NPC previously discussed. Figure A.30(e) and (f) show the converter three-phase output currents are sinusoidal despite the fact that the arm currents are not. Observe also, that the arm currents contain a dc component, the fundamental frequency component and low-order harmonics Figure A.30(g) shows the cell capacitor voltages in the three phases are tightly regulated around $\frac{1}{2}V_{dc}$, exploiting the switch combination of the voltage level 0 presented above. In this example, measurements of the cell capacitor voltages are updated per single carrier cycle ($1/2100 \approx 476$ μs) in an attempt to reduce the switching frequency at the cells.

Example A.10 Use the per-arm approach with SPWM to plot the key waveforms of a single-phase modular converter using the following parameters: number of cells per arm N = 2, cell capacitance = 2 mF, arm inductance L_{dc} = 3 mH, dc link voltage V_{dc} = 5 kV, RL

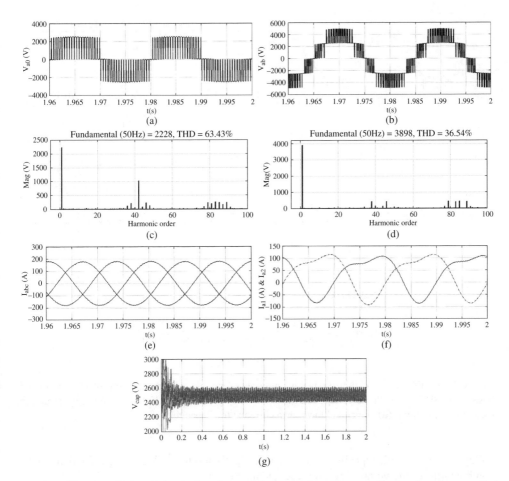

Figure A.30 Key results obtained when the modular converter with two cells per phase is controlled using per-phase approach. (a) Phase voltage relative to the supply mid-point; (b) Line voltage; (c) Phase voltage spectrum; (d) Line voltage spectrum; (e) Three-phase load currents; (f) Sample of the upper and lower arm current; (g) Voltage across the cell capacitors of the three phases.

load of 10 Ω and 23.5 mH per phase. Assume modulation index m = 0.9 and 2.1 kHz carrier frequency.

Solution. Figure A.31 shows the key results obtained when a single-phase modular converter is controlled using SPWM on a per-arm basis (two reference signals per phase, one for each arm). Observe that this approach produces more voltage levels per phase, and pushes all the harmonics away from the power frequency range. In addition, it places most of the harmonic energy around the 2nd carrier component (Figure A.31(a) and (b)). Figure A.31(c) shows that the per arm approach produces higher quality output phase current than the per-phase approach (a single reference signal per phase). The currents in the upper and lower arm are similar to those obtained with the per-phase approach (Figure A.31(d)). Figure A.31(e) shows that the

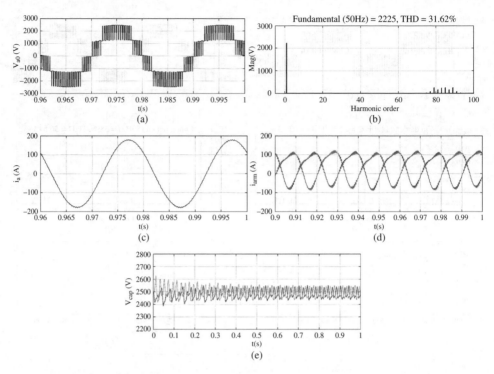

Figure A.31 Key results obtained when a single-phase modular converter is controlled on a per-arm basis using SPWM. (a) Phase voltage relative to the supply mid-point; (b) Phase voltage spectrum; (c) Output phase current; (d) Upper and lower arm currents; (e) Capacitor voltages.

voltage across the cell capacitors is maintained tightly around $^{1}/_{2}V_{dc}$ using redundant switch combinations of the voltage level 0 presented above.

Example A.11   Use the per-arm approach with staircase modulation to control a single-phase modular converter with the following parameters: Number of cells N = 16, dc link voltage V_{dc} = 40 kV, cell capacitance = 5 mF, arm inductance L_{dc} = 10 mH, RL load of 10 Ω and 23.8 mH, and 0.95 modulation index. The reference signals and capacitor voltage measurements are updated at 1.8 kHz.

Solution. Figure A.32 shows the key results obtained when a single-phase modular converter is controlled using staircase modulation, with two reference signals. Figure A.32(a) and (b) display phase voltage and its spectrum; observe that when number of cells is sufficiently large, staircase modulation produces a high quality output voltage as shown in Figure A.32(b), with no significant harmonic components that would require an ac filter. Figure A.32(c) and (d) show output phase current and arm currents. The half-bridge cell currents in Figure A.32(e) show that converter switches operate at reduced switching frequency when staircase modulation is used, hence lower switching losses are expected. Figure A.32 shows the voltage balance of the cell capacitors is maintained, with all capacitor voltages regulated around $\frac{1}{16}V_{dc}$.

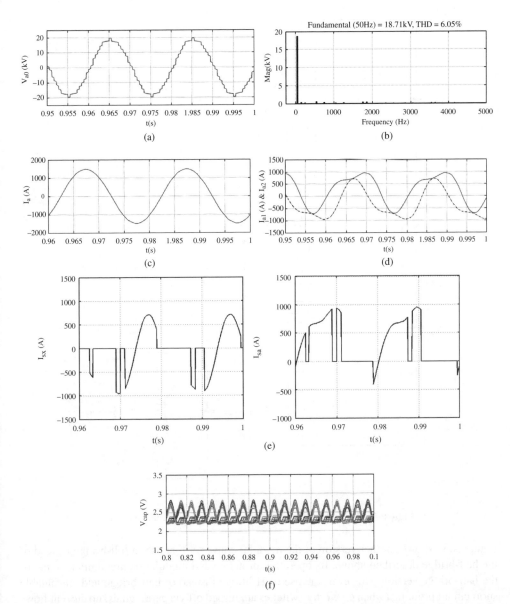

Figure A.32 Key results obtained when a single-phase modular converter with 16 cell per arm is controlled using staircase modulation, with two sinusoidal references per phase. (a) Phase voltage measured relative to the supply mid-point; (b) Phase voltage spectrum; (c) Output phase current; (d) Upper and lower arm currents; (e) Sample of the current waveforms in the converter switches that aim to illustrate the operation at reduced switching frequency; (f) Capacitor voltages.

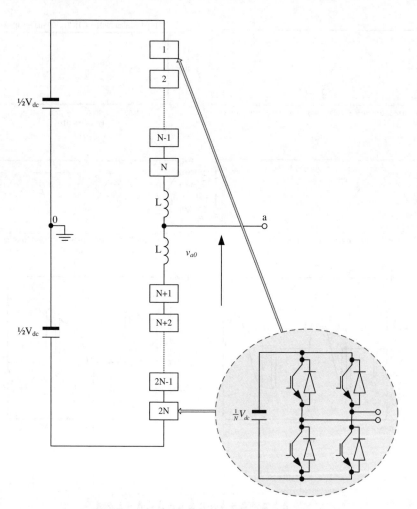

Figure A.33   Full-bridge modular converter with N cells per arm.

Figure A.33 shows another version of the modular converter that uses a full-bridge instead of the half-bridge described above. Its operating principle and modulation are similar to that of the half-bridge version. However, the use of H-bridge instead of half-bridge adds invaluable short circuit proofing (when converter switches are turned off via gate signals, no current flows between ac and dc side during a dc fault).

References

Abu-rub, H., Holtz, J., Rodriguez, J. and Baoming, G. (2010) Medium-voltage multilevel converters – state of the art, challenges, and requirements in industrial applications. *Industrial Electronics, IEEE Transactions on*, **57**, 2581–2596.

Aceiton, R., Weber, J. and Bernet, S. (2012) Level-Shifted PWM for a multilevel traction converter using a state composer. Energy Conversion Congress and Exposition (ECCE), 2012 IEEE, 15–20 September 2012, pp. 713–717.

Acuna, P.F., Moran, L.A., Weishaupt, C.A. and Dixon, L.W. (2011) An active power filter implemented with multilevel single-phase NPC converters. IECON 2011–37th Annual Conference on IEEE Industrial Electronics Society, 7–10 November 2011, pp. 4367–4372.

Adam, G.P. (2007) Quasi Two-level of five-level diode clamped converter. PhD in electrical engineering, power electronics, University of Strathclyde.

Adam, G.P., Ahmed, K.H., Finney, S.J. *et al.* (2013a) New Breed of Network Fault-Tolerant Voltage-Source-Converter HVDC Transmission System. *Power Systems, IEEE Transactions on*, **28**, 335–346.

Adam, G.P., Ahmed, K.H., Finney, S.J. and Williams, B.W. (2011a) Modular multilevel converter for medium-voltage applications. Electric Machines & Drives Conference (IEMDC), 2011 IEEE International, 15–18 May 2011, pp. 1013–1018.

Adam, G.P., Alajmi, B., Ahmed, K.H. *et al.* (2011b) New flying capacitor multilevel converter. Industrial Electronics (ISIE), 2011 IEEE International Symposium on, 27–30 June 2011, pp. 335–339.

Adam, G.P., Anaya-Lara, G.O. and Burt, G. (2012a) Statcom based on modular multilevel converter: Dynamic performance and transient response during ac network disturbances. Power Electronics, Machines and Drives (PEMD 2012), 6th IET International Conference on, 27–29 March 2012, pp. 1–5.

Adam, G.P., Anaya-Lara, O. and Burt, G. (2010a) Steady-state and transient performance of DC transmission systems based on HVDC technology. AC and DC Power Transmission, 2010. ACDC. 9th IET International Conference on, 19–21 October 2010, pp. 1–5.

Adam, G.P., Anaya-Lara, O., Burt, G. *et al.* (2009) Comparison between Two VSC-HVDC Transmission Systems Technologies: Modular and Neutral Point Clamped Multilevel Converter. IEEE 13th Annual conference of the Industrial Electronic Society IECON2009, 3–5 November 2009, Porto-Portugal.

Adam, G.P., Anaya-Lara, O., Burt, G. and Mcdonald, J. (2009) Transformerless STATCOM based on a five-level modular multilevel converter. Power Electronics and Applications. EPE '09. 13th European Conference on, 8–10 September 2009, pp. 1–10.

Adam, G.P., Anaya-Lara, O., Burt, G.M. *et al.* (2010b) Modular multilevel inverter: Pulse width modulation and capacitor balancing technique. *Power Electronics, IET*, **3**, 702–715.

Adam, G.P., Finney, S.J., Bell, K. and Williams, B.W. (2012b) Transient capability assessments of HVDC voltage source converters. Power and Energy Conference at Illinois (PECI), 2012 IEEE, 24–25 February 2012, pp. 1–8.

Adam, G.P., Finney, S.J., Ojo, O. and Williams, B.W. (2012c). Quasi-two-level and three-level operation of a diode-clamped multilevel inverter using space vector modulation. *Power Electronics, IET*, **5**, 542–551.

Adam, G.P., Finney, S.J. and Williams, B.W. (2013b). Hybrid converter with ac side cascaded H-bridge cells against H-bridge alternative arm modular multilevel converter: steady-state and dynamic performance. *Generation, Transmission & Distribution, IET*, **7**(3), 318–328.

Adamowicz, M., Strzelecki, R. and Krzeminski, Z. (2012) Hybrid high-frequency-SiC and line-frequency-Si based PEBB for MV modular power converters. IECON 2012–38th Annual Conference on IEEE Industrial Electronics Society, 25–28 October 2012, pp. 5197–5202.

Ahmed, N., Haider, A., Van Hertem, D. *et al.* (2011) Prospects and challenges of future HVDC SuperGrids with modular multilevel converters. Power Electronics and Applications (EPE 2011), Proceedings of the 2011–14th European Conference on, 30 August 2011–1 September 2011, pp. 1–10.

Ahmed, N., Ngquist, L., Norrga, S. and Nee, H.P. (2012) Validation of the continuous model of the modular multilevel converter with blocking/deblocking capability. AC and DC Power Transmission (ACDC 2012), 10th IET International Conference on, 4–5 December 2012, pp. 1–6.

Akagi, H. (2011a) Classification, terminology, and application of the modular multilevel cascade converter (MMCC). *Power Electronics, IEEE Transactions on*, **26**, 3119–3130.

Akagi, H. (2011b) New trends in medium-voltage power converters and motor drives. Industrial Electronics (ISIE), 2011 IEEE International Symposium on, 27–30 June 2011, pp. 5–14.

Akagi, H., Fujita, H., Yonetani, S. and Kondo, Y. (2008) A 6.6-kV transformerless STATCOM based on a five-level diode-clamped PWM converter: system design and experimentation of a 200-V 10-kVA laboratory model. *Industry Applications, IEEE Transactions on*, **44**, 672–680.

Akagi, H., Inoue, S. and Yoshii, T. (2007) Control and Performance of a Transformerless Cascade PWM STATCOM With Star Configuration. *Industry Applications, IEEE Transactions on*, **43**, 1041–1049.

Akagi, H. and Kitada, R. (2010) Control of a modular multilevel cascade BTB system using bidirectional isolated DC/DC converters. Energy Conversion Congress and Exposition (ECCE), 2010 IEEE, 12–16 September 2010, pp. 3549–3555.

Akagi, H. and Kitada, R. (2011) Control and Design of a Modular Multilevel Cascade BTB System Using Bidirectional Isolated DC/DC Converters. *Power Electronics, IEEE Transactions on*, **26**, 2457–2464.

Alam, M.K. and Khan, F.H. (2013) State space modeling and performance analysis of a multilevel modular switched-capacitor converter using pulse dropping switching technique. Applied Power Electronics Conference and Exposition (APEC), 2013 Twenty-Eighth Annual IEEE, 17–21 March 2013, pp. 1753–1758.

Alizadeh, O. and Farhangi, S. (2007) Voltage balancing technique with low switching frequency for cascade multilevel active front-end. Power Electronics, 2007. ICPE '07. 7th Internatonal Conference on, 22–26 October 2007, pp. 749–753.

Allebrod, S., Hamerski, R. and Marquardt, R. (2008) New transformerless, scalable Modular Multilevel Converters for HVDC-transmission. Power Electronics Specialists Conference, 2008. PESC 2008. IEEE, 15–19 June 2008, pp. 174–179.

Amankwah, E.K., Clare, J.C., Wheeler, P.W. and Watson, A.J. (2012a) Cell capacitor voltage control in a parallel hybrid modular multilevel voltage source converter for HVDC applications. Power Electronics, Machines and Drives (PEMD 2012), 6th IET International Conference on, 27–29 March 2012, pp. 1–6.

Amankwah, E.K., Clare, J.C., Wheeler, P.W. and Watson, A.J. (2012b) Multi carrier PWM of the modular multilevel VSC for medium voltage applications. Applied Power Electronics Conference and Exposition (APEC), 2012 Twenty-Seventh Annual IEEE, 5–9 February 2012, pp. 2398–2406.

Angquist, L., Antonopoulos, A., Siemaszko, D. *et al.* (2010) Inner control of Modular Multilevel Converters – An approach using open-loop estimation of stored energy. Power Electronics Conference (IPEC), 2010 International, 21–24 June 2010, pp. 1579–1585.

Angquist, L., Antonopoulos, A., Siemaszko, D. *et al.* (2011a) Open-loop control of modular multilevel converters using estimation of stored energy. *Industry Applications, IEEE Transactions on*, **47**, 2516–2524.

Angquist, L., Haider, A., Nee, H.P. and Hongbo, J. (2011b) Open-loop approach to control a Modular Multilevel Frequency Converter. Power Electronics and Applications (EPE 2011), Proceedings of the 2011–14th European Conference on, 30 August 2011–1 September 2011, pp. 1–10.

Antonopoulos, A., Angquist, L., Harnefors, L. *et al.* (2013) Global asymptotic stability of modular multilevel converters. *Industrial Electronics, IEEE Transactions on*, **61**(2), 603–612.

Antonopoulos, A., Angquist, L. and Nee, H.P. (2009) On dynamics and voltage control of the Modular Multilevel Converter. Power Electronics and Applications, 2009. EPE '09. 13th European Conference on, 8–10 September 2009, pp. 1–10.

Antonopoulos, A., Angquist, L., Norrga, S. *et al.* (2012) Modular multilevel converter ac motor drives with constant torque form zero to nominal speed. Energy Conversion Congress and Exposition (ECCE), 2012 IEEE, 15–20 September 2012, pp. 739–746.

Antonopoulos, A., Ilves, K., Ngquist, L. and Nee, H.P. (2010) On interaction between internal converter dynamics and current control of high-performance high-power AC motor drives with modular multilevel converters. Energy Conversion Congress and Exposition (ECCE), 2010 IEEE, 12–16 September 2010, pp. 4293–4298.

Araque, J.A., Diaz Rodriguez, J.L. and Garcia, A.P. (2012) Modular development of a single phase multilevel converter. Circuits and Systems (CWCAS), 2012 IEEE 4th Colombian Workshop on, 1–2 November 2012, pp. 1–6.

Azmi, S.A., Adam, G.P., Ahmed, K.H. *et al.* (2013) Grid interfacing of multimegawatt photovoltaic inverters. *Power Electronics, IEEE Transactions on*, **28**, 2770–2784.

Baier, C.R., Munoz, J.A., Espinoza, J.R. *et al.* (2011) Improving power quality in cascade multilevel converters based on single-phase non-regenerative power cells. IECON 2011–37th Annual Conference on IEEE Industrial Electronics Society, 7–10 November 2011, pp. 4192–4197.

Baker, A.C., Zaffanella, L.E., Anzivino, L.D. *et al.* (1988) Contamination performance of HVDC station post insulators. *Power Delivery, IEEE Transactions on*, **3**, 1968–1975.

Balzani, M., Reatti, A. and Salvadori, G. (2006) Design, Assembly and Testing of Modular Multilevel Converter with Multicarrier PWM Method. Research in Microelectronics and Electronics 2006, PhD, 2006, pp. 57–60.

Barbosa, P., Steimer, P., Steinke, J. *et al.* (2005) Active-neutral-point-clamped (ANPC) multilevel converter technology. Power Electronics and Applications, 2005 European Conference on, 2005, p. 10.

Barker, C.D. and Kirby, N.M. (2011) Reactive power loading of components within a modular multi-level HVDC VSC converter. Electrical Power and Energy Conference (EPEC), 2011 IEEE, 3–5 October 2011, pp. 86–90.

Barnklau, H., Gensior, A. and Bernet, S. (2012) Derivation of an equivalent submodule per arm for modular multilevel converters. Power Electronics and Motion Control Conference (EPE/PEMC), 2012 15th International, 4–6 September 2012, pp. LS2a.2-1–LS2a.2-5.

Barnklau, H., Gensior, A. and Rudolph, J. (2013) A model-based control scheme for modular multilevel converters. *Industrial Electronics, IEEE Transactions on*, **60**, 5359–5375.

Barros, J.D., Silva, J.F.A. and Jesus, E.G.A. (2013) Fast-predictive optimal control of NPC multilevel converters. *Industrial Electronics, IEEE Transactions on*, **60**, 619–627.

Baruschka, L. and Mertens, A. (2011a) Comparison of Cascaded H-Bridge and Modular Multilevel Converters for BESS application. Energy Conversion Congress and Exposition (ECCE), 2011 IEEE, 17–22 September 2011, pp. 909–916.

Baruschka, L. and Mertens, A. (2011b) A new 3-phase direct modular multilevel converter. Power Electronics and Applications (EPE 2011). Proceedings of the 2011–14th European Conference on, 30 August 2011–1 September 2011, pp. 1–10.

Bech, M.M., Blaabjerg, F. and Pedersen, J.K. (2000) Random modulation techniques with fixed switching frequency for three-phase power converters. *Power Electronics, IEEE Transactions on*, **15**, 753–761.

Beig, A.R. (2012) Synchronized SVPWM algorithm for the overmodulation region of a low switching frequency medium-voltage three-level VSI. *Industrial Electronics, IEEE Transactions on*, **59**, 4545–4554.

Ben-Brahim, L. (2008) A discontinuous PWM method for balancing the neutral point voltage in three-level inverter-fed variable frequency drives. *Energy Conversion, IEEE Transactions on*, **23**, 1057–1063.

Ben-Brahim, L. and Tadakuma, S. (2006) A novel multilevel carrier-based PWM-control method for GTO inverter in low index modulation region. *Industry Applications, IEEE Transactions on*, **42**, 121–127.

Ben-Sheng, C. and Hsu, Y.-Y. (2007) An analytical approach to harmonic analysis and controller design of a STATCOM. *Power Delivery, IEEE Transactions on*, **22**, 423–432.

Bergna, G., Berne, E., Egrot, P. *et al.* (2013) An energy-based controller for HVDC modular multilevel converter in decoupled double synchronous reference frame for voltage oscillation reduction. *Industrial Electronics, IEEE Transactions on*, **60**, 2360–2371.

Bergna, G., Berne, E., Egrot, P. *et al.* (2012a) A generalized power control approach in ABC frame for modular Multilevel Converters based on mathematical optimization. Energy Conference and Exhibition (ENERGYCON), 2012 IEEE International, 9–12 September 2012, pp. 158–165.

Bergna, G., Berne, E., Egrot, P. *et al.* (2012b) Modular Multilevel Converter-energy difference controller in rotating reference frame. Power Electronics and Motion Control Conference (EPE/PEMC), 2012 15th International, 4–6 September 2012, pp. LS2c.1-1–LS2c.1-4.

Bergna, G., Boyra, M. and Vivas, J.H. (2011) Evaluation and proposal of MMC-HVDC control strategies under transient and steady state conditions. Power Electronics and Applications (EPE 2011), Proceedings of the 2011–14th European Conference on, 30 August 2011–1 September 2011, pp. 1–10.

Bergna, G., Suul, J., Berne, E. *et al.* (2012c) Mitigating DC-side power oscillations and negative sequence load currents in Modular Multilevel Converters under unbalanced faults- first approach using resonant PI. IECON 2012 – 38th Annual Conference on IEEE Industrial Electronics Society, 25–28 October 2012, pp. 537–542.

Bergna, G., Vannier, J.C., Lefranc, P. *et al.* (2012d) Modular Multilevel Converter leg-energy controller in rotating reference frame for voltage oscillations reduction. Power Electronics for Distributed Generation Systems (PEDG), 2012 3rd IEEE International Symposium on, 25–28 June 2012, pp. 698–703.

Betanzos, J.D., Rodriguez, J.J. and Peralta, E. (2013) Space vector pulse width modulation for three-level NPC-VSI. *Latin America Transactions, IEEE (Revista IEEE America Latina)*, **11**, 759–767.

Beza, M. and Norrga, S. (2010) Three-level converters with selective Harmonic Elimination PWM for HVDC application. Energy Conversion Congress and Exposition (ECCE), 2010 IEEE, 12–16 September 2010, pp. 3746–3753.

Bierk, H., Al-Judi, A., Rahim, A. and Nowicki, E. (2009) Elimination of low-order harmonics using a modified SHE-PWM technique for medium voltage induction motor applications. Power & Energy Society General Meeting, 2009. PES '09. IEEE, 26–30 July 2009, pp. 1–8.

Bierk, H., Albakkar, A. and Nowicki, E. (2011) Harmonic reduction in the parallel arrangements of grid-connected voltage source inverters. Electric Power and Energy Conversion Systems (EPECS), 2011 2nd International Conference on, 15–17 Nov. 2011, pp. 1–6.

Bode, G.H. and Holmes, D.G. (2001) Improved current regulation for voltage source inverters using zero crossings of the compensated current errors. Industry Applications Conference, 2001. Thirty-Sixth IAS Annual Meeting. Conference Record of the 2001 IEEE, 30 September 2001–4 October 2001, pp. 1007–1014 vol.2.

Bor-Ren, L. and Ta-Chang, W. (2004) Space vector modulation strategy for an eight-switch three-phase NPC converter. *Aerospace and Electronic Systems, IEEE Transactions on*, **40**, 553–566.

Bowes, S.R. (in press) New sinusoidal pulse width modulation inverter. *Proc. IEE*, **122**, 1279–1285.

Bowes, S.R. (1975) New sinusoidal pulsewidth-modulated invertor. *Electrical Engineers, Proceedings of the Institution of*, **122**, 1279–1285.

Bowes, S.R. (1988) Comments, with reply, on "An algebraic algorithm for microcomputer-based (direct) inverter pulsewidth modulation" by Y.H. Kim and M. Ehsani. *Industry Applications, IEEE Transactions on*, **24**, 998–1004.

Bowes, S.R. (1990) Regular-sampled harmonic elimination/minimisation PWM techniques. Applied Power Electronics Conference and Exposition, 1990. APEC '90, Conference Proceedings 1990, Fifth Annual, 11–16 March 1990, pp. 532–540.

Bowes, S.R. (1993a) Advanced regular-sampled PWM control techniques for drives and static power converters. Industrial Electronics, Control, and Instrumentation, 1993. Proceedings of the IECON '93., International Conference on, 15–19 November 1993, vol. 2, pp. 662–669.

Bowes, S.R. (1993b) Efficient microprocessor real-time PWM drive control using regular-sampled harmonic minimisation techniques. Industrial Electronics, 1993. Conference Proceedings, ISIE'93 – Budapest, IEEE International Symposium on, 1993, pp. 211–218.

Bowes, S.R. (1993c) Novel real-time harmonic minimized PWM control for drives and static power converters. Applied Power Electronics Conference and Exposition, 1993. APEC '93. Conference Proceedings 1993, Eighth Annual, 7–11 March 1993, pp. 561–567.

Bowes, S.R. (1994) Novel real-time harmonic minimized PWM control for drives and static power converters. *Power Electronics, IEEE Transactions on*, **9**, 256–262.

Bowes, S.R. (1995) Advanced regular-sampled PWM control techniques for drives and static power converters. *Industrial Electronics, IEEE Transactions on*, **42**, 367–373.

Bowes, S.R. and Bird, B.M. (1975) Novel approach to the analysis and synthesis of modulation processes in power convertors. *Electrical Engineers, Proceedings of the Institution of*, **122**, 507–513.

Bowes, S.R. and Clare, J.C. (1988) Computer-aided design of PWM power-electronic variable-speed drives. *Electric Power Applications, IEE Proceedings B*, **135**, 240–260.

Bowes, S.R. and Clark, P.R. (1988) Transputer based optimal PWM control of inverter drives. Industry Applications Society Annual Meeting, 1988. Conference Record of the 1988 IEEE, 2–7 Oct. 1988. vol. 1, pp. 314–321.

Bowes, S.R. and Clark, P.R. (1990) Simple microprocessor implementation of new regular-sampled harmonic elimination PWM techniques. Industry Applications Society Annual Meeting, 1990. Conference Record of the 1990 IEEE, 7–12 October 1990. vol. 1, pp. 341–347.

Bowes, S.R. and Clark, P.R. (1992) Transputer-based harmonic-elimination PWM control of inverter drives. *Industry Applications, IEEE Transactions on*, **28**, 72–80.

Bowes, S.R. and Grewal, S. (1998a) A novel harmonic elimination PWM strategy. Power Electronics and Variable Speed Drives, 1998. Seventh International Conference on (Conf. Publ. No. 456), 21–23 September 1998, pp. 426–432.

Bowes, S.R. and Grewal, S. (1998b) Simplified harmonic elimination PWM control strategy. *Electronics Letters*, **34**, 325–326.

Bowes, S.R. and Grewal, S. (1999) Novel space vector based harmonic elimination inverter control. Industry Applications Conference, 1999. Thirty-Fourth IAS Annual Meeting. Conference Record of the 1999 IEEE, 1999, vol. 3, pp. 1616–1622.

Bowes, S.R., Grewal, S. and Holliday, D. (2000a) High frequency PWM technique for two and three level single-phase inverters. *Electric Power Applications, IEE Proceedings*, **147**, 181–191.

Bowes, S.R., Grewal, S. and Holliday, D. (2000b) Simplified ultrasonic regular-sampled PWM technique. *Electronics Letters*, **36**, 854–855.

Bowes, S.R. and Holliday, D. (2006) Comparison of pulse-width-modulation control strategies for three-phase inverter systems. *Electric Power Applications, IEE Proceedings*, **153**, 575–584.

Bowes, S.R. and Yen-Shin, L. (1997) The relationship between space-vector modulation and regular-sampled PWM. *Industrial Electronics, IEEE Transactions on*, **44**, 670–679.

Bradshaw, J., Madawala, U., Patel, N. and Vilathgamuwa, M. (2012) Bit-stream-based space vector modulators. *Power Electronics, IET*, **5**, 205–214.

Brando, G., Coppola, M., Dannier, A. *et al.* (2013) An analysis of modular multilevel converter for full frequency range operations. Ecological Vehicles and Renewable Energies (EVER), 2013 8th International Conference and Exhibition on, 27–30 March 2013, pp. 1–7.

Bruckner, T. and Holmes, D.G. (2003) Optimal pulse width modulation for three-level inverters. Power Electronics Specialist Conference, 2003. PESC '03. 2003 IEEE 34th Annual, 15–19 June 2003, vol. 1, pp. 165–170.

Buja, G.S. and Indri, G.B. (1977) Optimal pulsewidth modulation for feeding AC motors. *Industry Applications, IEEE Transactions on*, **IA-13**, 38–44.

Burgos, R., Lai, R., Pei, Y. *et al.* (2007) Space Vector Modulation for Vienna-Type Rectifiers Based on the Equivalence between Two- and Three-Level Converters: A Carrier-Based Implementation. Power Electronics Specialists Conference, 2007. PESC 2007. IEEE, 17–21 June 2007, pp. 2861–2867.

Burgos, R., Rixin, L., Yunqing, P. *et al.* (2008) Space vector modulator for vienna-type rectifiers based on the equivalence between two- and three-level converters: a carrier-based implementation. *Power Electronics, IEEE Transactions on*, **23**, 1888–1898.

Buschendorf, M., Weber, J. and Bernet, S. (2012) Comparison of IGCT and IGBT for the use in the modular multilevel converter for HVDC applications. Systems, Signals and Devices (SSD), 2012 9th International Multi-Conference on, 20–23 March 2012, pp. 1–6.

Busquets-Monge, S., Alepuz, S., Bordonau, J. and Peracaula, J. (2007) Voltage Balancing Control of Diode-Clamped Multilevel Converters with Passive Front-Ends. Industrial Electronics, 2007. ISIE 2007. IEEE International Symposium on, 4–7 June 2007, pp. 544–549.

Busquets-Monge, S., Rocabert, J., Crebier, J.C. and Peracaula, J. (2009) Diode-clamped multilevel converters with integrable gate-driver power-supply circuits. Power Electronics and Applications, 2009. EPE '09. 13th European Conference on, 8–10 September 2009, pp. 1–10.

Carnielutti, F., Pinheiro, H. and Rech, C. (2012) Generalized carrier-based modulation strategy for cascaded multilevel converters operating under fault conditions. *Industrial Electronics, IEEE Transactions on*, **59**, 679–689.

Carpaneto, M., Marchesoni, M. and Vaccaro, L. (2007) A New Cascaded Multilevel Converter Based on NPC Cells. Industrial Electronics, 2007. ISIE 2007. IEEE International Symposium on, 4–7 June 2007, pp. 1033–1038.

Carpento, M., Ferrando, G., Marchesoni, M. and Vaccaro, L.R. (2006) A new modular multilevel conversion structure with passive filter minimization. IEEE Industrial Electronics, IECON 2006 – 32nd Annual Conference on, 6–10 Nov. 2006, pp. 2432–2437.

Carrara, G., Casini, D., Gardella, S. and Salutari, R. (1993) Optimal PWM for the control of multilevel voltage source inverter. Power Electronics and Applications, 1993. Fifth European Conference on, 13–16 September 1993, vol. 4, pp. 255–259.

Carrara, G., Gardella, S., Marchesoni, M. *et al.* (1990) A new multilevel PWM method: a theoretical analysis. Power Electronics Specialists Conference, 1990. PESC '90 Record., 21st Annual IEEE, 1990, pp. 363–371.

Carrara, G., Gardella, S., Marchesoni, M. *et al.* (1992) A new multilevel PWM method: a theoretical analysis. *Power Electronics, IEEE Transactions on*, **7**, 497–505.

Carter, P.J. and Johnson, G.B. (1988) Space charge measurements downwind from a monopolar 500 kV HVDC test line. *Power Delivery, IEEE Transactions on*, **3**, 2056–2063.

Casadei, D., Dujic, D., Levi, E. *et al.* (2008) General modulation strategy for seven-phase inverters with independent control of multiple voltage space vectors. *Industrial Electronics, IEEE Transactions on*, **55**, 1921–1932.

Casadei, G., Teodorescu, R., Vlad, C. and Zarri, L. (2012) Analysis of dynamic behavior of Modular Multilevel Converters: Modeling and control. System Theory, Control and Computing (ICSTCC), 2012 16th International Conference on, 12–14 October 2012, pp. 1–6.

Cataliotti, A., Genduso, F., Raciti, A. and Galluzzo, G.R. (2007) Generalized PWM–VSI control algorithm based on a universal duty-cycle expression: theoretical analysis, simulation results, and experimental validations. *Industrial Electronics, IEEE Transactions on*, **54**, 1569–1580.

Ceballos, S., Pou, J., Gabiola, I. *et al.* (2006a) Fault-Tolerant Multilevel Converter Topology. Industrial Electronics, 2006 IEEE International Symposium on, 9–13 July 2006, pp. 1577–1582.

Ceballos, S., Pou, J., Robles, E. *et al.* (2008a) Three-level converter topologies with switch breakdown fault-tolerance capability. *Industrial Electronics, IEEE Transactions on*, **55**, 982–995.

Ceballos, S., Pou, J., Robles, E. *et al.* (2010) Performance evaluation of fault-tolerant neutral-point-clamped converters. *Industrial Electronics, IEEE Transactions on*, **57**, 2709–2718.

Ceballos, S., Pou, J., Sanghun, C. *et al.* (2011a) Analysis of voltage balancing limits in modular multilevel converters. IECON 2011 – 37th Annual Conference on IEEE Industrial Electronics Society, 7–10 November 2011, pp. 4397–4402.

Ceballos, S., Pou, J., Zaragoza, J. *et al.* (2006b) Efficient Modulation Technique for a Four-Leg Fault-Tolerant Neutral-Point-Clamped Inverter. IEEE Industrial Electronics, IECON 2006 – 32nd Annual Conference on, 6–10 November 2006, pp. 2090–2095.

Ceballos, S., Pou, J., Zaragoza, J. *et al.* (2008b) Efficient modulation technique for a four-leg fault-tolerant neutral-point-clamped inverter. *Industrial Electronics, IEEE Transactions on*, **55**, 1067–1074.

Ceballos, S., Pou, J., Zaragoza, J. *et al.* (2007) Soft-Switching Topology for a Fault-Tolerant Neutral-Point-Clamped Converter. Industrial Electronics, 2007. ISIE 2007. IEEE International Symposium on, 4–7 June 2007, pp. 3186–3191.

Ceballos, S., Pou, J., Zaragoza, J. *et al.* (2011b) Fault-tolerant neutral-point-clamped converter solutions based on including a fourth resonant leg. *Industrial Electronics, IEEE Transactions on*, **58**, 2293–2303.

Congzhe, G., Xinjian, J., Yongdong, L. *et al.* (2013) A DC-link voltage self-balance method for a diode-clamped modular multilevel converter with minimum number of voltage sensors. *Power Electronics, IEEE Transactions on*, **28**, 2125–2139.

Chang, G.W., Chia-Ming, Y. and Wei-Cheng, C. (2006) Meeting IEEE-519 current harmonics and power factor constraints with a three-phase three-wire active power filter under distorted source voltages. *Power Delivery, IEEE Transactions on*, **21**, 1648–1654.

Changjiang, Z., Arulampalam, A. and Jenkins, N. 2003. Four-wire dynamic voltage restorer based on a three-dimensional voltage space vector PWM algorithm. *Power Electronics, IEEE Transactions on*, **18**, 1093–1102.

Changjiang, Z., Ramachandaramurthy, V.K., Arulampalam, A. *et al.* (2001) Dynamic voltage restorer based on voltage-space-vector PWM control. *Industry Applications, IEEE Transactions on*, **37**, 1855–1863.

Chao, C., Adam, G.P., Finney, S.J. and Williams, B.W. (2012) Post – DC fault recharging of the H-bridge modular multilevel converter. AC and DC Power Transmission (ACDC 2012), 10th IET International Conference on, 4–5 December 2012, pp. 1–5.

Chaudhuri, T., Rufer, A. and Steimer, P.K. (2010) The common cross-connected stage for the 5L ANPC medium voltage multilevel inverter. *Industrial Electronics, IEEE Transactions on*, **57**, 2279–2286.

Cherix, N., Vasiladiotis, M. and Rufer, A. (2012) Functional modeling and Energetic Macroscopic Representation of Modular Multilevel Converters. Power Electronics and Motion Control Conference (EPE/PEMC), 2012 15th International, 4–6 September 2012, pp. LS1a-1.3-1–LS1a-1.3-8.

Ching-Tsai, P. and Jenn-Jong, S. (1999) A single-stage three-phase boost-buck AC/DC converter based on generalized zero-space vectors. *Power Electronics, IEEE Transactions on*, **14**, 949–958.

Chong, H., Huang, A.Q., Yu, L. and Bin, C. (2007) A Generalized Control Strategy of Per-Phase DC Voltage Balancing for Cascaded Multilevel Converter-based STATCOM. Power Electronics Specialists Conference, 2007. PESC 2007. IEEE, 17–21 June 2007, pp. 1746–1752.

Chunmei, F., Jun, L. and Agelidis, V.G. (2003) A novel voltage balancing control method for flying capacitor multilevel converters. Industrial Electronics Society, 2003. IECON '03. The 29th Annual Conference of the IEEE, 2–6 November 2003, vol. 2, pp. 1179–1184.

Chunmei, F., Jun, L. and Agelidis, V.G. (2007) Modified phase-shifted PWM control for flying capacitor multilevel converters. *Power Electronics, IEEE Transactions on*, **22**, 178–185.

Chunmei, F., Jun, L., Agelidis, V.G. and Green, T.C. (2006) A Multi-Modular System Based On Parallel-Connected Multilevel Flying Capacitor Converters Controlled with Fundamental Frequency SPWM. IEEE Industrial Electronics, IECON 2006 – 32nd Annual Conference on, 6–10 November 2006, pp. 2360–2365.

D'errico, L., Lidozzi, A., Serrao, V. and Solero, L. (2009) Multilevel converters for high fundamental frequency application. Power Electronics and Applications, 2009. EPE '09. 13th European Conference on, 8–10 September 2009, pp. 1–14.

da Silva, E.R.C., Dos Santos, E.C. and Jacobina, C.B. (2011) Pulsewidth modulation strategies. *Industrial Electronics Magazine, IEEE*, **5**, 37–45.

Dae-Wook, K., Byoung-Kuk, L., Jae-Hyun, J. *et al.* (2005) A symmetric carrier technique of CRPWM for voltage balance method of flying-capacitor multilevel inverter. *Industrial Electronics, IEEE Transactions on*, **52**, 879–888.

Dae-Wook, K., Yo-Han, L., Bum-Seok, S. *et al.* (2003) An improved carrier-based SVPWM method using leg voltage redundancies in generalized cascaded multilevel inverter topology. *Power Electronics, IEEE Transactions on*, **18**, 180–187.

Dahidah, M.S.A. and Agelidis, V.G. (2008) Selective harmonic elimination PWM control for cascaded multilevel voltage source converters: a generalized formula. *Power Electronics, IEEE Transactions on*, **23**, 1620–1630.

Dahidah, M.S.A., Konstantinou, G., Flourentzou, N. and Agelidis, V.G. (2010) On comparing the symmetrical and non-symmetrical selective harmonic elimination pulse-width modulation technique for two-level three-phase voltage source converters. *Power Electronics, IET*, **3**, 829–842.

Dahidah, M.S.A., Konstantinou, G.S. and Agelidis, V.G. (2011) SHE-PWM control for asymmetrical hybrid multilevel flying capacitor and H-bridge converter. Power Electronics and Drive Systems (PEDS), 2011 IEEE Ninth International Conference on, 5–8 December 2011, pp. 29–34.

Dahidah, M.S.A., Konstantinou, G.S. and Agelidis, V.G. (2012) Selective harmonic elimination pulse-width modulation seven-level cascaded H-bridge converter with optimised DC voltage levels. *Power Electronics, IET*, **5**, 852–862.

Dai, S.Z., Lujara, N. and Boon-Teck, O. (1992) A unity power factor current-regulated SPWM rectifier with a notch feedback for stabilization and active filtering. *Power Electronics, IEEE Transactions on*, **7**, 356–363.

Das, S. and Narayanan, G. 2012. Novel switching sequences for a space-vector-modulated three-level inverter. *Industrial Electronics, IEEE Transactions on*, **59**, 1477–1487.

Davoodnezhad, R., Holmes, D.G. and Mcgrath, B.P. (2012) Hysteresis current regulation of three phase flying capacitor inverter with balanced capacitor voltages. Power Electronics and Motion Control Conference (IPEMC), 2012 7th International, 2–5 June 2012, pp. 47–52.

de Alvarenga, M.B. and Pomilio, J.A. (2011) Modulation strategy for minimizing commutations and capacitor voltage balancing in symmetrical cascaded multilevel converters. Industrial Electronics (ISIE), 2011 IEEE International Symposium on, 27–30 June 2011, pp. 1875–1880.

de Leon Morales, J., Escalante, M.F. and Mata-Jimenez, M.T. (2007) Observer for DC voltages in a cascaded H-bridge multilevel STATCOM. *Electric Power Applications, IET*, **1**, 879–889.

Deschamps, E., Borges, A.R. and da Silva, A.S.G. (2003) The influence of gate signals differences on the voltage across the switches in a flying-capacitor DC-to-DC multilevel converter. Industrial Electronics, 2003. ISIE '03. 2003 IEEE International Symposium on, 9–11 June 2003, vol. 1, pp. 656–661.

Dieckerhoff, S., Bernet, S. and Krug, D. (2003) Evaluation of IGBT multilevel converters for transformerless traction applications. Power Electronics Specialist Conference, 2003. PESC '03. 2003 IEEE 34th Annual, 15–19 June 2003, vol. 4, pp. 1757–1763.

Dieckerhoff, S., Bernet, S. and Krug, D. (2005) Power loss-oriented evaluation of high voltage IGBTs and multilevel converters in transformerless traction applications. *Power Electronics, IEEE Transactions on*, **20**, 1328–1336.

Do-Hyun, J. (2007) PWM methods for two-phase inverters. *Industry Applications Magazine, IEEE*, **13**, 50–61.

Dordevic, O., Jones, M. and Levi, E. (2013a) A comparison of carrier-based and space vector PWM techniques for three-level five-phase voltage source inverters. *Industrial Informatics, IEEE Transactions on*, **9**, 609–619.

Dordevic, O., Levi, E. and Jones, M. (2013b) A vector space decomposition based space vector PWM algorithm for a three-level seven-phase voltage source inverter. *Power Electronics, IEEE Transactions on*, **28**, 637–649.

Dunford, W.G. and Van Wyk, J.D. (1992) Harmonic imbalance in asynchronous PWM schemes. *Power Electronics, IEEE Transactions on*, **7**, 480–486.

Duran, M.J., Prieto, J., Barrero, F. *et al.* (2013) Space-vector PWM with reduced common-mode voltage for five-phase induction motor drives. *Industrial Electronics, IEEE Transactions on*, **60**, 4159–4168.

Dyck, D.N., Gilbert, G. and Lowther, D.A. (2010) A performance model of an induction motor for transient simulation with a PWM drive. *Magnetics, IEEE Transactions on*, **46**, 3093–3096.

Eckel, H.G. and Runge, J. (2011) Comparison of the semiconductor losses in self commutated inverter topologies for HVDC. Power Electronics and Applications (EPE 2011), Proceedings of the 2011–14th European Conference on, 30 August 2011–1 September 2011, pp. 1–8.

Escalante, M.F. (2012) Least-squares estimation of capacitor voltages in flying capacitor multilevel converters. *Power Electronics, IET*, **5**, 1741–1747.

Fan, Z., Peng, F.Z. and Zhaoming, Q. (2004) Study of the multilevel converters in DC-DC applications. Power Electronics Specialists Conference, 2004. PESC 04. 2004 IEEE 35th Annual, 20–25 June 2004, vol. 2, pp. 1702–1706.

Fang, H., Yang, R., Yu, Y. and Xu, D. (2012) A study on the DC voltage control techniques of cascaded multilevel APF. Power Electronics and Motion Control Conference (IPEMC), 2012 7th International, 2–5 June 2012, pp. 2727–2731.

Fazel, S.S., Bernet, S., Krug, D. and Jalili, K. (2007) Design and comparison of 4-kV neutral-point-clamped, flying-capacitor, and series-connected H-bridge multilevel converters. *Industry Applications, IEEE Transactions on*, **43**, 1032–1040.

Feng, C. and Agelidis, V.G. (2004) A DSP-based controller design for multilevel flying capacitor converters. Applied Power Electronics Conference and Exposition, 2004. APEC '04. Nineteenth Annual IEEE, 2004, vol. 3, pp. 1740–1744.

Floricau, D., Floricau, E., Parvulescu, L. and Gateau, G. (2010a) Loss balancing for Active-NPC and Active-Stacked-NPC multilevel converters. Optimization of Electrical and Electronic Equipment (OPTIM), 2010 12th International Conference on, 20–22 May 2010, pp. 625–630.

Floricau, D., Gateau, G. and Leredde, A. (2010b) New active stacked NPC multilevel converter: operation and features. *Industrial Electronics, IEEE Transactions on*, **57**, 2272–2278.

Flourentzou, N. and Agelidis, V.G. (2007) Harmonic performance of multiple sets of solutions of SHE-PWM for a 2-level VSC topology with fluctuating DC-link voltage. Power Engineering Conference, 2007. AUPEC 2007. Australasian Universities, 9–12 December 2007, pp. 1–8.

Franquelo, L.G., Napoles, J., Guisado, R.C.P. *et al.* (2007) A flexible selective harmonic mitigation technique to meet grid codes in three-level PWM converters. *Industrial Electronics, IEEE Transactions on*, **54**, 3022–3029.

Franquelo, L.G., Rodriguez, J., Leon, J.I. *et al.* (2008) The age of multilevel converters arrives. *Industrial Electronics Magazine, IEEE*, **2**, 28–39.

Fukuda, S., Iwaji, Y. and Hasegawa, H. (1990) PWM technique for inverter with sinusoidal output current. *Power Electronics, IEEE Transactions on*, **5**, 54–61.

Fukuda, S., Matsumoto, Y. and Sagawa, A. (1999) Optimal-regulator-based control of NPC boost rectifiers for unity power factor and reduced neutral-point-potential variations. *Industrial Electronics, IEEE Transactions on*, **46**, 527–534.

Ghias, A.M.Y.M., Ciobotaru, M., Pou, J. and Agelidis, V.G. (2012a) Performance evaluation of a five-level flying capacitor converter with reduced DC bus capacitance under two different modulation schemes. Power Electronics for Distributed Generation Systems (PEDG), 2012 3rd IEEE International Symposium on, 25–28 June 2012, pp. 857–864.

Ghias, A.M.Y., Pou, J., Ciobotaru, M. and Agelidis, V.G. (2012b) Voltage balancing of a five-level flying capacitor converter using optimum switching transitions. IECON 2012 – 38th Annual Conference on IEEE Industrial Electronics Society, 25–28 October 2012, pp. 5006–5012.

Gonza, X., Lez, M.A. and Escalante, M.F. (2010) Traction system for an EV based on induction motor and 3-level NPC inverter multilevel converters. Power Electronics Congress (CIEP), 2010 12th International, 22–25 August 2010, pp. 73–77.

Gonzalez, S.A., Valla, M.I. and Christiansen, C.F. (2010) Five-level cascade asymmetric multilevel converter. *Power Electronics, IET*, **3**, 120–128.

Green, R.M. and Boys, J.T. (1982) PWM sequence selection and optimization: a novel approach. *Industry Applications, IEEE Transactions on*, **IA-18**, 146–151.

Grigoletto, F.B. and Pinheiro, H. (2011) Generalised pulse width modulation approach for DC capacitor voltage balancing in diode-clamped multilevel converters. *Power Electronics, IET*, **4**, 89–100.

Gruson, F., Le Moigne, P., Delarue, P. *et al.* (2013) A simple carrier-based modulation for the SVM of the matrix converter. *Industrial Informatics, IEEE Transactions on*, **9**, 947–956.

Hadiouche, D., Baghli, L. and Rezzoug, A. (2006) Space-vector PWM techniques for dual three-phase AC machine: analysis, performance evaluation, and DSP implementation. *Industry Applications, IEEE Transactions on*, **42**, 1112–1122.

Handley, P.G. and Boys, J.T. (1992) Practical real-time PWM modulators: an assessment. *Electric Power Applications, IEE Proceedings B*, **139**, 96–102.

Hava, A.M., Kerkman, R.J. and Lipo, T.A. (1997a) A high performance generalized discontinuous PWM algorithm. Applied Power Electronics Conference and Exposition, 1997. APEC '97 Conference Proceedings 1997. Twelfth Annual, 23–27 February 1997, vol. **2**, pp. 886–894.

Hava, A.M., Kerkman, R.J. and Lipo, T.A. (1997b) Simple analytical and graphical tools for carrier based PWM methods. Power Electronics Specialists Conference, 1997. PESC '97 Record., 28th Annual IEEE, 22–27 June 1997, vol. 2, pp. 1462–1471.

Hava, A.M., Kerkman, R.J. and Lipo, T.A. (1998) Carrier-based PWM-VSI over modulation strategies: analysis, comparison, and design. *Power Electronics, IEEE Transactions on*, **13**, 674–689.

Hava, A.M. and Ün, E. (2011) A high-performance PWM Algorithm for common-mode voltage reduction in three-phase voltage source inverters. *Power Electronics, IEEE Transactions on*, **26**, 1998–2008.

Hoevenaars, T., Ledoux, K. and Colosino, M. (in press) Interpreting IEEE Std 519 and Meeting its Harmonic Limits in VFD Applications, Mirus International Inc, #12, 6805 Invader Cres. Mississauga, ON L5T 2K6, Canada USA.

Holmes, D.G. (1995) The significance of zero space vector placement for carrier based PWM schemes. Industry Applications Conference, 1995. Thirtieth IAS Annual Meeting, IAS '95., Conference Record of the 1995 IEEE, 8–12 October 1995, vol. 3, pp. 2451–2458.

Holmes, D.G. (1996) The significance of zero space vector placement for carrier-based PWM schemes. *Industry Applications, IEEE Transactions on*, **32**, 1122–1129.

Holmes, D.G. (1998) A general analytical method for determining the theoretical harmonic components of carrier based PWM strategies. Industry Applications Conference, 1998. Thirty-Third IAS Annual Meeting. The 1998 IEEE, 12–15 October 1998, vol. 2, pp. 1207–1214.

Holmes, D.G. and Mcgrath, B.P. (1999) Opportunities for harmonic cancellation with carrier based PWM for two-level and multi-level cascaded inverters. Industry Applications Conference, 1999. Thirty-Fourth IAS Annual Meeting. Conference Record of the 1999 IEEE, 1999, vol. 2, pp. 781–788.

Holmes, D.G. and Mcgrath, B.P. (2001) Opportunities for harmonic cancellation with carrier-based PWM for a two-level and multilevel cascaded inverters. *Industry Applications, IEEE Transactions on*, **37**, 574–582.

Holmes, D.G. and Lipo, T.A. (2003) *Pulse Width Modulation for Power Converters: Principle and Practice*, Wiley IEEE Press.

Hosseini, S.H. and Sadeghi, M. (2011) Reduced Diode Clamped Multilevel Converter with a modified control method. Electrical and Electronics Engineering (ELECO), 2011 7th International Conference on, 1–4 December 2011, pp. I-302–I-306.

In-Dong, K., Eui-Cheol, N., Heung-Geun, K. and Jong-Sun, K. (2004) A generalized Undeland snubber for flying capacitor multilevel inverter and converter. *Industrial Electronics, IEEE Transactions on*, **51**, 1290–1296.

Ito, T., Kamaga, M., Sato, Y. and Ohashi, H. (2010) An investigation of voltage balancing circuit for DC capacitors in diode-clamped multilevel inverters to realize high output power density converters. Energy Conversion Congress and Exposition (ECCE), 2010 IEEE, 12–16 September 2010, pp. 3675–3682.

Jiang, Q., Holmes, D.G. and Giesner, D.B. (1991) A method for linearising optimal PWM switching strategies to enable their computation on-line in real-time. Industry Applications Society Annual Meeting, 1991. Conference Record of the 1991 IEEE, 28 September 1991–4 October 1991, vol. 1, pp. 819–825.

Jih-Sheng, L. and Fang Zheng, P. (1995) Multilevel converters-a new breed of power converters. Industry Applications Conference, 1995. Thirtieth IAS Annual Meeting, IAS '95., Conference Record of the 1995 IEEE, vol. 3, pp. 2348–2356.

Khajehoddin, S.A., Bakhshai, A. and Jain, P.K. (2006) A Current Flow Model for m-Level Diode-Clamped Multilevel Converters. IEEE Industrial Electronics, IECON 2006 – 32nd Annual Conference on, 6–10 November 2006, pp. 2477–2482.

Khazraei, M., Sepahvand, H., Corzine, K. and Ferdowsi, M. (2010) A generalized capacitor voltage balancing scheme for flying capacitor multilevel converters. Applied Power Electronics Conference and Exposition (APEC), 2010 Twenty-Fifth Annual IEEE, 21–25 February 2010, pp. 58–62.

Khazraei, M., Sepahvand, H., Corzine, K.A. and Ferdowsi, M. (2012) Active capacitor voltage balancing in single-phase flying-capacitor multilevel power converters. *Industrial Electronics, IEEE Transactions on*, **59**, 769–778.

Kouzou, A., Abu Rub, H., Iqbal, A. *et al.* (2011) Selective Harmonics Elimination for a three-level diode clamped five-phase inverter based on Particle Swarm Optimization. IECON 2011 – 37th Annual Conference on IEEE Industrial Electronics Society, 7–10 November 2011, pp. 3495–3500.

Leon, J.I., Franquelo, L.G., Portillo, R.C. and Prats, M.M. (2005) DC-link capacitors voltage balancing in multilevel four-leg diode-clamped converters. Industrial Electronics Society, 2005. IECON 2005. 31st Annual Conference of IEEE, 6–10 November 2005, pp. 6.

Leon, J.I., Portillo, R., Franquelo, L.G. *et al.* (2007) New Space Vector Modulation Technique for Single-Phase Multilevel Converters. Industrial Electronics, 2007. ISIE 2007. IEEE International Symposium on, 4–7 June 2007, pp. 617–622.

Marchesoni, M. and Tenca, P. (2002) Diode-clamped multilevel converters: a practicable way to balance DC-link voltages. *Industrial Electronics, IEEE Transactions on*, **49**, 752–765.

Mathew, J., Rajeevan, P.P., Mathew, K. *et al.* (2013) A multilevel inverter scheme with dodecagonal voltage space vectors based on flying capacitor topology for induction motor drives. *Power Electronics, IEEE Transactions on*, **28**, 516–525.

Mcgrat, B.P., Meynard, T.A., Gateau, G. and Holmes, D.G. (2005) Optimal Modulation of Flying Capacitor and Stacked Multicell Converters using a State Machine Decoder. Power Electronics Specialists Conference, 2005. PESC '05. IEEE 36th, 16–26 June 2005, pp. 1671–1677.

Mcgrath, B.P. and Holmes, D.G. (2002a) An analytical technique for the determination of spectral components of multilevel carrier-based PWM methods. *Industrial Electronics, IEEE Transactions on*, **49**, 847–857.

Mcgrath, B.P. and Holmes, D.G. (2002b) Sinusoidal PWM of multilevel inverters in the overmodulation region. Power Electronics Specialists Conference, 2002. pesc 02. 2002 IEEE 33rd Annual, 2002, vol. 2, pp. 485–490.

Mcgrath, B.P. and Holmes, D.G. (2007a) Analytical Determination of the Capacitor Voltage Balancing Dynamics for Three Phase Flying Capacitor Converters. Industry Applications Conference, 2007. 42nd IAS Annual Meeting. Conference Record of the 2007 IEEE, 23–27 September 2007, pp. 1974–1981.

Mcgrath, B.P. and Holmes, D.G. (2007b) Analytical Modelling of Voltage Balance Dynamics for a Flying Capacitor Multilevel Converter. Power Electronics Specialists Conference, 2007. PESC 2007. IEEE, 17–21 June 2007, pp. 1810–1816.

Mcgrath, B.P. and Holmes, D.G. (2007c) Natural Current Balancing of Multicell Current Source Converters. Power Electronics Specialists Conference, 2007. PESC 2007. IEEE, 17–21 June 2007, pp. 968–974.

Mcgrath, B.P. and Holmes, D.G. (2009a) Enhanced voltage balancing of a flying capacitor multilevel converter using Phase Disposition (PD) modulation. Energy Conversion Congress and Exposition, 2009. ECCE 2009. IEEE, 20–24 September 2009, pp. 3108–3115.

Mcgrath, B.P. and Holmes, D.G. (2009b) Natural capacitor voltage balancing for a flying capacitor converter induction motor drive. *Power Electronics, IEEE Transactions on*, **24**, 1554–1561.

Mcgrath, B.P., Holmes, D.G. and Lipo, T. (2003) Optimized space vector switching sequences for multilevel inverters. *Power Electronics, IEEE Transactions on*, **18**, 1293–1301.

Meili, J., Ponnaluri, S., Serpa, L. *et al.* (2006) Optimized Pulse Patterns for the 5-Level ANPC Converter for High Speed High Power Applications. IEEE Industrial Electronics, IECON 2006 – 32nd Annual Conference on, 6–10 November 2006, pp. 2587–2592.

Minshull, S.R., Bingham, C.M., Stone, D.A. and Foster, M.P. (2010) Compensation of nonlinearities in diode-clamped multilevel converters. *Industrial Electronics, IEEE Transactions on*, **57**, 2651–2658.

Mohzani, Z., Mcgrath, B.P. and Holmes, D.G. (2011) Natural balancing of the Neutral Point voltage for a three-phase NPC multilevel converter. IECON 2011 – 37th Annual Conference on IEEE Industrial Electronics Society, 7–10 November 2011, pp. 4445–4450.

Mohzani, Z., Mcgrath, B.P. and Holmes, D.G. (2012) The balancing properties of DC link compensation for 3-phase Neutral Point Clamped (NPC) Converter. Power Electronics and Motion Control Conference (IPEMC), 2012 7th International, 2–5 June 2012, pp. 574–579.

Mwinyiwiwa, B., Boon-Teck, O. and Wolanski, Z. (1998a) UPFC using multiconverter operated by phase-shifted triangle carrier SPWM strategy. *Industry Applications, IEEE Transactions on*, **34**, 495–500.

Mwinyiwiwa, B., Wolanski, Z. and Boon-Teck, O. (1998b) Microprocessor-implemented SPWM for multiconverters with phase-shifted triangle carriers. *Industry Applications, IEEE Transactions on*, **34**, 487–494.

Mwinyiwiwa, B., Wolanski, Z. and Boon-Teck, O. (1999) Current equalization in SPWM FACTS controllers at lowest switching rates. *Power Electronics, IEEE Transactions on*, **14**, 900–905.

Nademi, H., Das, A. and Norum, L. (2011) Nonlinear observer-based capacitor voltage estimation for sliding mode current controller in NPC multilevel converters. PowerTech, 2011 IEEE Trondheim, 19–23 June 2011, pp. 1–7.

Nami, A., Zare, F., Ledwich, G. and Ghosh, A. (2007) A new configuration for multilevel converters with diode clamped topology. Power Engineering Conference, 2007. IPEC 2007. International, 3–6 December 2007, pp. 661–665.

Newton, C. and Summer, M. (1998) Multi-level convertors a real solution to medium/high-voltage drives? *Power Engineering Journal*, **12**, 21–26.

Newton, C. and Sumner, M. (1998) A novel arrangement for balancing the capacitor voltages of a five level diode clamped inverter. Power Electronics and Variable Speed Drives, 1998. Seventh International Conference on (Conf. Publ. No. 456), 21–23 September 1998, pp. 465–470.

Newton, C. and Sumner, M. (1999) Novel technique for maintaining balanced internal DC link voltages in diode clamped five-level inverters. *Electric Power Applications, IEE Proceedings*, **146**, 341–349.

Newton, C., Sumner, M. and Alexander, T. (1997) Multi-level converters: a real solution to high voltage drives? Update on New Power Electronic Techniques (Digest No: 1997/091), IEE Colloquium on, 23 May 1997, pp. 3/1–3/5.

Ooi, B.T., Joos, G. and Xiaogang, H. (1999) Operating principles of shunt STATCOM based on 3-level diode-clamped converters. *Power Delivery, IEEE Transactions on*, **14**, 1504–1510.

Patel, H. and Agarwal, V. (2008) Control of a stand-alone inverter-based distributed generation source for voltage regulation and harmonic compensation. *Power Delivery, IEEE Transactions on*, **23**, 1113–1120.

Peng, F.Z., Wei, Q. and Dong, C. (2010) Recent advances in multilevel converter/inverter topologies and applications. Power Electronics Conference (IPEC), 2010 International, 21–24 June 2010, pp. 492–501.

Pongiannan, R.K., Paramasivam, S. and Yadaiah, N. (2011) Dynamically reconfigurable PWM controller for three-phase voltage-source inverters. *Power Electronics, IEEE Transactions on*, **26**, 1790–1799.

Steinke, J.K. (1992) Switching frequency optimal PWM control of three-level inverter. *IEEE Transaction on Power Electronics*, **7**, 487–496.

Zhang, Y., Adam, G.P., Lim, T.C. *et al.* (2010) Voltage Source Converter in High Voltage Applications: Multilevel versus Two-level Converters. IET, the 9th International Conference on AC and DC Power Transmission, 2010 London, UK.

B

Worked-out Examples

Exercise 1

Demonstrate that the average voltages (expressed in terms of the modulator signal assuming a $F_c \geq 10F_s$) generated by a three-phase VSC can be reduced to:

$$v_{vsc\_a} = v_{cd}\left(\frac{1}{3}m_a 1 - \frac{1}{6}m_b 1 - \frac{1}{6}m_c 1\right) = \frac{v_{cd}}{2}m_a 1$$

$$v_{vsc\_b} = v_{cd}\left(-\frac{1}{6}m_a 1 + \frac{1}{3}m_b 1 - \frac{1}{6}m_c 1\right) = \frac{v_{cd}}{2}m_b 1$$

$$v_{vsc\_c} = v_{cd}\left(-\frac{1}{6}m_a 1 - \frac{1}{6}m_b 1 + \frac{1}{3}m_c 1\right) = \frac{v_{cd}}{2}m_c 1$$

Demonstration:
  Knowing that

$$m_a = \cos(\omega t)$$

$$m_b = \cos\left(\omega t - \frac{2}{3}\pi\right)$$

$$m_b = \cos\left(\omega t + \frac{2}{3}\pi\right)$$

and using the trigonometric identity

$$\cos(\alpha \pm \beta) = \cos(\alpha)\cos(\beta) \mp \sin(\alpha)\sin(\beta)$$

Offshore Wind Energy Generation: Control, Protection, and Integration to Electrical Systems, First Edition.
Olimpo Anaya-Lara, David Campos-Gaona, Edgar Moreno-Goytia and Grain Adam.
© 2014 John Wiley & Sons, Ltd. Published 2014 by John Wiley & Sons, Ltd.
Companion Website: www.wiley.com/go/offshore_wind_energy_generation

For the a phase the following expression is obtained

$$v_{dc}\left[\frac{1}{3}\cos(\omega t)-\frac{1}{6}\cos\left(\omega t-\frac{2}{3}\pi\right)-\frac{1}{6}\sin\left(\omega t+\frac{2}{3}\pi\right)\right]$$

$$=v_{dc}\left[\frac{1}{3}\cos(\omega t)+\frac{1}{12}\cos(\omega t)-\frac{1}{12}\sin(\omega t)\sqrt{3}+\frac{1}{12}\cos(\omega t)+\frac{1}{12}\sin(\omega t)\sqrt{3}\right]$$

$$=\frac{v_{dc}}{2}\cos(\omega t).$$

For the b phase the following expression is obtained

$$v_{dc}\left[-\frac{1}{6}\cos(\omega t)+\frac{1}{3}\cos\left(\omega t-\frac{2}{3}\pi\right)-\frac{1}{6}\sin\left(\omega t+\frac{2}{3}\pi\right)\right]$$

$$=v_{dc}\left[-\frac{1}{6}\cos(\omega t)-\frac{1}{6}\cos(\omega t)+\frac{1}{6}\sin(\omega t)\sqrt{3}+\frac{1}{12}\cos(\omega t)+\frac{1}{12}\sin(\omega t)\sqrt{3}\right]$$

$$=v_{dc}\left[-\frac{1}{4}\cos(\omega t)+\frac{1}{4}\sin(\omega t)\sqrt{3}\right].$$

Knowing that $\cos(\frac{2}{3}\pi)=-\frac{1}{2}$ and $\sin(\frac{2}{3}\pi)=\frac{\sqrt{3}}{2}$ then the following expression is obtained

$$v_{dc}\left[-\frac{1}{4}\cos(\omega t)+\frac{1}{4}\sin(\omega t)\sqrt{3}\right]=v_{dc}\left[\frac{1}{2}\cos(\omega t)\cos\left(\frac{2}{3}\pi\right)+\frac{1}{2}\sin(\omega t)\sin\left(\frac{2}{3}\pi\right)\right]$$

$$=\frac{v_{dc}}{2}\cos\left(\omega t-\frac{2}{3}\pi\right).$$

For the c phase the following expression is obtained

$$v_{dc}\left[-\frac{1}{6}\cos(\omega t)-\frac{1}{6}\cos\left(\omega t-\frac{2}{3}\pi\right)+\frac{1}{3}\sin\left(\omega t+\frac{2}{3}\pi\right)\right]$$

$$=v_{dc}\left[-\frac{1}{6}\cos(\omega t)+\frac{1}{12}\cos(\omega t)-\frac{1}{12}\sin(\omega t)\sqrt{3}-\frac{1}{6}\cos(\omega t)-\frac{1}{6}\sin(\omega t)\sqrt{3}\right]$$

$$=v_{dc}\left[-\frac{1}{4}\cos(\omega t)-\frac{1}{4}\sin(\omega t)\sqrt{3}\right].$$

Knowing that $\text{co}(\frac{2}{3}\pi)=-\frac{1}{2}$ and $\sin(\frac{2}{3}\pi)=\frac{\sqrt{3}}{2}$ then the following expression is obtained

$$v_{dc}\left[-\frac{1}{4}\cos(\omega t)-\frac{1}{4}\sin(\omega t)\sqrt{3}\right]=v_{dc}\left[\frac{1}{2}\cos(\omega t)\cos\left(\frac{2}{3}\pi\right)-\frac{1}{2}\sin(\omega t)\sin\left(\frac{2}{3}\pi\right)\right]$$

$$=\frac{v_{dc}}{2}\cos\left(\omega t+\frac{2}{3}\pi\right).$$

Exercise 2

A DFIG-based wind turbine with terminal voltage of $V_s=(690+j0.0)$V and rotor voltage, $V_r'=(-50.43-j7.18)$V is working at 1686 rpm when subjected to a three-phase fault at time $t=0$.

The generator parameters are (all quantities referred to the stator):

Rotor resistance, $r'_r = 0.001309\ \Omega$

Stator resistance, $r_s = 0.001164\ \Omega$

Rotor self reactance, $X'_r = 0.9647\ \Omega$

Stator self reactance, $X_s = 0.963\ \Omega$

Mutual reactance, $X_m = 0.941\ \Omega$

Pole pairs, $p_g = 2$

(a) What is the transient current at the DFIG terminals immediately after the fault happening?

First calculate the slip speed

$$s_\omega = (\omega_s - p_g\omega_r)/\omega_s = \frac{\left[(2\pi 50) - (2)\,1686\dfrac{2\pi}{60}\right]}{2\pi 50} = -.124$$

Then calculate the stator transient inductance and the rotor and stator coupling factors

$$X_{ts} = X_s - \frac{X_m^2}{X'_r} = 0.963 - \frac{(0.941)^2}{.9641} = 0.04454$$

$$k_s = \frac{L_m}{L_s} = \frac{X_m}{X_s} = \frac{0.914}{0.936} = 0.9491$$

$$k_r = \frac{L_m}{L'_r} = \frac{X_m}{X'_r} = \frac{0.914}{0.9641} = 0.9480$$

Then, making use of Eq. (2.119), the following expression is obtained

$$|i_{s,0}| = \frac{\sqrt{2}}{iX_{ts}} \left[|v_s|e^{-t/T_s} - \left[(|v_s|k_r k_s + |v'_r|s_\omega^{-1}k_r)\,e^{j\omega_s t}\right]e^{-t/T_r}\right]$$

$$= \left|\frac{\sqrt{2}}{0.04454i}\left[690e^0 - \left[(690)(0.9480)(0.9491) + 50.9385e^{-3i}(-.124)^{-1}(0.9480)e^0\right]e^0\right]\right|$$

$$= \left|\frac{\sqrt{2}}{0.04454i}\left[690 - (620.825 + (385.536 - 54.9568i))\right]\right| = |-1744.75 - 10018.8i|$$

$$= 10169.5A$$

Exercise 3

A DFIG-based wind turbine has the following parameters are (all quantities referred to the stator):

Rotor resistance, $r'_r = 0.001309 \ \Omega$

Rotor self reactance, $X'_r = 0.9647 \ \Omega$

Stator self reactance, $X_s = 0.963 \ \Omega$

(a) What is the minimum value of the crowbar resistance needed to allow a transient DFIG speed deviation of ± 0.5 pu?.

$$S_{\omega max} = \frac{(r'_r + r_{cb})}{\sqrt{(r'_r + r_{cb})^2 + (X_s + X'_r)^2}} \rightarrow r_{cb} = -r_r + s_{\omega max}\sqrt{-\frac{X_s^2 + 2X_sX_r + X_r^2}{s^2_{\omega max}} - 1}$$

$$= -0.001309 + 0.5\sqrt{-\frac{3.716}{0.25 - 1}} = 1.111 \ \Omega$$

Exercise 4

A 1.27 MW DFIG-based wind turbine has the following parameters (all quantities referred to the stator):

Rotor resistance, $r'_r = 0.001309 \ \Omega$

Stator resistance, $r_s = 0.001164 \ \Omega$

Rotor reactance, $X'_r = 0.9647 \ \Omega$

Stator reactance, $X_s = 0.963 \ \Omega$

Mutual reactance, $X_m = 0.941 \ \Omega$

ac frequency $F_s = 50$ Hz

Lumped inertia constant $J = 3\ 716\ 063.08 \ \text{kg} \cdot \text{m}^2$

Damping coefficient $B_m = 504 \ \text{kNm/(rad/s)}$

(a) Calculate the PI controller parameters, using zero-pole cancellation, to attain a rotor current closed-loop controller time-response of 100 ms

$$\text{Rotor self inductance } L'_r = \frac{X'_r}{2\pi 50} = \frac{0.9647 \ \Omega}{314.159} = 0.003071H$$

$$\text{Stator self inductance } L_s = \frac{X_s}{2\pi 50} = \frac{0.936 \ \Omega}{314.159} = 0.002979H$$

$$\text{Mutual inductance } L_m = \frac{X_m}{2\pi 50} = \frac{0.941 \ \Omega}{314.159} = 0.002995H$$

Knowing that the rotor current transfer function is given by

$$\frac{i'_{dr}(s)}{v'_{dr}(s)} = \frac{i'_{qr}(s)}{v'_{qr}(s)} = \frac{i'_{r}(s)}{v'_{r}(s)} = P_r(s) = \frac{L_s}{s(L'_r L_s - L^2_m) + r'_r L_s}$$

and the rotor current controller given by

$$K_{ir}(s) = \frac{Kp_r s + Ki_r}{s}$$

then the following expression for the open loop gain controller is obtained

$$\ell_r(s) = K_r(s)P_r(s) = \left(\frac{Kp_r L_s}{(L'_r L_s - L^2_m)s}\right)\frac{\left(s + \dfrac{Ki_r}{Kp_r}\right)}{\left(s + \dfrac{r'_r L_s}{(L'_r L_s - L^2_m)}\right)}.$$

Making $Ki_r/Kp_r = r'_r L_s/(L'_r L_s - L^2_m)$ and $Kp_r L_s/(L'_r L_s - L^2_m) = \frac{1}{0.1}$ the stable pole of the plant is cancelled with the zero of the PI controller. Setting the response time of the current loop to 100 ms, the solution for the PI constants turns to be:

$$Kp_r = \frac{L'_r L_s - L^2_m}{0.1 L_s} = 0.000599$$

$$Ki_r = \frac{r'_r L_s}{(L'_r L_s - L^2_m)}\frac{(L'_r L_s - L^2_m)}{0.01 L_s}\frac{r'_r}{0.1} = 0.01309.$$

(b) Calculate the PI controller parameters, using zero-pole cancellation, to attain a rotor speed closed loop controller time-response of 2 s.

Knowing that the rotor speed transfer function is given by

$$P_\omega(s) = \frac{\omega_r(s)}{T_e(s)} = -\frac{1}{J + B_m}$$

and the rotor current controller given by

$$K_\omega(s) = \frac{Kp_\omega s + Ki_\omega}{s}$$

then the following expression for the open loop gain controller is obtained

$$K_\omega(s)P_\omega(s) = -\left(\frac{Kp_\omega}{Js}\right)\frac{\left(s + \dfrac{Ki_\omega}{Kp_\omega}\right)}{\left(s + \dfrac{B_m}{J}\right)}.$$

Making $Ki_\omega/Kp_\omega = B_m/J$ and $Kp_\omega/J = 1/2$ the stable pole of the plant is cancelled with the zero of the PI controller. Setting the response time of the rotor speed loop to 2 s, the solution for the PI constants turns to be:

$$Kp_\omega = \frac{J}{2} = 1.858031540 \cdot 10^6$$

$$Ki_\omega = \frac{B_m}{J}\frac{J}{2} = \frac{504 \cdot 10^3}{2} = 252 \cdot 10^3.$$

Exercise 5

A 2 MW DFIG-based wind turbine has the following parameters:

Stator terminal voltage $V_s = (690 + j0.0)$V

Back-to-back dc circuit capacitance: 10 000 μF

Grid side converter inductance: 0.02308 H

Grid side converter resistance: 0.0 002 308 Ω

(a) What is the value of the active damping loop to achieve an artificial time constant of the dc voltage plant of 200 ms?

Selecting G_{dc} to be 0.000 242 the following expression for Eq. (2.100) is obtained

$$-\frac{3 \cdot 690}{0.01s + 3 \cdot 690 \cdot 0.00024154} = -\frac{2070}{0.01s + 0.5} \rightarrow \left(-\frac{\frac{1}{0.5} \cdot 2070}{\frac{1}{0.5}}\right)\left(\frac{1}{0.02s + 1}\right).$$

(b) Calculate the dc voltage PI controller parameters, using zero-pole cancellation, to attain a time-response of 100 ms for the closed loop controller of the improved dc voltage plant of section a)

Knowing that the improved dc voltage transfer function is given by

$$M_{dc}(s) = -\frac{P_{dc}(s)}{1 + P_{dc}(s)G_{dc}} = -\frac{3v_d}{Cs + 3v_d G_{dc}}$$

and the voltage controller is given by

$$K_{dc}(s) = \frac{Kp_{dc}s + Ki_{dc}}{s} = \frac{Kp_{dc}}{s}\left(s + \frac{Ki_{dc}}{Kp_{dc}}\right)$$

then the following expression for the open loop gain controller is obtained

$$\ell_{dc}(s) = K_{dc}(s)M_{dc}(s) = \left(\frac{3Kp_{dc}v_d}{Cs}\right)\frac{\left(s + \frac{Ki_{dc}}{Kp_{dc}}\right)}{\left(s + \frac{3v_d G_{dc}}{C}\right)}.$$

Making $Ki_{dc}/Kp_{dc} = 3v_d G_{dc}/C$ and $3Kp_{dc}v_d/C = 1/\tau_{dc}$ the stable pole of the plant is cancelled with the zero of the PI controller. Setting the response time of the dc voltage controller loop to 100 ms, the solution for the PI constants turns to be:

$$Kp_{dc} = \frac{\left(\dfrac{C}{\tau_{dc}}\right)}{3v_d} = 4.830 \cdot 10^{-5}$$

$$Ki_{dc} = \frac{3v_d G_{dc} Kp_{dc}}{C} = 0.002415.$$

(c) Calculate the GSC current PI controller parameters, using zero-pole cancellation, to attain a GSC current closed loop controller time-response ten times smaller than the dc voltage controller time-response calculated above.

Knowing that the GSC plant is given by

$$P_{gsc}(s) = \frac{1}{Ls + r}$$

and the voltage controller is given by

$$K_{gsc}(s) = \frac{Kp_{gsc}s + Ki_{gsc}}{s} = \frac{Kp_{gsc}}{s}\left(s + \frac{Ki_{gsc}}{Kp_{gsc}}\right)$$

then the following expression for the open loop gain controller is obtained

$$\ell_{gsc}(s) = K_{gsc}(s)P_{gsc}(s) = \left(\frac{Kp_{gsc}}{Ls}\right)\frac{\left(s + \dfrac{Ki_\omega}{Kp_\omega}\right)}{\left(s + \dfrac{r}{L}\right)}.$$

Making $Ki_{gsc}/Kp_{gsc} = r/L$ and $Kp_{gsc}/L = 1/\tau_{gsc}$ the stable pole of the plant is cancelled with the zero of the PI controller. Setting the response time of the GSC current controller loop to 10 ms, the solution for the PI constants turns to be:

$$Kp_{gsc} = \left(\frac{L}{\tau_{gsc}}\right) = 2.308$$

$$Ki_{gsc} = \frac{rKp_{gsc}}{L} = 0.02308.$$

Exercise 6

A 2 MW DFIG-based wind turbine has the following parameters:

Stator terminal voltage $V_s = (690 + j0.0)\text{V}$

Back-to-back dc circuit capacitance (C): 10 000 μF

Back-to-back dc voltage in steady state (v_{dc}): 1200 V

Back-to-back dc voltage reference (v_{dc_ref}): 1200 V

Back-to-back power rating: 600 kW

Grid side converter inductance (L): 0.02308 H

Grid side converter resistance (r): 0.0 002 308 Ω

(a) What is the minimum dc voltage controller bandwidth α_{dc_min} required to allow a maximum v_{dc} deviation of $0.9v_{dc\_ref}$ during a sudden full power interruption of the back-to-back converter (i.e. 600 kW)

The maximum dc voltage during surge $v_{dc\,max\_step}^2$ must be less than:

$$v_{dc\,max\_step}^2 < \left[v_{dc}^2 - (0.9v_{dc\_ref})^2 \right] \Rightarrow v_{dc\,max\_step}^2 < 1.19v_{dc}^2.$$

Next, making use of Eq. (2.209)

$$\alpha_{dc\_min} = \frac{\pm 2P_{r\_max}}{\left(\pm v_{dc\,max\_step}^2 \right)(C)} e^{-1} \alpha_{dc} \geq \alpha_{dc\_min}$$

$$\alpha_{dc\_min} = \frac{\pm 2 \cdot 600000}{1.19 \cdot (1200)^2 (0.01)} e^{-1} = 70.02 \text{ rad.}$$

(b) Calculate the constants of the two degrees of freedom IMC dc voltage controller using the α_{dc_min} calculated on section a)

Using (2.204)

$$G_{dc} = -\frac{1}{3} \frac{\alpha_{dc}C}{v_d} = \frac{1}{3} \frac{70.02 \cdot 0.01}{690} = 0.000338.$$

Using (2.203)

$$Kp_{dc} = -\frac{\alpha_{dc}C}{3v_d} = -\frac{70.02 \cdot 0.01}{3 \cdot 690} = -0.000338$$

$$Ki_{dc} = \alpha_{dc}G_{dc} = 70.02 \cdot 0.000338 = 0.023685.$$

(c) Calculate the constants of the two degrees of freedom IMC grid side converter controller using a bandwidth 10 times larger than the α_{dc_min} calculated on section a).

Using (2.214)

$$G_{gsc} = \alpha_{gsc}L - r = 700.2 \cdot 0.02308 - 0.0002308 = 16.1604.$$

Using (2.213)

$$Kp_{gsc} = \alpha_{gsc}L = 16.1606$$

$$Ki_{gsc} = (r + G_{gsc}) \alpha_{gsc} = 11315.7.$$

Index

Offshore Wind Energy Generation: Control, Protection, and Integration to Electrical Systems, First Edition.
Olimpo Anaya-Lara, David Campos-Gaona, Edgar Moreno-Goytia and Grain Adam.
© 2014 John Wiley & Sons, Ltd. Published 2014 by John Wiley & Sons, Ltd.
Companion Website: www.wiley.com/go/offshore_wind_energy_generation